Data Structure Practice

for Collegiate Programming Contests and Education

Data Structure Practice

Practice

for Collegiate Programming Contests and Education

Yonghui Wu and Jiande Wang

CRC Press
Taylor & Francis Group
Boca Raton   London   New York

CRC Press is an imprint of the
Taylor & Francis Group, an **informa** business

Published with arrangement with the original publisher, Beijing Huazhang Graphics and Information Company.

CRC Press
Taylor & Francis Group
6000 Broken Sound Parkway NW, Suite 300
Boca Raton, FL 33487-2742

First issued in paperback 2020

© 2016 by Taylor & Francis Group, LLC
CRC Press is an imprint of Taylor & Francis Group, an Informa business

No claim to original U.S. Government works

Version Date: 20151203

ISBN 13: 978-0-367-57506-9 (pbk)
ISBN 13: 978-1-4822-1539-7 (hbk)

Visit the Taylor & Francis Web site at
http://www.taylorandfrancis.com

and the CRC Press Web site at
http://www.crcpress.com

Contents

SUMMARY OF SECTION I

SECTION II   EXPERIMENTS FOR LINEAR LISTS

SUMMARY OF SECTION II

SECTION III   EXPERIMENTS FOR TREES

SUMMARY OF SECTION III

Preface

Since the 1990s, the ACM International Collegiate Programming Contest (ACM-ICPC) has become a worldwide programming contest. Every year, more than 10,000 students and more than 1,000 universities participate in local contests, preliminary contests, and regional contests all over the world. In the meantime, programming contests' problems from all over the world can be gotten, analyzed, and solved by us. These contest problems can be used not only for programming contest training, but also for education.

In our opinion, not only a programming contestant's ability, but also a computer student's ability is based on his or her programming knowledge system and programming strategies for solving problems. The programming knowledge system can be summarized as a famous formula: algorithms + data structures = programs. It is also the foundation for the knowledge system of computer science and engineering. Strategies solving problems are strategies for data modeling and algorithm design. When data models and algorithms for problems are not standard, what strategies we should take to solve these problems?

Based on the ACM-ICPC, we published a series of books, not only for systematic programming contest training, but also for better polishing computer students' programming skill, using programming contests' problems: "Data Structure Experiment: For Collegiate Programming Contest and Education," "Algorithm Design Experiment: For Collegiate Programming Contest and Education," and "Programming Strategies Solving Problems" in Mainland China. And the traditional Chinese versions for "Data Structure Experiment: For Collegiate Programming Contest and Education" and "Programming Strategies Solving Problems" were also published in Taiwan.

"Data Structure Practice: For Collegiate Programming Contests and Education" is the English version for "Data Structure Experiment: For Collegiate Programming Contest and Education." There are 4 sections, 14 chapters, and 200 programming contest problems in this book. Section I, "Fundamental Programming Skills," focuses on experiments and practices for simple computing, simple simulation, and simple recursion, for students just learning programming languages. Section II, "Experiments for Linear Lists," Section III, "Experiments for Trees," and Section IV, "Experiments for Graphs," focus on experiments and practices for data structure.

Characteristics of the book are as follows:

1. The book's outlines are based on the outlines of data structures. Programming contest problems and their analyses and solutions are used as experiments. For each chapter, there is a "Problems" section to let students solve programming contests' problems, and hints for these problems are also shown.
2. Problems in the book are all selected from the ACM-ICPC regional and world finals programming contests, universities' local contests, and online contests, and from 1990 to now.

3. Not only analyses and solutions or hints to problems are shown, but also test data for most of problems are provided. Sources and IDs for online judges for these problems are also given. They can help readers better and more easily polish their programming skills.

Therefore, the book can be used not only as an experiment book, but also for systematic programming contests' training.

We appreciate Professors Steven Skiena, Rezaul Chowdhury, C. Jinshong Hwang, Ziliang Zong, Hongchi Shi, and Rudolf Fleischer. They provided us platforms in which English is the native language that improved our manuscript.

We appreciate our students Julaiti Alafate, Zheyun Yao, and Hao Zhang. They finished programs in the book.

The work is supported by the China Scholarship Council.

Online judge systems for problems in this book are as follows:

Online Judge Systems	Abbreviations	Website
Peking University Online Judge System	POJ	http://poj.org/
Zhejiang University Online Judge System	ZOJ	http://acm.zju.edu.cn/onlinejudge/
UVA Online Judge System	UVA	http://uva.onlinejudge.org/
		http://livearchive.onlinejudge.org/
Ural Online Judge System	Ural	http://acm.timus.ru/
SGU Online Judge System	SGU	http://acm.sgu.ru/

If you discover anything you believe to be an error, please contact us through Yonghui Wu's email: yhwu@fudan.edu.cn. Your help is appreciated.

Yonghui Wu
Jiande Wang

Authors

Yonghui Wu is associate professor at Fudan University. He acted as the coach of Fudan University Programming Contest teams from 2001 to 2011. Under his guidance, Fudan University qualified for Association for Computing Machinery International Collegiate Programming Contest (ACM-ICPC) World Finals every year and won three medals (bronze medal in 2002, silver medal in 2005, and bronze medal in 2010). Since 2012, he has published a series of books for programming contests and education. Since 2013, he has given lectures in Oman, Taiwan, and the United States for programming contest training. He is the chair of the ICPC Asia Programming Contest 1st Training Committee now.

Jiande Wang is a senior high school teacher and a famous coach for the Olympiad in Informatics in China. He has published 24 books for programming contests since the 1990s. Under his guidance, his students have won seven gold medals, three silver medals, and two bronze medals in the International Olympiad in Informatics for China.

FUNDAMENTAL PROGRAMMING SKILLS

I

Programming language is an introductory course of data structures and algorithms. This course enables students to program by programming languages. Programming languages, data structures, and algorithm designs are skills that computer students must polish. Therefore, polishing fundamental programming skills is the first section for this book. There are three chapters in Section I covering

1. Computing
2. Simulation
3. Recursion

These three chapters are not only a review of programming languages, but also an introductory course on data structure.

Chapter 1

Practice for Simple Computing

The pattern of a programming contest problem is input–process–output. A problem for simple computing is a problem whose process is simple. For such a problem, we should only consider optimizing the process and dealing with input and output correctly. The goals of Chapter 1 are as follows:

1. Students master C/C++ or Java programming language.
2. Students become familiar with online judge systems and programming environments.
3. Students begin to learn how to transfer a practical problem into a computing process, implement the computing process by a program, and debug the program to pass all test cases.

"God is in the details." In Chapter 1, problems are relatively simple. We should notice formats of input and output, precision, and time complexity. Therefore, the following topics will be discussed in this chapter:

1. Programming style
2. Multiple test cases
3. Precision of real numbers
4. Improving time complexity by dichotomy

Normally, a complex problem consists of several subproblems for simple computing. "Even the longest journey begins with a single step." Polishing programming skills should begin with solving simple computing problems.

1.1 Improving Programming Style

A program's writing style is not only for its visual sense, but also for examining the program and debugging its errors. A program's style also shows whether its programming idea is clear. It is hard to say which kind of programming style is good, but there are some rules for programing style. They are discussed in the following experiments.

1.1.1 Financial Management

Larry graduated this year and finally has a job. He's making a lot of money, but somehow never seems to have enough. Larry has decided that he needs to get a hold of his financial portfolio and solve his financial problems. The first step is to figure out what's been going on with his money. Larry has his bank account statements and wants to see how much money he has. Help Larry by writing a program to take his closing balance from each of the past 12 months and calculate his average account balance.

Input

The input will be 12 lines. Each line will contain the closing balance of his bank account for a particular month. Each number will be positive and displayed to the penny. No dollar sign will be included.

Output

The output will be a single number, the average (mean) of the closing balances for the 12 months. It will be rounded to the nearest penny, preceded immediately by a dollar sign, and followed by the end of the line. There will be no other spaces or characters in the output.

Sample Input	Sample Output
100.00	$1581.42
489.12	
12454.12	
1234.10	
823.05	
109.20	
5.27	
1542.25	
839.18	
83.99	
1295.01	
1.75	

Source: ACM Mid-Atlantic United States 2001.

IDs for online judges: POJ 1004, ZOJ 1048, UVA 2362.

Analysis

The problem's pattern, input–process–output, is very simple: First, the income of 12 months $a[0 .. 11]$ is input by a for statement *for*($i = 0$; $i < 12$; i++), and the total income

$$sum = \sum_{i=0}^{11} a[i]$$

is calculated. Then the average monthly income $avg = sum/12$ is calculated. Finally, avg is output in accordance with the problem's output format.

Program

```
#include<iostream>                              // Preprocessor Directive
using namespace std;                           // Using C++ Standard Library
int main()                                     // Main function
{
     double avg, sum=0.0, a[12]={0};           // Real variable avg and sum,
and real array a
     int i;                                    // Integer variable i
     for(i=0;i<12;i++){               // Input the income of 12 months a[0..11]
and summation
              cin>>a[i];
              sum+=a[i];
          }
     avg=sum/12;                               // Calculate the average monthly
income
     printf("$%.2f",avg);                      // Output the average monthly
income
     return 0;
   }
```

From the above program, we can get the following:

First, the input and output of the program must meet formats for input and output. In this problem, each input number will be positive and displayed to the penny, and the output will be rounded to the nearest penny, preceded immediately by a dollar sign and followed by an end of the line. If the program doesn't meet formats for input and output, it will be judged as the wrong answer.

Second, a program should be readable. The style of a program should be serration based on a logical level.

Finally, program annotations should be given.

1.2 Multiple Test Cases

The financial management problem (Section 1.1.1) has only one test case. In order to guarantee the correctness of a program, for most problems there are multiple test cases. In some circumstances, the number of test cases is given; in other circumstances, the number of test cases isn't given, but the mark of the input end is given.

1.2.1 Doubles

As part of an arithmetic competency program, your students will be given randomly generated lists of 2–15 unique positive integers and asked to determine how many items in each list are twice some other item in the same list. You will need a program to help you with the grading. This program should be able to scan the lists and output the correct answer for each one. For example, given the list

$$1\ 4\ 3\ 2\ 9\ 7\ 18\ 22$$

your program should answer 3, as 2 is twice 1, 4 is twice 2, and 18 is twice 9.

Input

The input file will consist of one or more lists of numbers. There will be one list of numbers per line. Each list will contain from 2 to 15 unique positive integers. No integer will be larger than 99. Each line will be terminated with the integer 0, which is not considered part of the list. A line with the single number –1 will mark the end of the file. The example input below shows three separate lists. Some lists may not contain any doubles.

Output

The output will consist of one line per input list, containing a count of the items that are double some other item.

Sample Input	Sample Output
1 4 3 2 9 7 18 22 0	3
2 4 8 10 0	2
7 5 11 13 1 3 0	0
–1	

Source: ACM Mid-Central United States 2003.

IDs for online judges: POJ 1552, ZOJ 1760, UVA 2787.

Analysis

There are multiple test cases for the problem. Therefore, a loop statement is used to deal with multiple test cases. The loop enumerates every test case. –1 marks the end of the input. Therefore, –1 is the end condition of the loop. In the loop statement, there are two steps:

1. A loop inputs a test case into array a and accumulates the number of elements n in the test case. 0 marks the end of the test case.
2. A double loop enumerates all pairs of $a[i]$ and $a[j]$ ($0 <= i < n - 1, i + 1 <= j < n$) in the test case and determines whether ($a[i]*2 == a[j] \parallel a[j]*2 == a[i]$) holds.

Program

```
#include <iostream>          // Preprocessor Directive
using namespace std;         // Using C++ standard library
int main()                   //Main function
{
    int i, j, n, count, a[20];      //  Integer variables i,
j, n, count and array a
```

```
    cin>>a[0];                          // Input the first element
    while(a[0]!=-1)                     // If it is not the end of input,
input a new test case
    {   n=1;                            // Input array a
        for( ; ; n++)
        {
          cin>>a[n];
          if (a[n]==0) break;
        }
        count=0;            // Determine how many items in each list are
twice some other item
        for (i=0; i<n-1; i++)                    // Enumerate all pairs
        {
          for (j=i+1; j<n; j++)
          {
            if (a[i]*2==a[j] || a[j]*2==a[i])    // Accumulation
            count++;
          }
        }
        cout<<count<<endl;              // Output the result
        cin>>a[0];                      // Input the first element of
next test case
    }
    return 0;
}
```

In this problem, the number of test cases and the size of a test case are unknown. Normally, a double-loop statement is used for the program structure: the outer loop is used to enumerate every test case, and the inner loop is used to deal with a test case.

In some problems, if the size of the test data is larger, all the test cases are dealt with by the same method, and the result area is known, its time complexity can be improved by an offline method. First, all solutions within the specified range are calculated and stored in a constant array. Then the program deals with the constant array directly for each test case. It can avoid duplication of computing.

1.2.2 Sum of Consecutive Prime Numbers

Some positive integers can be represented by a sum of one or more consecutive prime numbers. How many such representations does a given positive integer have? For example, the integer 53 has two representations 5 + 7 + 11 + 13 + 17 and 53. The integer 41 has three representations: 2 + 3 + 5 + 7 + 11 + 13, 11 + 13 + 17, and 41. The integer 3 has only one representation, which is 3. The integer 20 has no such representations. Note that summands must be consecutive prime numbers, so neither 7 + 13 nor 3 + 5 + 5 + 7 is a valid representation for the integer 20. Your mission is to write a program that reports the number of representations for the given positive integer.

Input

The input is a sequence of positive integers, each in a separate line. The integers are between 2 and 10,000, inclusive. The end of the input is indicated by a zero.

Output

The output should be composed of lines each corresponding to an input line, except the last zero. An output line includes the number of representations for the input integer as the sum of one or more consecutive prime numbers. No other characters should be inserted in the output.

Sample Input	Sample Output
2	1
3	1
17	2
41	3
20	0
666	0
12	1
53	2
0	

Source: ACM Japan 2005.

IDs for online judges: POJ 2739, UVA 3399.

Analysis

Because the program needs to deal with consecutive prime numbers for each test case, and the upper limit of prime numbers is 10,000, the offline method can be used to solve the problem.

First, all prime numbers less than 10,001 are obtained and stored in array *prime*[1 .. *total*] in ascending order.

Then we deal with the test cases one by one:

Suppose the input number is *n*; the sum of consecutive prime numbers is *cnt*; the number of representations for *cnt* == *n* is *ans*.

A double loop is used to get the number of representations for *n*:

■ The outer loop *i*: *for*(int *i* = 0; *n* >= *prime*[*i*]; *i*++) enumerates all possible minimum *prime*[*i*].
■ The inner loop *j*: *for*(int *j* = *i*; *j* < *total* && *cnt* < *n*; *j*++), *cnt* += *prime*[*j*], is to calculate the sum of consecutive prime numbers. If *cnt* ≥ *n*, then the loop ends, and if *cnt* == *n*, then the number of representations is *ans*++.

When the outer loop ends, *ans* is the solution to the test case.

Program

```
#include <iostream>           // Preprocessor Directive
using namespace std;          // Using C++ Standard Library
const int maxp = 2000, n = 10000;  // Set the size of prime array and
the upper limit of  prime numbers
int prime[maxp], total = 0;       // Initialization
```

```
bool isprime(int k)                   // Determine whether k is a
prime number or not
{
    for (int i = 0; i < total; i++)
        if (k % prime[i] == 0)
            return false;
    return true;
}
int main(void)                        // Main Function
{
    for (int i = 2; i <= n; i++)         //  get all prime numbers
less than 10001
        if (isprime(i))
            prime[total++] = i;
    prime[total] = n + 1;
    int m;
    cin >> m;                         // Input the first positive integer
    while (m) {
        int ans = 0;                       // Initialization
        for (int i = 0; m >= prime[i]; i++) { // Enumerate the least
prime number
            int cnt = 0;                   // Calculate the sum of
consecutive prime numbers
            for (int j = i; j < total && cnt < m; j++)
                cnt += prime[j];
            if (cnt == m)                  // if cnt==n, then ++ans
                ++ans;
        }
        cout << ans << endl;           // Output the result
        cin >> m;                      // Input the next positive integer
    }
    return 0;
}
```

1.3 Precision of Real Numbers

In some cases, we need to deal with real numbers and real arithmetics to solve problems, such as judging whether two real numbers are equal, and so on. For a programming language, precision of real numbers is limited. And sometime programs are required to meet requirements for accuracy errors of real numbers. If the program can't deal with such details well, it will lead to the wrong answer even though its algorithm is correct.

1.3.1 I Think I Need a Houseboat

Fred Mapper is considering purchasing some land in Louisiana to build his house on. In the process of investigating the land, he learned that the state of Louisiana is actually shrinking by 50 square miles each year, due to erosion caused by the Mississippi River. Since Fred is hoping to live in this house for the rest of his life, he needs to know if his land is going to be lost to erosion.

After doing more research, Fred has learned that the land that is being lost forms a semicircle. This semicircle is part of a circle centered at (0, 0), with the line that bisects the circle being the *X*

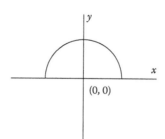

Figure 1.1 The land that is being lost forms a semicircle.

axis. Locations below the X axis are in the water. The semicircle has an area of 0 at the beginning of year 1. (The semicircle is illustrated in Figure 1.1.).

Input

The first line of input will be a positive integer indicating how many data sets will be included (N).

Each of the next N lines will contain the X and Y Cartesian coordinates of the land Fred is considering. These will be floating-point numbers measured in miles. The Y coordinate will be nonnegative. (0, 0) will not be given.

Output

For each data set, a single line of output should appear. This line should take the form of

Property N: This property will begin eroding in year Z.

where N is the data set (counting from 1) and Z is the first year (start from 1) this property will be within the semicircle AT THE END OF YEAR Z. Z must be an integer. After the last data set, this should print out "END OF OUTPUT."

Sample Input	Sample Output
2	Property 1: This property will begin eroding in year 1.
1.0 1.0	Property 2: This property will begin eroding in year 20.
25.0 0.0	END OF OUTPUT.

Source: ACM Mid-Atlantic United States 2001.

Note: No property will appear exactly on the semicircle boundary: it will be either inside or outside. This problem will be judged automatically. Your answer must match exactly, including the capitalization, punctuation, and white space. This includes the periods at the ends of the lines. All locations are given in miles.

IDs for online judges: POJ 1005, ZOJ 1049, UVA 2363.

Analysis

The number of test cases n is given. Therefore, a *for* repetition statement is used to deal with all test cases. Each test case contains only X and Y Cartesian coordinates. The ith test case (X_i, Y_i) and the center of the circle (0, 0) constitute the semicircle that will be eroded. Each year 50 square miles of land is eroded. And the number of years is an integer. When (X_i, Y_i) is in water, the number of years must be the least integer that is greater than

$$\frac{\text{Area of the semicircle}}{50}$$

and function *ceil*(x) is used to round up the fare.

Program

```
#include <stdio.h>                    // Preprocessor Directive
#include <math.h>
#define M_PI 3.14159265
int num_props;                        // The number of test cases
float x, y;                            // X and Y Cartesian coordinates
int i;
double calc;                          // The area of the semicircle/50
int years;                            // The number of years
int main( )                  // Main function
{
    scanf("%d", &num_props);          // Input the number of test cases
    for (i = 1; i <= num_props; i++)
    {
        scanf("%f %f", &x, &y);        // Input the i-th test case
        calc = (x*x + y*y)* M_PI / 2 / 50; // Calculate the area of the
semi-circle/50
        years = ceil(calc);
        printf("Property %d: This property will begin eroding in year %d.n",
i, years); //Output
    }
    printf("END OF OUTPUT.n");
}
```

In real arithmetic, sometimes we need to determine whether real number x and real number y are equal. Using $y - x == 0$ as the condition may result in error of precision. The method avoiding error of precision is to set a constant of precision *delta*. If $|y - x| < delta$, then we can judge x and y are equal. The hangover problem (Section 1.4.1) shows such an example.

1.4 Improving Time Complexity by Dichotomy

In some cases, the data area of a problem is an ordered interval. Dichotomy is used to divide the interval into two subintervals and then determine if the process of computation is in the left subinterval or the right subinterval. If the solution isn't obtained, then repeat the above steps. For a problem whose time complexity is O(n), if dichotomy can be used to solve it, its time complexity can be improved to O($\log_2(n)$).

Dichotomy is used in many algorithms, such as binary search, recursive halving method, quick sort, merge sort, binary search tree, and segment tree. Among these methods, binary search and recursive halving method are relatively simple algorithms.

The idea for binary search is as follows: Suppose the data area is an interval in ascending order. The search begins by comparing x with the number in the middle of the interval. If x equals this number, the search terminates. If x is smaller than the number, then we need only search in the left half; if x is greater than the number, then we need only search in the right half. We repeat the above steps until the search ends.

1.4.1 Hangover

How far can you make a stack of cards overhang a table? If you have one card, you can create a maximum overhang of half a card length. (We're assuming that the cards must be perpendicular to the table.) With two cards, you can make the top card overhang the bottom one by half a card length, and the bottom one overhang the table by a third of a card length, for a total maximum overhang of $1/2 + 1/3 = 5/6$ card lengths. In general, you can make n cards overhang by $1/2 + 1/3 + 1/4 + \ldots + 1/(n + 1)$ card lengths, where the top card overhangs the second by $1/2$, the second overhangs the third by $1/3$, the third overhangs the fourth by $1/4$, and so on, and the bottom card overhangs the table by $1/(n + 1)$. This is illustrated in Figure 1.2.

Input

The input consists of one or more test cases, followed by a line containing the number 0.00 that signals the end of the input. Each test case is a single line containing a positive floating-point number c whose value is at least 0.01 and at most 5.20; c will contain exactly three digits.

Output

For each test case, output the minimum number of cards necessary to achieve an overhang of at least c card lengths. Use the exact output format shown in the examples.

Sample Input	Sample Output
1.00	3 card(s)
3.71	61 card(s)
0.04	1 card(s)
5.19	273 card(s)
0.00	

Source: ACM Mid-Central United States 2001.

IDs for online judges: POJ 1003, UVA 2294.

Analysis

The problem's data area is little. Therefore, first lengths that cards achieve are calculated, and the length is at most 5.20 card lengths. Suppose the *total* is the number of cards and *len[i]* is the length

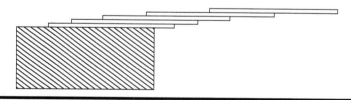

Figure 1.2   A stack of cards overhangs a table.

that i cards achieve. That is, $len[i] = len[i - 1] + 1/(i + 1)$, where $i \geq 1$ and $len[0] = 0$. Obviously, array len is in ascending order.

Because elements of len and x are real numbers, the accuracy error must be controlled. Suppose $delta = 1e - 8$, and function $zero(x)$ marks x is a positive real number, a negative real number, or a zero. Function $zero(x)$ is defined as follows:

$$zero(x) = \begin{cases} 1 & x > delta \\ -1 & x < -delta \\ 0 & otherwise \end{cases}$$

Initially $len[0] = 0$. Array len can be obtained through the following loop:

```
for (total=1; zero(len[total-1]-5.20)<0; total++)
    len[total]=len[total-1]+1.0/double(total+1);
```

After array len is obtained, the program inputs the first test data x and enters the loop of $while(zero(x))$. In each loop, dichotomy is used to get the minimum number of cards necessary to achieve an overhang of at least x card lengths, and then the next test data x is input. The loop terminates when $x = 0.00$.

The procedure of dichotomy is as follows:

The initial interval $[l, r] = [1, total]$ and $mid = [(l + r)/2]$. If $zero(len[mid] - x) < 0$, then search the right half $(l = mid)$; otherwise, search the left half $(r = mid)$. Repeat the above steps in interval $[l, r]$ until $l + 1 \geq r$. r is the minimum number of cards.

Program

```
#include <iostream>          // Preprocessor Directive
using namespace std;         // Using C++ Standard Library
const int maxn = 300;         // Size of array len
const double delta = 1e-8;     // Set the accuracy error
int zero(double x)            // In the area of accuracy error delta, if x
is a negative real number less than 0, then return -1; if x is a positive
real number larger than 0, then return 1; and if x is 0, then return 0.
{
    if (x < -delta)
        return -1;
    return x > delta;
}
int main(void)              // Main Function
{
    double len[maxn];        // Define array len and the length of len
    int total;
    len[0] = 0.0;            // Calculate array len, and len[i] is the length
that i cards achieve
    for (total = 1; zero(len[total - 1] - 5.20) < 0; total++)
        len[total] = len[total - 1] + 1.0 / double(total + 1);
    double x;
    cin >> x;                // Input the first test case x
    while (zero(x)) {        // Using dichotomy to get the minimum number
of cards necessary to achieve an overhang of at least x card lengths.
```

```
        int l, r;
        l = 0;              // Set left pointer l and right pointer r for
the interval
        r = total;
        while (l + 1 < r) {
            int mid = (l + r) / 2;
            if (zero(len[mid] - x) < 0)      // If the middle value is
less than x, then search the right half, else search the left half.
                l = mid;
            else
                r = mid;
        }
        cout << r << " card(s)" << endl;     // Output the minimum number
of cards
        cin >> x;                            //Input the next test case
    }
    return 0;
}
```

Dichotomy can be used not only in a data search, but also in function calculation. Suppose there are variables x_1, x_2, and x_3 and function $x_1 = f(x_2, x_3)$ holds. The recursive halving method can be used to calculate x_3 when x_1 and x_2 are known. The method is as follows:

Halving is to halve the data area of a problem (such as the data area of x_3), and the property of the problem (such as $x_1 = f(x_2, x_3)$) is not changed. Suppose the size of data area for the problem is n. We can first make use of some methods to change the original problem into c subproblems with half of the data area (c is a constant, is related to the problem, and is not related to the data area), and then solve the problem by solving subproblems whose size of data area is $n/2$. Properties for these subproblems are the same as those for the original problem, but the size of the data area for these subproblems is smaller.

Recursion is to repeat the above halving steps. A problem whose size of data area is $n/2$ is changed into c subproblems whose size is $n/4$, and so on. Repeat the above steps until the subproblems can be solved easily.

1.4.2 Humidex

The humidex is a measurement used by Canadian meteorologists to reflect the combined effect of heat and humidity. It differs from the heat index used in the United States in using dew point rather than relative humidity.

When the temperature is 30°C (86°F) and the dew point is 15°C (59°F), the humidex is 34 (note that humidex is a dimensionless number, but the number indicates an approximate temperature in Celsius). If the temperature remains 30°C and the dew point rises to 25°C (77°F), the humidex rises to 42.3.

The humidex tends to be higher than the U.S. heat index at equal temperature and relative humidity.

The current formula for determining the humidex was developed by J.M. Masterton and F.A. Richardson of Canada's Atmospheric Environment Service in 1979.

According to the Meteorological Service of Canada, a humidex of at least 40 causes "great discomfort" and above 45 is "dangerous." When the humidex hits 54, heat stroke is imminent.

The record humidex in Canada occurred on June 20, 1953, when Windsor, Ontario, hit 52.1. (The residents of Windsor would not have known this at the time, since the humidex had yet to be invented.) More recently, the humidex reached 50 on July 14, 1995, in both Windsor and Toronto.

The humidex formula is as follows:

humidex = temperature + h
$h = (0.5555)*(e - 10.0)$
$e = 6.11*exp$ [5417.7530*((1/273.16) − (1/(dewpoint + 273.16)))]

where $exp(x)$ is 2.718281828 raised to the exponent x.

While humidex is just a number, radio announcers often announce it as if it were the temperature, for example, "It's 47° out there ... with the humidex." Sometimes weather reports give the temperature and dew point, or the temperature and humidex, but rarely do they report all three measurements. Write a program that, given any two of the measurements, will calculate the third.

You may assume that for all inputs, the temperature, dew point, and humidex are all between −100°C and 100°C.

Input

Input will consist of a number of lines. Each line except the last will consist of four items separated by spaces: a letter, a number, a second letter, and a second number. Each letter specifies the meaning of the number that follows it and will be either T, indicating temperature; D, indicating dew point; or H, indicating humidex. The last line of input will consist of the single letter E.

Output

For each line of input except the last, produce one line of output. Each line of output should have the form:

T number D number H number

where the three numbers are replaced with the temperature, dew point, and humidex. Each value should be expressed rounded to the nearest tenth of a degree, with exactly one digit after the decimal point. All temperatures are in degrees Celsius.

Sample Input	Sample Output
T 30 D 15	T 30.0 D 15.0 H 34.0
T 30.0 D 25.0	T 30.0 D 25.0 H 42.3
E	

Source: Waterloo Local Contest, July 14, 2007.

ID for online judge: POJ 3299.

Analysis

Based on the humidex formula *humidex = temperature + h*, *h* is proportional to the dew point. If the dew point and temperature (or humidex) are known, the value of *h* can be inferred, and the humidex or temperature can be calculated by the humidex formula. If temperature and humidex are known, the recursive halving method can be used to calculate the dew point.

The program sets an initial value 0 to the dew point and enters a loop: the initial value of the increment for dew point is 100; halve the increment value each time as the loop is performed. If the humidex obtained from the formula is larger than the announced humidex, then the value of the dew point decreases an increment (i.e., $h\searrow$; decrease the humidex to be close to the announced humidex); otherwise, the value of the dew point increases an increment (i.e., $h\nearrow$; increase the humidex to be close to the announced humidex). The repetition condition is the increment value greater than 0.0001. When the loop ends, the dew point is the answer.

Program

```
#include <stdio.h>                    // Preprocessor Directive
#include <math.h>
#include <assert.h>
char a,b;                            //Two characters for Test Mark
double A,B,temp,hum,dew;
double dohum(double tt, double dd){ // Calculate humidex based on
temperature tt and dew point dd
   double e = 6.11 * exp (5417.7530 * ((1/273.16) - (1/(dd+273.16))));
   double h = (0.5555)*(e - 10.0);
   return tt + h;                   // Return humidex
}
double dotemp( ){        // Calculate temperature based on dew point dew
and humidex hum
   double e = 6.11 * exp (5417.7530 * ((1/273.16) - (1/(dew+273.16))));
   double h = (0.5555)*(e - 10.0);
   return hum - h;                  // Return temperature
}
double dodew( ){                //Calculate dew point based on
temperature temp and humidex hum
   double x = 0;                 // Initialization of dew point and
increment
   double delta=100;
//Loop; Halve the increment value each time the loop is performed. If
humidex gotten from the formula is larger than the announced humidex,
then the value of dew point decrease an increment; otherwise the value of
dew point increase an increment. Repeat the procedure until the increment
value delta<=0.0001.
   for (delta=100; delta>.00001; delta *=.5) {
      if (dohum(temp, x)>hum) x -= delta;
      else x += delta;
   }
   return x;                       //Return dew point x
}
int main( )                            // main function
{     //Loop: each loop inputs two measurements and loop-end condition
is 'E'.

   while (4 == scanf(" %c %lf %c %lf",&a,&A,&b,&B) && a != 'E'){
      temp = hum = dew = -99999;  // Initialization of temperature,
humidex and dew point
         if (a == 'T') temp = A;   // The first measurement is temperate.
         if (a == 'H') hum = A;    // The first measurement is humidex.
         if (a == 'D') dew = A;    // The first measurement is dew point.
         if (b == 'T') temp = B;   // The second measurement is temperate.
         if (b == 'H') hum = B;    // The second measurement is humidex.
         if (b == 'D') dew = B;    // The second measurement is dew point.
```

```
        if (hum == -99999) hum = dohum(temp, dew);// Calculate humidex
based on temperate and dew point.
        if (dew == -99999) dew=dodew( );   // Calculate dew point based on
temperate and humidex.
        if (temp == -99999) temp = dotemp( );   // Calculate temperate
based on humidex and dew point.
        printf("T %0.1lf D %0.1lf H %0.1lfn",temp, dew,hum);  //Output
temperate, dew point and humidex.
    }
    assert(a == 'E');                       //  Loop-end condition is 'E'.
}
```

1.5 Problems

1.5.1 Sum

Your task is to find the sum of all integer numbers lying between 1 and N inclusive.

Input

The input consists of a single integer N that is not greater than 10,000 by its absolute value.

Output

Write a single integer number that is the sum of all integer numbers lying between 1 and N inclusive.

Sample Input	Sample Output
–3	–5

Source: ACM 2000, Northeastern European Regional Programming Contest (test tour).

ID for online judge: Ural 1068.

Hint

Based on the summation formula of arithmetic progression $s = 1 + 2 + ... N$, if N is an integer larger than 0, then $s = [(1 + N)/2]*N$; otherwise, $s = [(1 – N)/2]*N + 1$.

1.5.2 Specialized Four-Digit Numbers

Find and list all four-digit numbers in decimal notation that have the property that the sum of their four digits equals the sum of their digits when represented in hexadecimal (base 16) notation and also equals the sum of their digits when represented in duodecimal (base 12) notation.

For example, the number 2991 has the sum of (decimal) digits $2 + 9 + 9 + 1 = 21$. Since 2991 = $1*1728 + 8*144 + 9*12 + 3$, its duodecimal representation is 1893_{12}, and these digits also sum up to 21. But in hexadecimal, 2991 is BAF_{16}, and $11 + 10 + 15 = 36$, so 2991 should be rejected by your program.

The next number (2992), however, has digits that sum to 22 in all three representations (including $BB0_{16}$), so 2992 should be on the listed output. (We don't want decimal numbers with fewer than four digits—excluding leading zeros—so that 2992 is the first correct answer.)

Input

There is no input for this problem.

Output

Your output is to be 2992 and all larger four-digit numbers that satisfy the requirements (in strictly increasing order), each on a separate line, with no leading or trailing blanks, ending with a new-line character. There are to be no blank lines in the output. The first few lines of the output are shown below:

Sample Input	Sample Output
	2992
	2993
	2994
	2995
	2996
	2997
	2998
	2999
	...

Source: ACM Pacific Northwest 2004.

IDs for online judges: POJ 2196, ZOJ 2405, UVA 3199.

Hint

First, function *calc*(*k*, *b*) is designed to calculate and return the sum of digits of number *k* represented in base *b*. Then every number *i* in [2992 ... 9999] is enumerated: if $calc(i, 10) == calc(i, 12) == calc(i, 16)$, then output *i*.

1.5.3 Quicksum

A checksum is an algorithm that scans a packet of data and returns a single number. The idea is that if the packet is changed, the checksum will also change, so checksums are often used for detecting transmission errors, validating document contents, and in many other situations where it is necessary to detect undesirable changes in data.

For this problem, you will implement a checksum algorithm called quicksum. A quicksum packet allows only uppercase letters and spaces. It always begins and ends with an uppercase letter. Otherwise, spaces and letters can occur in any combination, including consecutive spaces.

A quicksum is the sum of the products of each character's position in the packet times the character's value. A space has a value of zero, while letters have a value equal to their position in the alphabet. So, A = 1, B = 2, and so on, through Z = 26. Here are example quicksum calculations for the packets "ACM" and "MID CENTRAL":

ACM: 1*1 + 2*3 + 3*13 = 46
MID CENTRAL: 1*13 + 2*9 + 3*4 + 4*0 + 5*3 + 6*5 + 7*14 + 8*20 + 9*18 + 10*1 + 11*12
= 650

Input

The input consists of one or more packets followed by a line containing only # that signals the end of the input. Each packet is on a line by itself, does not begin or end with a space, and contains from 1 to 255 characters.

Output

For each packet, output its quicksum on a separate line in the output.

Sample Input	Sample Output
ACM	46
MID CENTRAL	650
REGIONAL PROGRAMMING	4690
CONTEST	49
ACN	75
A C M	14
ABC	15
BBC	
#	

Source: ACM Mid-Central United States 2006.

IDs for online judges: POJ 3094, ZOJ 2812, UVA 3594.

Hint
Function *value(c)* is implemented as follows: if character $c ==$ '_', then return 0; otherwise, return the corresponding value of c: $c - $ 'A' + 1.

The process is a loop. Each loop inputs a test case and calculates its *Quicksum*.

First, the location of character c and *Quicksum* are initialized 0, and string s is initialized NULL. Repeatedly input character c and add c into s until c is EOF or '\n'. If s is "#," the program ends.

$$Quicksum = \sum_{i=0}^{s.size-1} (i+1) * value(s[i])$$

1.5.4 A Contesting Decision

Judging a programming contest is hard work, with demanding contestants, tedious decisions, and monotonous work—not to mention the nutritional problems of spending 12 hours with only donuts, pizza, and soda for food. Still, it can be a lot of fun.

Software that automates the judging process is a great help, but the notorious unreliability of some contest software makes people wish that something better were available. You are part of a group trying to develop better, open-source, contest management software, based on the principle of modular design.

Your component is to be used for calculating the scores of programming contest teams and determining a winner. You will be given the results from several teams and must determine the winner.

Scoring

There are two components to a team's score. The first is the number of problems solved. The second is penalty points, which reflect the amount of time and incorrect submissions made before the problem is solved. For each problem solved correctly, penalty points are charged equal to the time at which the problem was solved plus 20 minutes for each incorrect submission. No penalty points are added for problems that are never solved.

So if a team solved problem 1 on their second submission at 20 minutes, they are charged 40 penalty points. If they submit problem 2 three times, but do not solve it, they are charged no penalty points. If they submit problem 3 once and solve it at 120 minutes, they are charged 120 penalty points. Their total score is two problems solved with 160 penalty points.

The winner is the team that solves the most problems. If teams tie for solving the most problems, then the winner is the team with the fewest penalty points.

Input

For the programming contest your program is judging, there are four problems. You are guaranteed that the input will not result in a tie between teams after counting penalty points.

```
Line 1: < nTeams >
Line 2: n+1 < Name > < p1Sub > < p1Time > < p2Sub > < p2Time > ...
< p4Time >
```

The first element on the line is the team name, which contains no white space. Following that, for each of the four problems, is the number of times the team submitted a run for that problem and the time at which it was solved correctly (both integers). If a team did not solve a problem, the time will be zero. The number of submissions will be at least one if the problem was solved.

Output

The output consists of a single line listing the name of the team that won, the number of problems they solved, and their penalty points.

Sample Input	Sample Output
4	Penguins 3 475
Stars 2 20 5 0 4 190 3 220	
Rockets 5 180 1 0 2 0 3 100	
Penguins 1 15 3 120 1 300 4 0	
Marsupials 9 0 3 100 2 220 3 80	

Source: ACM Mid-Atlantic 2003.

IDs for online judges: POJ 1581, ZOJ 1764, UVA 2832.

Hint

Suppose the name of the winner is *wname*, the number of problems that winner solved is *wsol*, and the winner's penalty points is *wpt*; the name of the current team is *name*, the number of problems that the current team solved is *sol*, and the current team's penalty points is *pt*. The submission number of the current problem is *sub*, and the time at which of the current problem is solved is *time*.

If the problem is solved (*time* > 0), then we accumulate the number of problems the current team solved (++*sol*) and compute the current team's penalty points *pt* (*pt* += (*sub* − 1)*20 + *time*).

After we deal with a team's case, if the number of problems the current team solved is the most, or the current team and other teams all solved the most number of problems, and the current team is with the fewest penalty points, that is, (*sol* > *wsol* || (*sol* == *wsol* && *wpt* > *pt*)) holds, then the current team is set as winner, and its team name, the number of solved problems, and its penalty points are recorded, that is, *wname* = *name*, *wsol* = *sol*, *wpt* = *pt*.

Obviously, after we deal with all teams' cases, *wname*, *wsol*, and *wpt* are solutions to the problem.

1.5.5 Dirichlet's Theorem on Arithmetic Progressions

If *a* and *d* are relatively prime positive integers, the arithmetic sequence beginning with *a* and increasing by *d*, that is, *a*, *a* + *d*, *a* + 2*d*, *a* + 3*d*, *a* + 4*d*, …, contains infinitely many prime numbers. This fact is known as Dirichlet's theorem on arithmetic progressions, which had been conjectured by Johann Carl Friedrich Gauss (1777–1855) and was proved by Johann Peter Gustav Lejeune Dirichlet (1805–1859) in 1837.

For example, the arithmetic sequence beginning with 2 and increasing by 3, that is,

2, 5, 8, 11, 14, 17, 20, 23, 26, 29, 32, 35, 38, 41, 44, 47, 50, 53, 56, 59, 62, 65, 68, 71, 74, 77, 80, 83, 86, 89, 92, 95, 98, ….

contains infinitely many prime numbers:

2, 5, 11, 17, 23, 29, 41, 47, 53, 59, 71, 83, 89, …

Your mission, should you choose to accept it, is to write a program to find the *n*th prime number in this arithmetic sequence for given positive integers *a*, *d*, and *n*.

Input

The input is a sequence of data sets. A data set is a line containing three positive integers *a*, *d*, and *n* separated by a space. *a* and *d* are relatively prime. You may assume $a \le 9307$, $d \le 346$, and $n \le 210$.

The end of the input is indicated by a line containing three zeros separated by a space. It is not a data set.

Output

The output should be composed of as many lines as the number of the input data sets. Each line should contain a single integer and should never contain extra characters.

The output integer corresponding to a data set *a*, *d*, *n* should be the *n*th prime number among those contained in the arithmetic sequence beginning with *a* and increasing by *d*.

For your information, it is known that the result is always less than 10^6 (1 million) under this input condition.

Sample Input	Sample Output
367 186 151	92809
179 10 203	6709
271 37 39	12037
103 230 1	103
27 104 185	93523
253 50 85	14503
1 1 1	2
9075 337 210	899429
307 24 79	5107
331 221 177	412717
259 170 40	22699
269 58 102	25673
0 0 0	

Source: ACM Japan 2006, Domestic.

ID for online judge: POJ 3006.

Hint

A test case consists of integers a, d, and n in an arithmetic sequence, and the end of the input is indicated by a line containing 0 0 0. Therefore, a *while* repetition statement is used for test cases. After the first a, d, and n are input, the program enters the *while*(a || d || n) loop. In the loop body, the steps are as follows:

1. Initialize the number of prime numbers *cnt* 0.
2. Construct an arithmetic sequence with n prime numbers through the loop statement *for*($m = a$; *cnt* < n; m += d). The control variable m is initialized with a, and the loop-continuation condition is *cnt* < n. In each loop, if m is a prime number, then *cnt*++, and d is added to control variable m.
3. Output the nth prime number $m - d$. (Because of the *for* loop, output $m - d$.)
4. Input a, d, and n for the next arithmetic sequence.

1.5.6 The Circumference of the Circle

To calculate the circumference of a circle seems to be an easy task—provided you know its diameter. But what if you don't?

You are given the Cartesian coordinates of three noncollinear points in the plane.

Your job is to calculate the circumference of the unique circle that intersects all three points.

Input

The input file will contain one or more test cases. Each test case consists of one line containing six real numbers, $x_1, y_1, x_2, y_2, x_3, y_3$, representing the coordinates of the three points. The diameter of the circle determined by the three points will never exceed 1 million. Input is terminated by the end of the file.

Output

For each test case, print one line containing one real number telling the circumference of the circle determined by the three points. The circumference is to be printed accurately rounded to two decimals. The value of π is approximately 3.141592653589793.

Sample Input	Sample Output
0.0 –0.5 0.5 0.0 0.0 0.5	3.14
0.0 0.0 0.0 1.0 1.0 1.0	4.44
5.0 5.0 5.0 7.0 4.0 6.0	6.28
0.0 0.0 –1.0 7.0 7.0 7.0	31.42
50.0 50.0 50.0 70.0 40.0 60.0	62.83
0.0 0.0 10.0 0.0 20.0 1.0	632.24
0.0 –500000.0 500000.0 0.0 0.0	3141592.65
500000.0	

Source: Ulm Local Contest 1996.

IDs for online judges: POJ 2242, ZOJ 1090.

Hint
The key to the problem is to find the center of a circle that intersects all three points. Suppose the Cartesian coordinates of three points are (x_0, y_0), (x_1, y_1), and (x_2, y_2), and the Cartesian coordinates of the center of the circle are (x_m, y_m). There are two solutions.

Determinant

Calculate the Cartesian coordinates of the center of the circle that intersects all three points:

$$x_m = \frac{x_1 + x_2}{2} + (y_2 - y_1) * \frac{\begin{vmatrix} y_1 - y_0 & \dfrac{x_2 - x_0}{2} \\ x_0 - x_1 & \dfrac{y_2 - y_0}{2} \end{vmatrix}}{\begin{vmatrix} y_1 - y_0 & y_1 - y_2 \\ x_0 - x_1 & x_2 - x_1 \end{vmatrix}}, \quad y_m = \frac{y_1 + y_2}{2} + (x_1 - x_2) * \frac{\begin{vmatrix} y_1 - y_0 & \dfrac{x_2 - x_0}{2} \\ x_0 - x_1 & \dfrac{y_2 - y_0}{2} \end{vmatrix}}{\begin{vmatrix} y_1 - y_0 & y_1 - y_2 \\ x_0 - x_1 & x_2 - x_1 \end{vmatrix}}$$

Based on this, the radius of the unique circle

$$r = \sqrt{(x_m - x_0)^2 + (y_m - y_0)^2}$$

and $2\pi r$ is the circumference of the unique circle that intersects all three points: (x_0, y_0), (x_1, y_1), and (x_2, y_2).

Theorem 1.1

The Cartesian coordinates of the center of the unique circle that intersects all three points (x_0, y_0), (x_1, y_1), and (x_2, y_2) are (x_m, y_m).

Proof

Suppose the Cartesian coordinate of the center of a circle is $P = (x_m, y_m)$; the perpendicular bisectors from P to $\overline{AB}$ and $\overline{BC}$ are $\overline{PN}$ and $\overline{PM}$, respectively. The point of intersection of $\overline{PN}$ and $\overline{AB}$ is N, and the point of intersection of $\overline{PM}$ and $\overline{BC}$ is M. Obviously, the Cartesian coordinates of M are $[(x_1 + x_2)/2, (y_1 + y_2)/2])$, and point $(y_2 - y_1, x_2 - x_1)$ is on $\overline{PM}$ (Figure 1.3).

Because $\overline{PM} \perp \overline{BC}$,

$$\frac{y_m - \dfrac{y_1 + y_2}{2}}{x_m - \dfrac{x_1 + x_2}{2}} * \frac{y_2 - y_1}{x_2 - x_1} = -1$$

Suppose

$$k = \frac{y_m - \dfrac{y_1 + y_2}{2}}{x_2 - x_1} = \frac{x_m - \dfrac{x_1 + x_2}{2}}{y_2 - y_1}$$

Now we need to prove

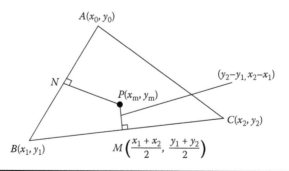

Figure 1.3   Three points and the center of the circle.

$$k = \frac{\begin{vmatrix} y_1 - y_0 & \dfrac{x_2 - x_0}{2} \\[2ex] x_0 - x_1 & \dfrac{y_2 - y_0}{2} \end{vmatrix}}{\begin{vmatrix} y_1 - y_0 & y_1 - y_2 \\ x_0 - x_1 & x_2 - x_1 \end{vmatrix}} \quad (*)$$

Because $\overline{PN} \perp \overline{AB}$,

$$\frac{y_m - \dfrac{y_0 + y_2}{2}}{x_m - \dfrac{x_0 + x_2}{2}} * \frac{y_1 - y_0}{x_1 - x_0} = -1$$

Because

$$x_m = \frac{x_1 + x_2}{2} + (y_2 - y_1) * k,$$

and

$$y_m = \frac{y_1 + y_2}{2} + (x_1 - x_2) * k$$

$$\frac{(x_1 - x_2)k + \dfrac{y_2 - y_0}{2}}{(y_2 - y_1)k + \dfrac{x_2 - x_0}{2}} * \frac{y_1 - y_0}{x_1 - x_0} = -1$$

holds. Therefore, (*) holds.

Elementary Geometry

Suppose $a = |\overline{AB}|, b = |\overline{BC}|, c = |\overline{CA}|$, and

$$p = \frac{a + b + c}{2}.$$

Based on Heron's formula,

$$s = \sqrt{p(p-a)(p-b)(p-c)}$$

the formula for calculating the area of a triangle

$$s = \frac{a * b * \sin(\angle ab)}{2}$$

and sine theorem

$$\frac{a}{\sin(\angle bc)} = \frac{b}{\sin(\angle ac)} = \frac{c}{\sin(\angle ab)} = \text{diameter of circumcircle } d$$

we can calculate the diameter of circumcircle $d = (a{*}b{*}c)/(2{*}s)$ and the circumference of circumcircle $l = d{*}\pi$.

1.5.7 Vertical Histogram

Write a program to read four lines of uppercase (i.e., all CAPITAL LETTERS) text input (no more than 72 characters per line) from the input file and print a vertical histogram that shows how many times each letter (but not blanks, digits, or punctuation) appears in the all-uppercase input. Format your output exactly as shown.

Input

Lines 1–4: Four lines of uppercase text, no more than 72 characters per line.

Output

Lines 1–?: Several lines with asterisks and spaces followed by one line with the uppercase alphabet separated by spaces. Do not print unneeded blanks at the end of any line. Do not print any leading blank lines.

Sample Input	Sample Output
THE QUICK BROWN FOX JUMPED OVER THE LAZY DOG. THIS IS AN EXAMPLE TO TEST FOR YOUR HISTOGRAM PROGRAM. HELLO!	<pre>*
*
*   *
*   *   *   *
*   *   *   *
*   *   *   *
*   *   *   *   *   *
*   *   *   *   *   *   *   *   *   *
*   *   *   *   * *   *   *   *   *   *   *   *
*   * * * * * *   * * * *   * * * *   * *
* *
A B C D E F G H I J K L M N O P Q R S T U V W X Y Z</pre>

Source: USACO, February 2003, Orange.

ID for online judge: POJ 2136.

Hint

The sequence of drawing a vertical histogram is from top to bottom and left to right. From top to bottom means dealing with every line in the frequency's descending order. From left to right means dealing with letters in the current line in the ordinal number's ascending order.

Suppose *cnt* is the frequency array for letters, where *cnt*[0] is the number of 'A', ..., *cnt*[25] is the number of 'Z', and

$$Maxc = \max_{0 \le i \le 25} \{cnt[i]\}$$

that is, *Maxc* is the height of the highest "pillar" in the vertical histogram. The algorithm is as follows:

1. Input a test case and count array *cnt*;
2. Getting the value of *Maxc*;
3. From the highest "pillar" in the vertical histogram, draw the vertical histogram from top to bottom. A repetition statement *for* (int *i* = 1; *i* ≤ *Maxc*; *i*++) implements it as follows:
   - Find the right boundary for the current line. That is, in array *cnt*, from 25 to 0, find the first letter whose serial number is *l*1–1, where *cnt*[*l*1–1] > *Maxc*–*i*.
   - For letters whose serial numbers are in [0..*l*1–1], if *cnt*[*j*] > *Maxc* – *i* (0 ≤ *j* ≤ l1 – 1), then output '*_'; otherwise, output '_ _'.
4. Output the last line 'A_B_..... _Z'.

1.5.8 Ugly Numbers

Ugly numbers are numbers whose only prime factors are 2, 3, or 5. The sequence 1, 2, 3, 4, 5, 6, 8, 9, 10, 12, ... shows the first 10 ugly numbers. By convention, 1 is included.

Given the integer *n*, write a program to find and print the *n*th ugly number.

Input

Each line of the input contains a positive integer *n* ($n \le 1500$). Input is terminated by a line with $n = 0$.

Output

For each line, output the *n*th ugly number. Don't deal with the line with $n = 0$.

Sample Input	Sample Output
1	1
2	2
9	10
0	

Source: New Zealand 1990, Division I.

IDs for online judges: POJ 1338, UVA 136.

Hint

An offline method is used to solve the problem. The first 1500 ugly numbers are calculated and stored in array $a[1 .. 1500]$.

Suppose the upper limit for the largest ugly number $limit = 1,000,000,000$. The outer loop (control variable i) enumerates multiples of 2. For each time, $i \leftarrow i*2$ is performed. The loop-continuation condition is $i < limit$. The middle loop (control variable j) enumerates multiples of 3. For each time, $j \leftarrow j*3$ is performed. The loop-continuation condition is $i*j < limit$. The inner loop (control variable k) enumerates multiples of 5. For each time, ugly number $i*j*k$ is stored in array a and $k \leftarrow k*5$ is performed. The loop-continuation condition is $i*j*k < limit$.

Then array a is sorted such that $a[x]$ is the xth large ugly number ($1 \leq x \leq 1500$).

1.5.9 Number Sequence

A single positive integer i is given. Write a program to find the digit located in the position i in the sequence of number groups $S_1 S_2 \dots S_k$. Each group S_k consists of a sequence of positive integer numbers ranging from 1 to k, written one after another.

For example, the first 80 digits of the sequence are as follows:
1121231234123451234561234567123456781234567891234567891012345678910111234567 8910

Input

The first line of the input file contains a single integer t ($1 \leq t \leq 10$), the number of test cases, followed by one line for each test case. The line for a test case contains the single integer i ($1 \leq i \leq 2147483647$).

Output

There should be one output line per test case containing the digit located in the position i.

Sample Input	Sample Output
2	2
8	2
3	

Source: ACM Tehran 2002, First Iran Nationwide Internet Programming Contest.

IDs for online judges: POJ 1019, ZOJ 1410.

Hint

First, two functions are implemented. The first function is to calculate the length for the first j groups (i.e., the number of digits for the first j groups) and is stored in an array. The second function is to return the digit located in the position l in a group S_m. Then dichotomy is used to find the group S_n containing the digit located in the position i. Finally, in the group S_n, the digit located in the position i is returned.

Chapter 2

Simple Simulation

In the real world, there are many problems that we can solve by simulating their processes. Such problems are called simulation problems. For these problems, solution procedures and rules are showed in problem descriptions. Programs must simulate procedures or implement rules based on descriptions.

Normally there are two kinds of simulations: stochastic simulation and process simulation.

Problems for stochastic simulation show or imply probabilities. Programmers make use of random functions and round functions to set the random value for a range, making the random value meet the probability as a parameter. Then programmers design the algorithm by simulating the mathematical model. Because of uncertainty, there are fewer problems for stochastic simulation in programming contests.

Problems for process simulation require programmers to design parameters for mathematical models, and to observe changes of states caused by parameters. Programmers design algorithms based on process simulation. Programs depend entirely on authenticity and correctness of the process simulation without any uncertainties.

This chapter focuses on process simulation. There are three kinds of process simulation:

1. Simulation of direct statement
2. Simulation by sieve method
3. Simulation by construction

2.1 Simulation of Direct Statement

For problems for simulation of direct statement, programmers are required to solve them by strictly following rules showed in the problems' descriptions. Programmers must read such problems carefully and simulate processes based on descriptions. A problem for simulation of direct statement gets harder as the number of rules increases. It causes the amount of code to grow and become more illegible.

2.1.1 Speed Limit

Bill and Ted are taking a road trip. But the odometer in their car is broken, so they don't know how many miles they have driven. Fortunately, Bill has a working stopwatch, so they can record their speed and the total time they have driven. Unfortunately, their record-keeping strategy is a little odd, so they need help computing the total distance driven. You are to write a program to do this computation.

For example, if their log shows

Speed in Miles per Hour	Total Elapsed Time in Hours
20	2
30	6
10	7

this means they drove 2 hours at 20 miles per hour, then $6 - 2 = 4$ hours at 30 miles per hour, then $7 - 6 = 1$ hour at 10 miles per hour. The distance driven is then $(2)(20) + (4)(30) + (1)(10) = 40 + 120 + 10 = 170$ miles. Note that the total elapsed time is always since the beginning of the trip, not since the previous entry in their log.

Input

The input consists of one or more data sets. Each set starts with a line containing an integer n, $1 \leq n \leq 10$, followed by n pairs of values, one pair per line. The first value in a pair, s, is the speed in miles per hour, and the second value, t, is the total elapsed time. Both s and t are integers, $1 \leq s \leq 90$ and $1 \leq t \leq 12$. The values for t are always in strictly increasing order. A value of -1 for n signals the end of the input.

Output

For each input set, print the distance driven, followed by a space, followed by the word *miles*.

Sample Input	Sample Output
3	170 miles
20 2	180 miles
30 6	90 miles
10 7	
2	
60 1	
30 5	
4	
15 1	
25 2	
30 3	
10 5	
−1	

Source: ACM Mid-Central United States 2004.

IDs for online judges: POJ 2017, ZOJ 2176, UVA 3059.

Analysis

This is a simple problem of direct statement. We can simulate the stopwatch's running to compute the total distance driven: if the last total elapsed time in hours is z, the current speed in miles per hour is x, and the current total elapsed time in hours is y, then the current distance driven is $(y - z)*x$, and we add it to the total distance driven.

Program

```cpp
#include <iostream>                       // Preprocessor Directive
using namespace std;                      // Using C++ Standard Library
int main()                                //Main Function
{
    int n, i, x, y, z, ans;
    // Multiple test cases are dealt with by a while loop statement
    while (cin >> n, n > 0)
    {
        ans = z = 0;
        // Simulate the stopwatch to calculate
        for (i = 0; i < n; i++)                       // Input and calculate the
current data set
        {
            cin >> x >> y;                    //Input the speed and the total
elapsed time
            ans += (y - z) * x;               // Accumulate the distance driven
            z = y;                            //Record the total elapsed time
        }
        cout << ans << " miles" << endl;   //Output the distance driven for
the current data set
    }
    return 0;
}
```

2.1.2 Ride to School

Many graduate students of Peking University are living on Wanliu Campus, which is 4.5 kilometers from the main campus—Yanyuan. Students in Wanliu have to either take a bus or ride a bike to go to school. Due to the bad traffic in Beijing, many students choose to ride a bike.

We may assume that all the students except "Charley" ride from Wanliu to Yanyuan at a fixed speed. Charley is a student with a different riding habit—he always tries to follow another rider to avoid riding alone. When Charley gets to the gate of Wanliu, he will look for someone who is setting off to Yanyuan. If he finds someone, he will follow that rider, or if not, he will wait for someone to follow. On the way from Wanliu to Yanyuan, at any time if a faster student surpasses Charley, he will leave the rider he is following and speed up to follow the faster one.

We assume the time that Charley gets to the gate of Wanliu is zero. Given the set-off time and speed of the other students, your task is to give the time when Charley arrives at Yanyuan.

Input

There are several test cases. The first line of each case is $N(1 \leq N \leq 10,000)$ representing the number of riders (excluding Charley). $N = 0$ ends the input. The following N lines are information of N different riders, in such format:

$$V_i \ [\text{TAB}] \ T_i$$

V_i is a positive integer ≤ 40, indicating the speed of the ith rider (kilometers per hour). T_i is the set-off time of the ith rider, which is an integer and counted in seconds. In any case, it is ensured that there always exists a nonnegative T_i.

Output

The output is one line for each case: the arrival time of Charley. Round up (ceiling) the value when dealing with a fraction.

Sample Input	Sample Output
4	780
20 0	771
25 −155	
27 190	
30 240	
2	
21 0	
22 34	
0	

Source: ACM Beijing 2004, Preliminary.

IDs for online judges: POJ 1922, ZOJ 2229.

Analysis

There is no mathematical formula to solve the problem. We can calculate the arrival time of Charley by simulating each student leaving from Wanliu to Yanyuan. For each test case, the time that Charley gets to the gate of Wanliu is zero. From it we calculate the arrival time of each student. Obviously, the earliest arrival time is the arrival time of Charley.

Suppose *min* is the earliest arrival time for the first $i - 1$ riders, the speed of the ith rider is v, and the set-off time of the ith rider is t. Then the time when the ith rider arrives at Yanyuan is $x = t + (4.5*3600)/v$. If $x < min$, then *min* is adjusted as x. Obviously, after the arrival time of all riders is calculated, *min* is the arrival time of Charley.

There is a trap in the test data. If T_i is a negative integer, we should neglect it. It doesn't affect the arrival time of Charley.

Program

```cpp
#include <iostream>              // Preprocessor Directive
#include <cmath>
using namespace std;            // Using C++ Standard Library
int main()                      //Main function
{
    const double DISTANCE = 4.50;     //The distance between Yanyuan and Wanliu
    while(true)                       //A while statement deals with test cases
    {
        int n;                        //The number of riders except Charley
        scanf("%d", &n);
```

```
        if (n == 0) break;              //Input ends
        double v, t, x, min = 1e100;    //min is initialized 10^100
        for(int i = 0; i < n; ++i)       // A while statement deals
with riders
        {
            scanf("%lf%lf", &v, &t);    // the speed and the set off time
of the i-th rider
// Calculate time x when the i-th rider arrives at Yanyuan. If x<min,
then min is adjusted to x.
            if (t >= 0 && (x = DISTANCE * 3600 / v + t) < min)
            min = x;
        }
        printf("%.0lf\n", ceil(min));   //Output the arrival time of
Charley
    }
    return 0;
}
```

2.2 Simulation by Sieve Method

The simulation by sieve method is to get constraints in the problem description, and such constraints constitute a sieve. And then all possible solutions are put in the sieve to filter out solutions that do not meet constraints from time to time. Finally, solutions settling in the sieve are solutions to the problem. The structure and idea for the simulation by sieve method is concise and clear, but also blind. Therefore, maybe the time efficiency is not good. The key to the simulation by sieve method is to find the constraints. Any errors and omissions will lead to failure. Because filtering rules do not need complex algorithm design, such problems are usually simple simulation problems.

2.2.1 Self-Numbers

In 1949, the Indian mathematician D.R. Kaprekar discovered a class of numbers called self-numbers. For any positive integer n, define $d(n)$ to be n plus the sum of the digits of n. (The d stands for digitadition, a term coined by Kaprekar.) For example, $d(75) = 75 + 7 + 5 = 87$. Given any positive integer n as a starting point, you can construct the infinite increasing sequence of integers n, $d(n)$, $d(d(n))$, $d(d(d(n)))$, …. For example, if you start with 33, the next number is $33 + 3 + 3 = 39$, the next is $39 + 3 + 9 = 51$, the next is $51 + 5 + 1 = 57$, and so on, and you generate the sequence

$$33, 39, 51, 57, 69, 84, 96, 111, 114, 120, 123, 129, 141, …$$

The number n is called a generator of $d(n)$. In the sequence above, 33 is a generator of 39, 39 is a generator of 51, 51 is a generator of 57, and so on. Some numbers have more than one generator; for example, 101 has two generators, 91 and 100. A number with no generators is a self-number. There are 13 self-numbers less than 100: 1, 3, 5, 7, 9, 20, 31, 42, 53, 64, 75, 86, and 97.

Input

There is no input for this problem.

Output

Write a program to output all positive self-numbers less than 10,000 in increasing order, one per line.

Sample Input	Sample Output
	1
	3
	5
	7
	9
	20
	31
	42
	53
	64
	|
	a lot more numbers
	|
	9903
	9914
	9925
	9927
	9938
	9949
	9960
	9971
	9982
	9993

Source: ACM Mid-Central United States 1998.

IDs for online judges: POJ 1316, ZOJ 1180, UVA 640.

Analysis

The simulation by sieve method is used to solve the problem. Suppose the sieve is array g, where $g[y] = x$ means y is a number in ascending sequence for x. Based on $d(x) = x +$ the sum of the digits of x, a subprogram *generate_sequence*(x) is to generate the ascending sequence $[d(x), d(d(x)), d(d(d(x))), ...]$ for x. Suppose x is the generation number for every number in the sequence:

$$g[d(x)] = g[d(d(x))] = g[d(d(d(x)))] = ... = x$$

If a number is in ascending sequence for x, it is not a self-number and should be sieved from sieve g. The process will repeat until the generated number ≥ 1000 or the generated number has been generated before ($g[x] \neq x$). If x has been generated, it is not a self-number.

The algorithm is as follows:

First, $g[i]$ is initialized as i ($1 \leq i \leq 1000$). Then *generate_sequence*(1) ... *generate_sequence*(1000) are called to calculate $g[1..1000]$. Finally, numbers left in the sieve, that is, $g[x]==x$, are self-numbers.

Program

```c
#include <stdio.h>           // Preprocessor Directive
#define N 10000              // All positive self-numbers less
than 10000
unsigned g[N];               //Array g is the sieve
unsigned sum_of_digits (unsigned n)   //Calculate the sum of the digits of n
{
  if (n < 10)
    return n;
  else
    return (n % 10) + sum_of_digits (n / 10);
}
void generate_sequence (unsigned n)   //Construct ascending sequence
for n.
{
  while (n < N)                        //n=10000 is the end condition
  {
    unsigned next=n+sum_of_digits(n);  //Calculate d[n]
    if (next >= N || g[next] != next)  //If d[n]>=N or d[n] is not a
self-number, return;
      return;
    g[next] = n;                       //put d[n] into ascending sequence for n.
    n = next;
  }
}
int main ( )
{
  unsigned n;
  for (n = 1; n < N; ++n)              //Initialization
    g[n] = n;
  for (n = 1; n < N; ++n)              //Calculate g[1..1000]
    generate_sequence (n);
  for (n = 1; n < N; ++n)              //Output all self-numbers
    if (g[n] == n)
      printf ("%un", n);
}
```

2.3 Construction Simulation

Construction simulation is a kind of relatively complex simulation method. It requires a complete and accurate mathematical model to represent and solve a problem. We need to design parameters of the model and calculate simulation results. Because such mathematical models represent objects and their relationships accurately, the efficiencies are relatively high.

2.3.1 Bee

In Africa, there is a very special species of bee. Every year, the female bees of the species give birth to one male bee, while the male bees give birth to one male bee and one female bee, and then they die.

Now scientists have accidentally found one "magical" female bee of the special species to the effect that she is immortal, but still able to give birth once a year as all the other female bees. The scientists would like to know how many bees there will be after N years. Write a program that helps them find the number of male bees and the total number of all bees after N years.

Input

Each line of input contains an integer N (≥ 0). Input ends with a case where $N = -1$. (This case should *not* be processed.)

Output

Each line of output should have two numbers, the first one the number of male bees after N years and the second one the total number of bees after N years. (The two numbers will not exceed 2^{32}.)

Sample Input	Sample Output
1	1 2
3	4 7
−1	

ID for online judge: UVA 11000.

Analysis

From the description of bees' breeding, it is a problem of process simulation. Because bees' breeding is based on rules, the corresponding mathematical model can be constructed. Therefore, it is also a problem of construction simulation.

There is only one integer for a test case, and −1 marks the end of input. After the first test case is input, there is a *while* repetition statement *while*($n > -1$). In the loop body, the calculation process is as follows:

1. Initialize the number of female bees a as 1 and the number of male bees b as 0. Because of the size of the operation, the type of a and b is long long.
2. Making a series of recurrences for i from 0 to $n - 1$. After $i + 1$ years, the number of female bees is the number of last year's male bees + 1, and the number of male bees is the number of last year's bees. Therefore, formulas are as follows:
   $c = 1 + b, d = a + b, a = c, b = d$
3. Output the number of male bees a and the number of bees $a + b$ after N years.
4. Input the next test case.

Program

```
#include <iostream>                          // Preprocessor Directive
using namespace std;                         // Using C++ Standard Library
int main(void)
{
    int n;
    cin >> n;                                //The number of years
    while (n > -1) {
        // Initialize the number of female bees a 1, and the number of
male bees b 0.
        long long a = 1;
        long long b = 0;
        for (int i = 0; i < n; i++) {        // a series of recurrences
            long long c, d;
            c = 1 + b;      //Calculate the number of female bees and the
number of male bees
```

```
            d = a + b;
            a = c;
            b = d;
        }
        // Output the number of male bees a and the number of bees a+b
after N years
        cout << b << ' ' << a + b << endl;
        cin >> n;                        //The next number of years
    }
    return 0;
}
```

The key to "construction simulation" is to find a mathematic model. Sometimes there are several mathematic models. We should select a suitable mathematic model based on its simulation efficiency and complexity of the program.

2.4 Problems

2.4.1 *Gold Coins*

The king pays his loyal knight in gold coins. On the first day of his service, the knight receives one gold coin. On each of the next two days (the second and third days of service), the knight receives two gold coins. On each of the next three days (the fourth, fifth, and sixth days of service), the knight receives three gold coins. On each of the next four days (the seventh, eighth, ninth, and tenth days of service), the knight receives four gold coins. This pattern of payments will continue indefinitely: after receiving N gold coins on each of N consecutive days, the knight will receive $N + 1$ gold coins on each of the next $N + 1$ consecutive days, where N is any positive integer.

Your program will determine the total number of gold coins paid to the knight in any given number of days (starting from day 1).

Input

The input contains at least 1, but no more than 21 lines. Each line of the input file (except the last one) contains data for one test case of the problem, consisting of exactly one integer (in the range 1 .. 10000), representing the number of days. The end of the input is signaled by a line containing the number 0.

Output

There is exactly one line of output for each test case. This line contains the number of days from the corresponding line of input, followed by one blank space and the total number of gold coins paid to the knight in the given number of days, starting with day 1.

Sample Input	Sample Output
10	10 30
6	6 14
7	7 18

(Continued)

Sample Input	Sample Output
11	11 35
15	15 55
16	16 61
100	100 945
10000	10000 942820
1000	1000 29820
21	21 91
22	22 98
0	

Source: ACM Rocky Mountain 2004.

IDs for online judges: POJ 2000, ZOJ 2345, UVA 3045.

Hint

The rule that the king pays his loyal knight in gold coins is showed in the problem description. We partition n days into p intervals. The ith interval is i days, and i gold coins are received every day $(1 \le i \le p, [(1 + p)/2]p \le n, [(2 + p)/2](p + 1) > n)$. Suppose n is the total number of days; *ans* is the number of received gold coins; i is the number of current days; j is the number of the current interval, that is, the interval in which the king pays his loyal knight the same number of gold coins every day; and k is the number of remaining days.

A double loop is used to calculate the total number of gold coins.
The outer loop enumerates every interval j: *for*(int $i = 0$, $j = 1$; $i \le n$; j++).
The inner loop calculates the total number of received gold coins in interval j: int $k = j$, while $(k$-- && ++$i \le n)$ *ans* += j.
Finally, *ans* is the total number of gold coins paid to the knight in the given number of days.

2.4.2 The 3n + 1 Problem

Problems in computer science are often classified as belonging to a certain class of problems (e.g., NP, unsolvable, recursive). In this problem you will analyze a property of an algorithm whose classification is not known for all possible inputs.
Consider the following algorithm:

```
1. input n
2. print n
3. if n = 1 then STOP
4. if n is odd then n <-- 3n + 1
5. else n <-- n/2
6. GOTO 2
```

Given the input 22, the following sequence of numbers will be printed: 22 11 34 17 52 26 13 40 20 10 5 16 8 4 2 1.
It is conjectured that the algorithm above will terminate (when a 1 is printed) for any integral input value. Despite the simplicity of the algorithm, it is unknown whether this conjecture is true.

It has been verified, however, for all integers n such that $0 < n < 1,000,000$ (and, in fact, for many more numbers than this).

Given an input n, it is possible to determine the number of numbers printed before the 1 is printed. For a given n this is called the cycle length of n. In the example above, the cycle length of 22 is 16.

For any two numbers i and j you are to determine the maximum cycle length over all numbers between i and j.

Input

The input will consist of a series of pairs of integers i and j, one pair of integers per line. All integers will be less than 10,000 and greater than 0.

You should process all pairs of integers and for each pair determine the maximum cycle length over all integers between and including i and j.

Output

For each pair of input integers i and j you should output i, j, and the maximum cycle length for integers between and including i and j. These three numbers should be separated by at least one space with all three numbers on one line and with one line of output for each line of input. The integers i and j must appear in the output in the same order in which they appeared in the input and should be followed by the maximum cycle length (on the same line).

Sample Input	Sample Output
1 10	1 10 20
100 200	100 200 125
201 210	201 210 89
900 1000	900 1000 174

Source: Duke Internet Programming Contest 1990.

IDs for online judges: POJ 1207, UVA 100.

Hint

It is a problem for classical simulation of direct statement; steps for the algorithm are shown in the problem description. If the input pair of integers are a and b, then the interval is [$min(a, b)$, $max(a, b)$]. A double loop is used to solve the problem.

The outer loop is a repetition statement $for(n = min(a, b); n \leq max(a, b); n++)$. It enumerates every number n in the interval ($min(a, b) \leq n \leq max(a, b)$).

The inner loop is a repetition statement $for(i = 1, m = n; m > 1; i++)$. It calculates cycle length of n (if ($m\%2 == 0$) $m /= 2$; else, $m = 3*m + 1$).

Obviously, the maximum cycle length over all numbers in [$min(a, b)$, $max(a, b)$] is the solution to the problem.

2.4.3 *Pascal Library*

Pascal University, one of the oldest in the country, needs to renovate its library building, because after all these centuries the building has started to show the effects of supporting the weight of the enormous amount of books it houses.

To help in the renovation, the alumni association of the university decided to organize a series of fund-raising dinners, for which all alumni were invited. These events proved to be a huge success, and several were organized during the past year. (One of the reasons for the success of this initiative seems to be the fact that students that went through the Pascal system of education have fond memories of that time and would love to see a renovated Pascal library.)

The organizers maintained a spreadsheet indicating which alumni participated in each dinner. Now they want your help to determine whether any alumnus or alumna took part in all of the dinners.

Input

The input contains several test cases. The first line of a test case contains two integers, N and D, indicating, respectively, the number of alumni and the number of dinners organized ($1 \leq N \leq$ 100 and $1 \leq D \leq 500$). Alumni are identified by integers from 1 to N. Each of the next D lines describes the attendees of a dinner and contains N integers X_i indicating if the alumnus or alumna i attended that dinner ($X_i = 1$) or not ($X_i = 0$). The end of input is indicated by $N = D = 0$.

Output

For each test case in the input your program must produce one line of output, containing either the word "yes," in case there exists at least one alumnus or alumna that attended all dinners, or the word "no" otherwise.

Sample Input	Sample Output
3 3	Yes
1 1 1	No
0 1 1	
1 1 1	
7 2	
1 0 1 0 1 0 1	
0 1 0 1 0 1 0	
0 0	

Source: ACM South America 2005.

IDs for online judges: POJ 2864, UVA 3470.

Hint

Suppose *yes* is the mark that there exists at least one alumnus or alumna that attended all dinners. Array *att* shows whether an alumni or alumna attends dinners or not, where $att[j] == 1$ represents alumni or alumna j attending all dinners so far.

First, input the number of alumni n and the number of dinners d for the first test case. Then there is a loop *while*($n \parallel d$). In the loop body, the process is as follows:

1. Initially, suppose all alumni attend dinners, that is, $att[0] = att[1] = \ldots att[n-1] = 1$.
2. A double loop is used to enumerate all dinners that alumni attend. The outer loop enumerates dinner j ($0 \leq j \leq d-1$). The inner loop enumerates alumni i ($0 \leq i \leq n-1$). In the loop

body, based on case k that alumnus or alumna j attends dinner i, calculate whether alumnus or alumna j attends dinners or not, that is, $att[j] = att[j]$ & k.

3. Calculate whether there exists at least one alumnus that attended all dinners, that is,

$$yes = \bigcup_{0 \leq i \leq n-1} att[i]$$

4. If yes == true, then output "yes;" else, output "no."
5. Input the next test case.

2.4.4 *Calendar*

Most of us have a calendar on which we scribble details of important events in our lives—visits to the dentist, the Regent 24-hour book sale, programming contests, and so on. However, there are also the fixed dates—partner's birthdays, wedding anniversaries, and the like—and we also need to keep track of these. Typically we need to be reminded of when these important dates are approaching—the more important the event, the further in advance we wish to have our memories jogged.

Write a program that will provide such a service. The input will specify the year for which the calendar is relevant (in the range 1901–1999). Bear in mind that, within the range specified, all years that are divisible by 4 are leap years and hence have an extra day (February 29) added. The output will specify "today's'" date, a list of forthcoming events and an indication of their relative importance.

Input

The first line of input will contain an integer representing the year (in the range 1901–1999). This will be followed by a series of lines representing anniversaries or days for which the service is requested.

An anniversary line will consist of the letter 'A'; three integer numbers (D, M, P) representing the date, the month, and the importance of the event; and a string describing the event, all separated by one or more spaces. P will be a number between 1 and 7 (both inclusive) and represents the number of days before the event that the reminder service should start. The string describing the event will always be present and will start at the first nonblank character after the priority.

A date line will consist of the letter 'D' and the date and month as above.

All anniversary lines will precede any date lines. No line will be longer than 255 characters in total. The file will be terminated by a line consisting of a single #.

Output

Output will consist of a series of blocks of lines, one for each date line in the input. Each block will consist of the requested date followed by the list of events for that day and as many following days as necessary.

The output should specify the date of the event (D and M), right justified in fields of width 3, and the relative importance of the event. Events that happen today should be flagged as shown below, events that happen tomorrow should have P stars, events that happen the day after tomorrow should have $P--1$ stars, and so on. If several events are scheduled for the same day, order them by relative importance (number of stars).

If there is still a conflict, order them by their appearance in the input stream. Follow the format used in the example below. Leave one blank line between blocks.

Sample Input	Sample Output
1993	Today is: 20 12
A 23 12 5 Partner's birthday	20 12 *TODAY* Unspecified anniversary
A 25 12 7 Christmas	23 12 *** Partner's birthday
A 20 12 1 Unspecified anniversary	25 12 *** Christmas
D 20 12	
#	

Source: New Zealand Contest 1993.

ID for online judge: UVA 158.

Hint

It is a problem of classical process simulation. The simulation is to directly implement the problem description.

Suppose e is a linear list for events, where $e[i].month$ and $e[i].day$ are the date of event i representing the month and the date, respectively; $e[i].level$ is the relative importance of event i; $e[i].index$ is the order of input; and $e[i].a$ is the string describing event i.

Based on the problem description, we deal with the input as follows:

1. Input the year and determine whether the year is a leap year.
2. Repeat deal with the input until '#'.

If the input is a letter 'A', then add up to the number of anniversaries n and input the nth anniversary date ($e[n].month$, $e[n].day$), the relative importance $e[n].level$, the string describing the event $e[n].a$, and the order of input $e[n].index = n$.

If the input is a letter 'D', do the following:

1. If it is the first time that 'D' is input, based on the order that the anniversary date is the first key, relative importance is the second key, and the order of input is the third key, sort events $e[1 .. n]$.
2. If it is not the first time that 'D' is input, then input the service date (*month*, *day*) and initialize date counter $cnt - 1$. Then the program enters a loop until cnt exceeds 7:
   - If it is today ($cnt == -1$), then store the event of the anniversary date (*month*, *day*) into s, and sort s based on the order of input. Output the date and the string describing the event, and the relative importance of the event is "TODAY".
   - If it is not today($cnt \neq -1$), search the event $e[i]$ of the anniversary date (*month*, *day*) in events e ($e[i].month == month$ && $e[i].day == day$, $1 \leq i \leq n$), and calculate the number of days before the event that the reminder service should start, $num = e[i].level - cnt$. If $num \leq 0$, the event is past; else, output the reminder service (num '*' and 8-num blanks) and the string describing the event $e[i].a$.
   - Accumulate the number of days ($cnt++$). If it exceeds the reminder service ($cnt == 7$), then break the loop; else, get the next date (*month*, *day*).

2.4.5 Manager

One of the programming paradigms in parallel processing is the producer–consumer paradigm that can be implemented using a system with a "manager" process and several "client" processes. The clients can be producers, consumers, and so on. The manager keeps a trace of client processes.

Each process is identified by its cost, which is a strictly positive integer in the range 1 .. 10,000. The number of processes with the same cost cannot exceed 10,000. The queue is managed according to three types of requests, as follows:

- *a x*: Add to the queue the process with the cost *x*.
- *r*: Remove a process, if possible, from the queue according to the current manager policy.
- *p i*: Enforce the policy *i* of the manager, where *i* is 1 or 2. The default manager policy is 1.
- *e*: Ends the list of requests.

There are two manager policies:

1. Remove the minimum cost process.
2. Remove the maximum cost process.

The manager will print the cost of a removed process only if the ordinal number of the removed process is in the removal list.

Your job is to write a program that simulates the manager process.

Input

The input is from the standard input. Each data set in the input has the following format:

- The maximum cost of the processes.
- The length of the removal list.
- The removal list—the list of ordinal numbers of the removed processes that will be displayed. For example, 1 4 means that the cost of the first and fourth removed processes will be displayed.
- The list of requests, each on a separate line.

Each data set ends with an *e* request. The data sets are separated by empty lines.

Output

The program prints on standard output the cost of each process that is removed, provided that the ordinal number of the remove request is in the list and the queue is not empty at that moment. If the queue is empty, the program prints –1. The results are printed on separate lines. An empty line separates the results of different data sets.

An example is given in the following:

Sample Input	Sample Output
5	2
2	5
1 3	
a 2	
a 3	

(*Continued*)

Sample Input	Sample Output
r	
a 4	
p 2	
r	
a 5	
r	
e	

Source: ACM Southeastern Europe 2002.

IDs for online judges: POJ 1281, UVA 2514.

Hint

Suppose *minp* is the minimum cost of the processes, $minp = 1$; *mapx* is the maximum cost of the processes; *print* is the removal sign list of the processes, where $print[k] ==$ true means process k is removed; *plen* is the number of processes that should be removed; *np* is the number of processes that have been removed; *cnt* stores the number of processes for each cost, where $cnt[k]$ is the number of processes costing k; *req* is the types of requests ('a', 'r', 'p', 'e'); and *condition* is manager policy (1 or 2).
 The format for each test case is as follows:

> The maximum cost of the processes *maxp*
> The length of the removal list *plen*
> *plen* removed processes
> The list of requests ('ax', 'r', 'pi', 'e'), and 'e' marks the end
> *mapx* $== 0$ marks the end of test cases

 Obviously, repetition statement *while*(cin >> *maxp*) constitutes the main program, and its procedure is as follows:

1. Input the length of the removal list *plen* and *plen* removed processes, and set these processes' *print* signs true.
2. Initialize *np* 0. Input the first request. Repetition statement *while*(*req* ! = 'e') deals with requests one by one. The procedure is as follows:
   a. If *req* is 'a', then add the process with the cost x into the queue, $cnt[x]++.$
   b. If *req* is 'r', then remove a process, if possible, from the queue according to the current manager policy:
      • If *condition* $== 1$, then remove the minimum cost process. k is enumerated in ascending order from *minp* to *maxp*. The process for the first $cnt[k] \neq 0$ will be removed, and $cnt[k]$--.
      • If *condition* $== 2$, then remove the maximum cost process. k is enumerated in descending order from *maxp* to *minp*. The process for the first $cnt[k] \neq 0$ will be removed, and $cnt[k]$--.
      The number of removed processes is *np*++. If (*print*[*np*] $==$ true), output the cost of the process k.
   c. If *req* is 'p', then change *condition* (1 or 2).
   d. Input the next request *req*.

Chapter 3

Simple Recursion

The programming technique of a program calling itself is called recursion. There are two kinds of recursions: direct recursion and indirect recursion. We can use the Droste effect to illustrate recursion (Figure 3.1):

> The Droste effect ... is the effect of a picture appearing within itself, in a place where a similar picture would realistically be expected to appear. The appearance is recursive: the smaller version contains an even smaller version of the picture, and so on. (http://en.wikipedia.org/wiki/Droste_effect)

Recursion is to change a large and complex problem into a smaller problem similar to the original problem and solve it. Therefore, a small amount of program code can implement repeated calculation, making the program more concise and clear.

A stack is used to implement a recursive process. When a program calls itself, it stores the point of return and pushes local variables of the current layer into a stack, and when it backtracks, it returns to the point of the current layer and pops local variables of the current layer from the stack. Recursive algorithms are normally concise. If a recursive process can't reach the end or its recursion times are too many, it will cause stack overflow. For example, a recursive process is as follows:

$$f(n) = \begin{cases} 1 & n = 1 \\ n + f(n-2) & n > 1 \end{cases}$$

Obviously, if n is an even number, then $f(n)$ can't reach f(1), and the program will run out of limit.

Recursive algorithms are normally used to solve three kinds of problems:

1. Functions' definitions are recursive (such as factorial, or Fibonacci function).
2. Solutions to problems are recursive (such as backtrack).
3. Definitions of data structures are recursive (such as tree traversal, or graph search).

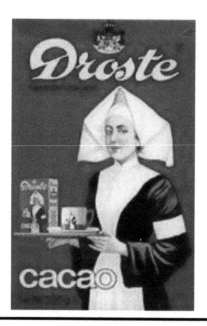

Figure 3.1   Droste effect. (From http://en.wikipedia.org/wiki/Droste_effect.)

3.1 Calculation of Recursive Functions

Definitions and calculations of factorial functions, power functions, and the Fibonacci sequence are recursive. For example, the recursive definition of factorial function $n!$ is as follows:

$$fac(n) = n! = \begin{cases} 1 & n = 0 \\ n * fac(n-1) & n \geq 1 \end{cases}$$

Based on the recursive definition, we can use recursive function $fac(n)$ to solve it.

```
int fac(int n);
{ if (n==0) return 1;         // end condition of recursion
  if (n>=1) return n*fac(n-1); // recursion
}
```

Obviously, the advantage of a recursive program is concise and readable, but its efficiency is relatively lower. For example, the recursive process of $fac(3)$ is shown in Figure 3.2.

In the program, $fac(0) = 1$ is called the end condition of recursion. The process $fac(3) \rightarrow fac(2) \rightarrow fac(1) \rightarrow fac(0)$ is a recursive process, and the process $fac(0) \rightarrow fac(1) \rightarrow fac(2) \rightarrow fac(3)$ is a back substitution process ($fac(0) = 1$ is back substitution to $fac(1)$, $fac(1)$ is back substitution to $fac(2)$, ..., until we calculate $fac(3) = 6$).

Similarly, function fib for the Fibonacci sequence is defined as follows:

$$fib(n) = \begin{cases} n & n = 0,1 \\ fib(n-1) + fib(n-2) & n > 1 \end{cases}$$

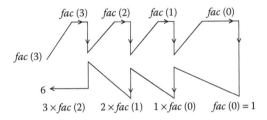

Figure 3.2 The recursive process of *fac*(3).

Based on the above recursive format, recursion function *fib*(*n*) is shown as follows:

```
int fib(int n);
{ if ((n==1)||(n==0)) return n;      // the end condition of recursion
  if (n>1) return fib(n-1)+fib(n-2);  // The recursive steps
}
```

From above examples, we can get following hints for recursion:

1. The recursive process is used to solve recursive functions. A recursive process can be directly implemented by a program based on the definition of recursive functions.
2. For a complex problem, if it can be decomposed into several relatively simple subproblems, these subproblems' solutions are the same or similar, and the original problem can be solved as long as these subproblems are solved, then it is a recursive solution. For example, before we calculate 4!, we calculate 3!, and the result of 3! is back substitution and we can calculate 4! (4! = 4*3!). The decomposition and solution strategy are called divide and conquer.
3. When the subproblems can be solved directly, the decomposition terminates. Such subproblems are called the end conditions of recursion. If the recursive function can't reach the end conditions, then the program will fail due to stack overflow. For example, the end condition for the factorial is *fac*(0) = 1, and the end conditions for the Fibonacci sequence are *fib*[0] = 0 and *fib*[1] = 1.

3.2 Solving Problems by Recursive Algorithms

If the initial status and the goal status are shown in the problem description, and rules and constraints for the expended status are the same, then a recursive algorithm can be used to find a solution from the initial status to the goal status. When a recursive algorithm is used, the following should be noted.

How to represent a status with value parameters or local variables in decomposition should be noticed to recover its original status in backtracking. If storage of these parameters is large (such as an array) and initial values should be passed by the main program, in order to avoid memory overflow, these variables must be set as global variables, and before backtracking, these variables are required to restore their original values before recursion.

The end conditions of recursion should be determined.

The search area and constraints should be determined, that is, in what situation the end conditions can't be reached and in what area search the recursive process can continue.

The above recursive algorithm is called backtracking. The essence of backtracking is the same as the essence of the depth-first search (DFS) in Section IV. DFS is used in graph traversal. Both methods make use of the recursive strategy of depth search.

3.2.1 Red and Black

There is a rectangular room, covered with square tiles. Each tile is colored either red or black. A man is standing on a black tile. From a tile, he can move to one of four adjacent tiles. But he can't move on red tiles; he can move only on black tiles. Write a program to count the number of black tiles that he can reach by repeating the moves described above.

Input

The input consists of multiple data sets. A data set starts with a line containing two positive integers W and H; W and H are the numbers of tiles in the x and y directions, respectively. W and H are not more than 20.

There are H more lines in the data set, each of which includes W characters. Each character represents the color of a tile as follows:

'.': A black tile
'#': A red tile
'@': A man on a black tile (appears exactly once in a data set)

The end of the input is indicated by a line consisting of two zeros.

Output

For each data set, your program should output a line that contains the number of tiles the man can reach from the initial tile (including itself).

Sample Input	Sample Output
6 9	45
...#.	59
.....#	6
......	13
......	
......	
......	
#@...#	
.#..#.	
11 9	

Sample Input	Sample Output
.#.........	
.#.#######.	
.#.#.....#.	
.#.#.###.#.	
.#.#..@#.#.	
.#.#####.#.	
.#.......#.	
.#########.	
....	
11 6	
..#..#..#..	
..#..#..#..	
..#..#..###	
..#..#..#@.	
..#..#..#..	
..#..#..#..	
7 7	
..#.#..	
..#.#..	
###.###	
...@...	
###.###	
..#.#..	
..#.#..	
0 0	

Source: ACM Japan 2004, Domestic.

ID for online judge: POJ 1979.

Analysis

The recursive method (backtracking) can be used to count the number of black tiles that the man can reach. Suppose n and m are the numbers of tiles in the x direction and y direction, respectively; ans is the number of black tiles that the man reach, and its initial value is 0; map represents the rectangular room covered with square tiles, where $map[i][j]$ is a character that represents the tile whose positions in the x direction and y direction are i and j, respectively; *visited* is visited marks for the man, where *visited*[i][j] == true means the man has reached the tile whose position is (i, j). The recursive function is *search*(i, j), where the following apply:

Status: The man's current position is (i, j). Obviously the position before the recursion is the position of the initial tile.

End condition of recursion: If the current position is out of the rectangular room ($i < 0$||$i >= n$||$j < 0$||$j >= m$) or can't pass (*map*[i][j] == '#'), or the position has been visited (*visited*[i][j] == true), then backtrack; otherwise, for the current position (i, j), *visited*[i] [j] = true, the number of tiles that the man has reached increases 1 (++*ans*), and then continue recursion.

Search area: For the current position (i, j), recursively search the four neighbor positions (*search*(i − 1, j); search(i + 1, j); search(i, j − 1); and search(i, j + 1);).

Program

```
#include <iostream>
#include <string>
#include <cstring>
using namespace std;
const int maxn = 20 + 5, maxm = 20 + 5;    // the upper limit of
numbers of cow and column
    int n, m, ans;    // numbers of cow and column, the number of black
tiles which the man reach
    string map[maxn];      // the rectangular room covered with square tiles
    bool visited[maxn][maxm];        // the man has reached the tile
    void search(int row, int col)        // recursively count the number of
black tiles which the man can reach from [row, col]
    {
        if (row < 0 || row >= n || col < 0 || col >= m || map[row][col] ==
'#' || visited[row][col])
        // backtrack condition
            return;
        visited[row][col] = true; //      visit mark
        ++ans;         // accumulation of the number of black tiles which
the man reach
        search(row - 1, col);            // recursions for four adjacent tiles
        search(row + 1, col);
        search(row, col - 1);
        search(row, col + 1);
    }
    int main(void)
    {
        cin >> m >> n;                // The room size of the first test case
        while (n || m) {
            int row, col;
            for (int i = 0; i < n; i++) {        // Input the current room
                cin >> map[i];
                for (int j = 0; j < m; j++)
                    if (map[i][j] == '@') {    // The position where the
man stands on
                        row = i;
                        col = j;
```

```
                }
            }
            memset(visited, false, sizeof(visited)); //Initialization
            ans = 0;
            search(row, col);              // Recursion calculation
            cout << ans << endl;    // Output the number of black tiles
                                    // which the man reach
            cin >> m >> n;          // Input the size of the room in the
                                    // next case
        }
        return 0;
    }
```

3.3 Solving Recursive Datum

When we construct a mathematical model for a problem, sometimes we find its data structure is in a recursive form. For example, tree traversal and depth-first traversal for a graph are defined recursively and will be discussed in Sections III and IV, respectively. An easy and interesting example is shown as follows.

3.3.1 Symmetric Order

In your job at Albatross Circus Management (yes, it's run by a bunch of clowns), you have just finished writing a program whose output is a list of names in nondescending order by length (so that each name is at least as long as the one preceding it). However, your boss does not like the way the output looks and instead wants the output to appear more symmetric, with the shorter strings at the top and bottom and the longer strings in the middle. His rule is that each pair of names belongs on opposite ends of the list, and the first name in the pair is always in the top part of the list. In the first example set below, Bo and Pat are the first pair, Jean and Kevin the second pair, and so forth.

Input

The input consists of one or more sets of strings, followed by a final line containing only the value 0. Each set starts with a line containing an integer, n, which is the number of strings in the set, followed by n strings, one per line, sorted in nondescending order by length. None of the strings contain spaces. There is at least 1 and no more than 15 strings per set. Each string is at most 25 characters long.

Output

For each input set print "SET n" on a line, where n starts at 1, followed by the output set as shown in the sample output.

Sample Input	Sample Output
7	SET 1
Bo	Bo

(Continued)

Sample Input	Sample Output
Pat	Jean
Jean	Claude
Kevin	Marybeth
Claude	William
William	Kevin
Marybeth	Pat
6	SET 2
Jim	Jim
Ben	Zoe
Zoe	Frederick
Joey	Annabelle
Frederick	Joey
Annabelle	Ben
5	SET 3
John	John
Bill	Fran
Fran	Cece
Stan	Stan
Cece	Bill
0	

Source: ACM Mid-Central United States 2004.

IDs for online judges: POJ 2013, ZOJ 2172.

Analysis

The list of names in nondescending order by length is s[1] … s[n]. The format of output is symmetric, with the shorter strings at the top and bottom, and the longer strings in the middle. Lengths of names in the upper half part are ascending, and lengths of names in the lower half part are descending. There are two solutions to the problem: nonrecursive method and recursive method.

1. Nonrecursive method: The input consists of one or more sets of strings sorted in nondescending order by length. The output is symmetric, with the shorter strings at the top and bottom and the longer strings in the middle. So the upper half of the output is as follows:

    s[1]
    s[3]

$s[5]$

$\ldots$

$s[n]$, if n is an odd number, or $s[n-1]$, if n is an even number

That is, the for statement *for*(int $i = 1$; $i \leq n$; i += 2) cout << $s[i]$ << endl; will implement the upper half of the output.

The lower half of the output is as follows:

$s[n - (n\%2)]$

$s[n - (n\%2) - 2]$

$s[n - (n\%2) - 4]$

$\ldots$

$s[2]$

That is, the for statement *for*(int $i = n - (n\%2)$; $i > 1$; i -= 2) cout << $s[i]$ << endl; will implement the lower half of the output.

2. Recursive method: First the program outputs $s[1]s[3] \ldots s[n]$, if n is odd; or the program outputs $s[1]s[3] \ldots s[n-1]$, if n is even; Then the program outputs $s[n]s[n-2] \ldots s[2]$, if n is even or $s[n-1]s[n-3] \ldots s[2]$ if n is odd.

n strings are divided into $\lceil n/2 \rceil$ groups, and each group contains two adjacent strings $s[1]$ $s[2]$, $s[3]s[4]$, $\ldots$, and so on.

If $s[k]s[k+1]$ $(1 \leq k < n)$ is in a group, first input the first parameter $s[k]$ and output it, and then input the second parameter $s[k+1]$ and push it into a stack. After the last group is input, elements in the stack are popped and output. It can be implemented by recursive function *print*(n).

The process for *print*(n) is as follows.

Input and output the first string of the current group $s[k]$; n--;

If $n > 0$, then input the second string of the current group $s[k+1]$ and push it into a stack, n--. If $n > 0$, then recursively call *print*(n) until $n == 0$.

In backtrack, $s[n-1]s[n-3] \ldots s[2]$ if n is even; or $s[n]s[n-2] \ldots s[2]$, if n is odd; are popped from the stack and output.

Program

```
#include <iostream>
using namespace std;
void print(int n)              //n strings are inputted, and are outputted
in symmetric order
{
    string s;                  // the current string
    cin >> s;                  // Input and output the first string
    cout << s << endl;
    if (--n) {                 // Input 2nd string and push it into a stack
through recursion
        cin >> s;
        if (--n)
            print(n);
        cout << s << endl;     //Backtracking, elements popped from
stack and output
    }
```

```
}
int main(void)
{
    int n, loop = 0;
    cin >> n;                                    //number of strings
    while (n) {
        cout << "SET " << ++loop << endl;
        print(n);                                // recursive function
        cin >> n;                                // number of strings in the next set
    }
    return 0;
}
```

3.4 Problems

3.4.1 Fractal

A fractal is an object or quantity that displays self-similarity, in a somewhat technical sense, on all scales. The object need not exhibit exactly the same structure at all scales, but the same "type" of structures must appear on all scales.

A box fractal is defined as below.

- ◾ A box fractal of degree 1 is simply

  X
- ◾ A box fractal of degree 2 is

  X X

   X

  X X
- ◾ If using $B(n - 1)$ to represent the box fractal of degree $n - 1$, then a box fractal of degree n is defined recursively as follows:

$$B(n - 1) \qquad B(n - 1)$$
$$B(n - 1)$$
$$B(n - 1) \qquad B(n - 1)$$

Your task is to draw a box fractal of degree n.

Input

The input consists of several test cases. Each line of the input contains a positive integer n that is no greater than 7. The last line of input is a negative integer -1 indicating the end of input.

Output

For each test case, output the box fractal using the 'X' notation. Please notice that 'X' is an uppercase letter. Print a line with only a single dash after each test case.

Sample Input	Sample Output
1	(see below)

```
Sample Input        Sample Output

1                   X
2                   —
3                   X X
4                    X
-1                  X X
                    —
                    X X    X X
                     X      X
                    X X    X X
                       X X
                        X
                       X X
                    X X    X X
                     X      X
                    X X    X X
                    —
                    X X    X X              X X    X X
                     X      X                X      X
                    X X    X X              X X    X X
                       X X                     X X
                        X                       X
                       X X                     X X
                    X X    X X              X X    X X
                     X      X                X      X
                    X X    X X              X X    X X
                          X X    X X
                           X      X
                          X X    X X
                             X X
                              X
                             X X
                          X X    X X
                           X      X
                          X X    X X
                    X X    X X              X X    X X
                     X      X                X      X
                    X X    X X              X X    X X
                       X X                     X X
                        X                       X
                       X X                     X X
                    X X    X X              X X    X X
                     X      X                X      X
                    X X    X X              X X    X X
                    —
```

Source: ACM Shanghai 2004, Preliminary.

IDs for online judges: POJ 2083, ZOJ 2423.

Hint

The size of a box fractal of degree n is 3^{n-1}, $n \geq 1$. That is, a box fractal of degree n is a square whose length of each side is 3^{n-1}. Recursive function $print(n, x, y)$ produces a square of degree n whose top left corner is at (x, y):

1. The end condition of recursion: If $n == 1$, then output a 'X' at (x, y).
2. If $n > 1$, then a box fractal of degree $n - 1$ is a square whose length of a side is 3^{n-2}, and output five box fractals of degree $n - 1$ at the top left corner, top right corner, center, low left corner, and low right corner, respectively.
   - For the box fractal of degree $n - 1$ at the top left corner, the coordinate at the top left corner is (x, y), and $print(n - 1, x, y)$ is called recursively to produce the box fractal.
   - For the box fractal of degree $n - 1$ at the top right corner, the coordinate at the top left corner is $(x, y + m)$, and $print(n - 1, x, y + m)$ is called recursively to produce the box fractal.
   - For the box fractal of degree $n - 1$ at the center, the coordinate at the top left corner is $(x + m, y + m)$, and $print(n - 1, x + m, y + m)$ is called recursively to produce the box fractal.
   - For the box fractal of degree $n - 1$ at the low left corner, the coordinate at the top left corner is $(x + 2*m, y)$, and $print(n - 1, x + 2*m, y)$ is called recursively to produce the box fractal.
   - For the box fractal of degree $n - 1$ at the low right corner, the coordinate at the top left corner is $(x + 2*m, y + 2*m)$, and $print(n - 1, x + 2*m, y + 2*m)$ is called recursively to produce the box fractal.

Obviously, $print(n, 0, 0)$ is called to produce a box fractal of degree n.

3.4.2 Sticks

George took sticks of the same length and cut them randomly until all parts became at most 50 units long. Now he wants to return the sticks to their original state, but he forgot how many sticks he had originally and how long they were originally. Please help him and design a program that computes the smallest possible original length of those sticks. All lengths expressed in units are integers greater than zero.

Input

The input contains blocks of two lines. The first line contains the number of stick parts after cutting; there are at most 64 sticks. The second line contains the lengths of those parts separated by the space. The last line of the file contains zero.

Output

The output should contain the smallest possible length of original sticks, one per line.

Sample Input	Sample Output
9	6
5 2 1 5 2 1 5 2 1	5

Sample Input	Sample Output
4	
1 2 3 4	
0	

Source: ACM Central Europe 1995.

IDs for online judges: POJ 1011, UVA 307.

Hint
Based on the problem description, given n sticks $sticks[0 .. n-1]$ (old sticks), the program computes the smallest possible original length of those sticks. Suppose the smallest possible original length of those sticks is len, and $sticks[0 .. n-1]$ are sorted in descending order. Obviously, properties for len are as follows.
    First,

$$sum = \sum_{i=0}^{n-1} sticks[i]$$

must be divided by len, that is, $sum \% len == 0$ and len must be a divisor for sum. Then, len must be larger than or equal to the length of each old stick, that is, $len \geq sticks[0]$.
    Second, if there are at least two sticks, then $sticks[0] \leq len \leq (sum/2)$. If the suitable length can't be found in the interval, then n old sticks must be obtained from only one stick. That is, len is sum.
    Therefore, the problem is how to find len in the interval $[sticks[0], (sum/2)]$.
    A recursive Boolean function $dfs(i, l, t)$ is to calculate len, where i is the serial number of the old stick that will be cut, l is the remainder length of the current stick, and t is the sum of lengths of the remainder sticks. If $dfs(0, len, sum)$ returns true, then len is the smallest original length of those sticks; otherwise, repeat len++ and $dfs(0, len, sum)$ until len is found or $len > (sum/2)$.
    Suppose array $used[ ]$ represents whether old sticks are used or not. Initially, $used[ ]$ is zero.
    The recursive boundary is $l == 0$; that is, the current stick is cut from sticks whose length is len, and $t -= len$. If $l == 0$ if $t == 0$, then n old sticks are all cut and the function returns true; if $t \neq 0$, then the first unused old stick i is searched in descending order of length. Old stick i is the longest old stick for unused old sticks, and it must be used ($used[i] = 1$). Therefore, the serial number of searched old sticks is $i + 1$, and $dfs(i + 1, len - sticks[i], t)$ is called. If the function returns true, len is the solution.
    If the remainder length of the current stick $l \neq 0$, then old stick j after old stick i is searched in descending order of length ($i \leq j \leq n - 1$).
    If the length of old stick $j - 1$ is the same as the length of old stick j, and old stick $j - 1$ is unused ($j > 0$ && ($sticks[j] == sticks[j-1]$ &&! $used[j-1]$)), then old stick j needn't to be tested, and old stick $j + 1$ is tried.
    If old stick j is unused and l is longer than or equal to its length ($!used[j]$ && $l >= sticks[j]$), then old stick j is cut ($l -= sticks[j]$; $used[j] = 1$), and $dfs(j, l, t)$ is called. If the function returns false, and if the length of old stick j is l before cut, the remainder sticks can't be cut, then break.

Program

```cpp
#include <iostream>
#include <algorithm>
using namespace std;
int sticks[65];                              // given n old sticks
int used[65];                                // whether a stick is used or not
int n,len;                                   // n: the number of old sticks
bool dfs(int i, int l, int t)      //determine whether len is the solution
{
    if (l==0)                                // old stick i is cut from a stick
whose length is len
    {
        t-=len;;                     // the sum of lengths of remainder sticks
        if (t==0) return true;                    //if n are all cut
        for (i=0; used[i]; ++i);                   // in descending
order, the first unused old stick i
        used[i]=1;                            // old stick i is used
        if(dfs(i+1, len-sticks[i], t))return true; // if from old stick
i+1, len can be determined
        used[i]=0; t+=len;            // recover parameter before recursion
    }
    else
    {
        for (int j=i; j<n; ++j)                // in descending order, search
old stick j after old stick i
        {
            if          (j>0&&(sticks[j]==sticks[j-1]&&!used[j-1]))
continue;   // If the length of old stick j-1 is the same as the length of
old stick j, and old stick j-1 is unused
            if (!used[j]&&l>=sticks[j])    // If old stick j is unused and l
is longer than or equal to its length
            {
                l-=sticks[j]; used[j]=1;
                if (dfs(j,l,t))return true;          //from old stick j, len
can be determined
                l+=sticks[j]; used[j]=0;
                if (sticks[j]==l) break;            // If the length of old
stick j is l before cut, the remainder sticks can't be cut, then break
            }
        }
    }
    return false;
}
bool cmp(const int a, const int b)
{
    return a>b;
}
int main()
{
    while (cin>>n&&n)          // the number of sticks parts after cutting
    {
        int sum=0;
        for(int i=0;i<n;++i)          // input the lengths of those parts
```

```
            {
                cin>>sticks[i]; sum+=sticks[i];
                used[i]=0;
            }
            sort(sticks,sticks+n,cmp);                //sort n old sticks in
descending order of length
            bool flag=false;
            for(len=sticks[0];len<=sum/2;++len)  // from the longest old
stick, search the smallest possible length of original sticks (at least 2
sticks)
            {
                if(sum%len==0)                        // if len is a divisor for sum
                {
                    if(dfs(0,len,sum))                //if len is the smallest
possible length of original sticks
                    {
                        flag=true;
                        cout<<len<<endl;              // output the smallest possible
length of original sticks
                        break;
                    }
                }
            }
            if(!flag) cout<<sum<<endl;            // if the suitable length can't
be found in [sticks[0],sum/2], then n old sticks must be gotten from one
stick
        }
    return 0;
}
```

SUMMARY OF SECTION I

Section I is for students who just learn programming languages. It is not only the review course for programming languages, but also the elementary course for data structure. Section I focused on experiments for simple computing, simple simulation, and simple recursion.

Experiments for simple computing allow students to understand that the pattern of a programming contest problem is input–process–output. Students should improve their programming style to make programs readable and meet requirements for input and output.

A simulation program is to simulate a process or implement some rules shown in the problem description. Experiments for simple simulation not only help students be familiar with this kind of problem, but also improve students' coding ability.

Recursion means a subprogram calls itself directly or indirectly. In Section I, experiments for calculation of recursive functions, solving problems by recursive algorithms, and solving recursive data were shown. In Sections III and IV, some data structures are recursively defined, and some algorithms are recursive algorithms.

In Section I, some programming strategies, such as dichotomy, were also shown. It will also be discussed in the future.

There are three kinds of data structures: linear list, tree, and graph. Their experiments will be shown in Sections II to IV, respectively.

EXPERIMENTS FOR LINEAR LISTS

A linear list consists of a finite ordered set of data elements. The type of all data elements in the set is the same. Each data element contains one or more items. Such a data structure is simple and commonly used. Its features are as follows:

Uniformity: The type of data elements in a linear list is the same. For example, the string is a linear structure; in a string, each data element is a single character. In a student score list, each data element consists of a student name, a student ID, scores of several subjects, and so on. And such items represent a student's attributes. Therefore, a student score list is also a linear list.

Order: If a linear list isn't null, in it there exist the first and the last data element; if there exist other elements, for each element there are only one direct predecessor and one direct successor. For the first data element, there isn't a direct predecessor, and for the last data element, there isn't a direct successor. For example, for characters in a string and data elements in a student score list, there exist such relationships.

Based on storage mode, this section discusses three kinds of linear lists:

1. Linear lists accessed directly
2. Application of sequential access
3. Generalized list using indexes

Chapter 4

Linear Lists Accessed Directly

In linear lists accessed directly, an element can be accessed directly without visiting its predecessor or successor. An array is one of these kinds of data structure.

An array is a set of data elements with the same type stored in a continuous area and with fixed length. A one-dimensional array (or single-dimension array) is a typical linear list accessed directly. Arrays can also have more than one dimension, such as two-dimensional arrays that can be used to represent matrices. Such an array is called a multidimensional array. In an array, indexes of data elements indirectly show their memory addresses, and data elements can be accessed directly. Therefore, time complexity for the access of one element in the array is O(1). In this sense, the storage structure of an array is a structure of direct access. For example, a string is a direct access structure. In a string, any character can be accessed directly.

Linear lists accessed directly (arrays) are used mostly, such as calculation of date, calculation of high precision, representation and access of polynomial, and calculation of numerical matrices.

4.1 Application of Arrays 1: Calculation of Dates

Date is represented by year, month, and day. Problems for date type can make use of arrays as data structures. Normally there are two kinds of storage modes:

1. A linear list (array) whose data element is a structure containing year, month, and day
2. Three integer arrays that record years, months, and days, respectively

Dates are stored in a linear list. As a linear list, it is finite (the number of date elements is finite), ordered (date elements are listed one by one in a permutation), and uniform (the type of all date elements is the same). Elements can be directly accessed. Therefore, the linear list for the date element is a typical linear list accessed directly.

The calculation of date and the conversion of calendar are based on a linear list. Months and days are generally represented by English words. Therefore, months and days are stored in arrays of strings, and the indexes also correspond to months and days. Two examples are shown as follows.

4.1.1 Calendar

A calendar is a system for measuring time, from hours and minutes, to months and days, and finally to years and centuries. The terms *hour, day, month, year,* and *century* are all units of time measurements of a calendar system.

According to the Gregorian calendar, which is the civil calendar in use today, years evenly divisible by 4 are leap years, with the exception of centurial years that are not evenly divisible by 400. Therefore, the years 1700, 1800, 1900, and 2100 are not leap years, but 1600, 2000, and 2400 are leap years.

Given the number of days that have elapsed since January 1, 2000 AD, your mission is to find the date and the day of the week.

Input

The input consists of lines each containing a positive integer, which is the number of days that have elapsed since January 1, 2000 AD. The last line contains an integer −1, which should not be processed.

You may assume that the resulting date won't be after the year 9999.

Output

For each test case, output one line containing the date and the day of the week in the format of "YYYY-MM-DD DayOfWeek," where "DayOfWeek" must be one of "Sunday," "Monday," "Tuesday," "Wednesday," "Thursday," "Friday," or "Saturday."

Sample Input	Sample Output
1730	2004-09-26 Sunday
1740	2004-10-06 Wednesday
1750	2004-10-16 Saturday
1751	2004-10-17 Sunday
−1	

Source: ACM Shanghai 2004, Preliminary.

IDs for online judges: POJ 2080, ZOJ 2420.

Analysis
First, two functions are designed as follows:

1. *days_of_year(year)*: Calculate the number of days in *year*. If *year* is a leap year, the number of days in *year* is 366; otherwise, the number of days in *year* is 365.
2. *days_of_month(month, year)*: Calculate the number of days in *month, year*. If *month* == 2 and *year* is a leap year, the number of days is 29; else, the number of days is 28. If *month* == 1, 3, 5, 7, 8, 10, or 12, the number of days is 31. If *month* == 4, 6, 9, or 11, the number of days is 30.

Then, we use January 1, 2000 (Saturday), as the benchmark. Suppose *year, month,* and *day* are variables; *wstr* is a string array storing the day of the week, that is, *wstr*[0 .. 6] = {"Saturday",

"Sunday", "Monday", "Tuesday", "Wednesday", "Thursday", "Friday"}. Initially *year* = 2000, *month* = 1, and *day* = 1. Suppose *n* is the number of days that have elapsed since January 1, 2000 AD. The steps finding the date and the day of the week are as follows:

Step 1 is to calculate the day of the week: Because January 1, 2000 (Saturday) is the benchmark, and *wstr*[0 .. 6] = {"Saturday", "Sunday", "Monday", "Tuesday", "Wednesday", "Thursday", "Friday"}, obviously *wstr*[*n* % 7] is the day of the week.

Step 2 is to calculate *year*: While *n* ≥ *days_of_year*(*year*), repeat statements *n*– = *days_of_year*(*year*); and ++*year*. When the loop ends, *year* is calculated, and *n* is the number of days in *year*.

Step 3 is to calculate *month* and *day*: While *n* ≥ *days_of_month*(*month*, *year*), repeat statements *n*– = *days_of_month*(*month*, *year*); ++*month*. When the loop ends, *month* is calculated. And the statement *day* += *n* is to calculate *day*.

Program

```cpp
#include <iostream>           //Preprocessor Directive
using namespace std;         //Using C++ Standard Library
const char wstr[][20] = {"Saturday", "Sunday", "Monday", "Tuesday",
"Wednesday", "Thursday", "Friday"};       //String array for the day of
the week
int days_of_year(int year)   // Return the number of days of year
{
    if (year % 100 == 0)
        return year % 400 == 0 ? 366 : 365;
    return year % 4 == 0 ? 366 : 365;
}
int days_of_month(int month, int year) // Return the number of days of
month in year
{
    if (month == 2)
        return days_of_year(year) == 366 ? 29 : 28;
    int d;
    switch (month) {
        case 1: case 3: case 5: case 7: case 8:
        case 10: case 12:
            d = 31;
            break;
        default:
            d = 30;
    }
    return d;
}
int main(void)
{
    int n;
    cin >> n;                //Input the first test case
    while (n >= 0) {
        int year, month, day, week;
        week = n % 7; // use January 1, 2000 (Saturday) as the benchmark
and the beginning of a week
        year = 2000;
        month = 1;
        day = 1;
```

```
        while (n) {
            if (n >= days_of_year(year)) { // Calculate the year
                n -= days_of_year(year);
                ++year;
            } else if (n >= days_of_month(month, year)) { // Calculate
the month
                n -= days_of_month(month, year);
                ++month;
            } else { // Calculate the day
                day += n;
                n = 0;
            }
        }
    //Output the date and the day of the week
        cout << year << '-' << (month < 10 ? "0" : "") << month << '-'
            << (day < 10 ? "0" : "") << day << ' ' << wstr[week] <<
endl;
        cin >> n; //Input the next test case
    }
    return 0;
}
```

4.1.2 *What Day Is It?*

The calendar now in use evolved from the Romans. Julius Caesar codified a calendar system that came to be known as the Julian calendar. In this system, all months have 31 days, except for April, June, September, and November, which have 30 days, and February, which has 28 days in non–leap years and 29 days in leap years. Also, in this system, leap years happened every 4 years. That is because the astronomers of ancient Rome computed the year to be 365.25 days long, so that after every 4 years, one needed to add an extra day to keep the calendar on track with the seasons. To do this, they added an extra day (February 29) to every year that was a multiple of 4.

> **Julian rule:** Every year that is a multiple of 4 is a leap year, that is, has an extra day (February 29).

In 1582, Pope Gregory's astronomers noticed that the year was not 365.25 days long, but closer to 365.2425. Therefore, the leap year rule would be revised to the following:

> **Gregorian rule:** Every year that is a multiple of 4 is a leap year, unless it is a multiple of 100 that is not a multiple of 400.

To compensate for how the seasons had shifted against the calendar up until that time, the calendar was actually shifted 10 days: the day following October 4, 1582, was declared to be October 15.

England and its empire (including the United States) didn't switch to the Gregorian calendar system until 1752, when the day following September 2 was declared to be September 14. (The delay was caused by the poor relationship between Henry VIII and the pope.)

Write a program that converts dates in the United States using a calendar of the time and outputs weekdays.

Input

The input will be a series of positive integers greater than zero, three integers per line, which represent dates, one date per line. The format for a date is "month day year," where month is a number between 1 (which indicates January) and 12 (which indicates December), day is a number between 1 and 31, and year is positive number.

Output

The output will be the input date and name of the weekday on which the given date falls in the format shown in the sample. An invalid date or nonexistent date for the calendar used in the United States at the time should generate an error message indicating a invalid date. The input will end with three zeros.

Sample Input	Sample Output
11 15 1997	November 15, 1997, is a Saturday.
1 1 2000	January 1, 2000, is a Saturday.
7 4 1998	July 4, 1998, is a Saturday.
2 11 1732	February 11, 1732, is a Friday.
9 2 1752	September 2, 1752, is a Wednesday.
9 14 1752	September 14, 1752, is a Thursday.
4 33 1997	4/33/1997 is an invalid date.
0 0 0	

Source: ACM Pacific Northwest 1997.

IDs for online judges: ZOJ 1256, UVA 602.

Analysis

The problem is a simulation problem. That is, you are asked to solve the problem following some rules. Generally, these rules are given in the problem description.

Because months and days of the week are strings in the output, string arrays should be defined first:

```
const  char wstr[ ] [maxs]        // A string array represents the day of
the week
={"Sunday", "Monday", "Tuesday", "Wednesday", "Thursday", "Friday",
"Saturday"};
const  char mstr[ ] [maxs]        // A string array represents months
= {"", "January", "February", "March", "April", "May", "June", "July",
"August", "September", "October", "November", "December"};
```

The indexes of arrays correspond to months and days of the week.

Suppose *year*, *month*, and *day* are variables representing the current date; *old* is a Boolean variable representing whether the current date is before September 2, 1752, or not, that is, $old = ((year < 1752)$ $|| (year == 1752$ && $month < 9) || (year == 1752$ && $month == 9$ && $day <= 2))$. If *old* is true and *year* can be divided by 4, then *year* is a leap year, or if *year* can be divided by 4, but can't be divided by 100, or can be divided by 400, *year* is a leap year.

Four functions are designed based on the Boolean variable *old*:

isLeap(year, old): determine whether *year* is a leap year or not.
days_of_year(year, old): Calculate the number days in *year*.
days_of_month(month, year, isLeap(year, old)): Calculate the number of days in *month*, *year*.
valid(month, day, year, old): Determine whether *month*, *day*, *year* is a valid date or not. If (*year* ≥ 1) && (1 ≤ *month* ≤ 12) && (1 ≤ *day* ≤ days_of_month(*month*, *year*, isLeap(*year*, *old*)) && (the date is not from September 3 to 13, 1752), it returns true.

The main algorithm is based on the above functions. Repeat inputting the current date—*year*, *month*, and *day*;—and for each date, the processes are as follows:

Determine whether the current date is before September 2, 1752, or not. The result is set to *old*.

Determine whether the current date is valid or not by *valid* (*month*, *day*, *year*, *old*). If it is invalid, then output the invalid date's message; else, calculate the total number of days from 0 AD to the current date:

$$sum = \sum_{i=1}^{year-1} day\_of\_year(i, old) + \sum_{i=1}^{month-1} day\_of\_month(i, year, isleap(year, old)) + day$$

If the current date is after September 2, 1752, the day of the week is (*sum* % 7); else, the day of the week is ((*sum* + 5)% 7). Output the transferred date.

Repeat the above processes until the test case is three zeros.

Program

```
#include <iostream>          // Preprocessor Directive
#include <cstdio>
#include <cstring>
using namespace std;         // Using C++ Standard Library
const int maxs = 20;         // The size of string array
                    // String array representing days of the week
const char wstr[ ][maxs] = {"Sunday", "Monday", "Tuesday", "Wednesday",
"Thursday", "Friday", "Saturday"};
                    // String array representing months
const char mstr[ ][maxs] = {"", "January", "February", "March",
"April","May", "June", "July", "August", "September", "October",
"November", "December"};
bool isLeap(int year, bool old = false)    //Determine whether year is a
leap year or not
{
    if (old)                             // The date is before September 2, 1752
        return year % 4 == 0 ? true : false;
```

```
    return (year % 100 == 0 ? (year % 400 == 0 ? true : false) : (year %
4 == 0 ? true : false));
}
int days_of_month(int month, int year, bool leap)    // Return the number
of days in month, year
{
    if (month == 2)
        return leap ? 29 : 28;
    int d;
    switch (month) {
        case 1: case 3: case 5: case 7: case 8:
        case 10: case 12:
            d = 31;
            break;
        default:
            d = 30;
    }
    return d;
}
int days_of_year(int year, bool old)                // Return the number of
days in year
{
    return isLeap(year, old) ? 366 : 365;
}
int getNum(char s[ ], const char ss[ ][maxs], int tot)  // Return the
position of s in ss. If there is no s in ss, return -1.
{
    int i = 0;
    while (i < tot && strcmp(s, ss[i]))
        ++i;
    return i < tot ? i : -1;
}
bool valid(int month, int day, int year, bool old)       // If year>=1
and month∈{1..12} and day∈{1.. the number of days from 0 A.D. to month,
year} and the current date is not from September 3 to 13, 1752; then
return true; else return false
{
    if (year < 1)
        return false;
    if (month < 0 || month > 12)
        return false;
    if (day < 1 || day > days_of_month(month, year, isLeap(year, old)))
        return false;
    if (year == 1752 && month == 9 && 3 <= day && day <= 13)
        return false;
    return true;
}
bool isOld(int month, int day, int year)  // If the current date is
before September 2, 1752, return true; else return false
{
    return year < 1752 || (year == 1752 && month < 9) ||
        (year == 1752 && month == 9 && day <= 2);
}
int main(void)                              // Main function
```

```
{
    int month, day, year;
    cin >> month >> day >> year;          // Input date
    while (month || day || year) {
        bool old = isOld(month, day, year); // Determine whether the date
is before September 2, 1752 or not
        if (!valid(month, day, year, old)) {  // If the date is invalid
            cout << month << '/' << day << '/' << year
                 << " is an invalid date." << endl;
        } else {                           // Accumulate the number of days
from 0 A.D.
            int sum = 0;
            for (int yy = 1; yy < year; yy++)
                sum += days_of_year(yy, old);
            for (int mm = 1; mm < month; mm++)
                sum += days_of_month(mm, year, isLeap(year, old));
            sum += day;
            int week = sum % 7;            // Calculate the day of the week
            if (old)                       // If the date is before
September 2, 1752
                week = (week + 5) % 7;
            cout << mstr[month] << ' ' << day << ", " << year // Output
the transferred date and the day of the week
                 << " is a " << wstr[week] << endl;
        }
        cin >> month >> day >> year;       // Input the next date
    }
    return 0;
}
```

4.2 Application of Arrays 2: Calculation of High-Precision Numbers

In programming languages, range and precision for integer and real are limited. If the range and precision of integer or real in a problem is out of the range and precision for integer or real in programming languages, calculation of high-precision numbers should be used. For calculation of high-precision numbers, there are two fundamental problems: representation of high-precision numbers and fundamental calculations of high-precision numbers.

1. Representation of high-precision numbers: A high-precision number can be represented by an array. Numbers are separated by decimal digits, and each decimal digit is sequentially stored in an array. In the program, first a string is used to store a high-precision number, and in the string a character stores a digit. Then the string is conversed to the corresponding decimal number and stored in an array. For example, for a long positive integer, the program segment is as follows:

```
int a[100]={0};        // Array a is used to store a long positive
integer, one digit is stored in one element. Initial values are 0.
int n;                 // n is the number of digits for the long integer
string s;              // String s is used to receive the integer
```

```
cin>>s;                    // Input the integer into s
   n=s.length( );               // Calculate the number of digits
   for (i=0; i<n; i++)     // Array a stores the integer from right to
   left, and one element stores one digit.
      a[i]=s[n-i-1]-'0';
```

From the above discussion, a high-precision number is stored in an integer array in the order of decimal digits. The array's indexes correspond to digit numbers of the high-precision number so that one digit of the high-precision number is stored in one element in the array.

2. Fundamental calculations of high-precision numbers: Fundamental calculations of high-precision numbers are +, −, *, and /.
   - Addition and subtraction of high-precision numbers: Rules for addition and subtraction of high-precision numbers are the same as rules of arithmetic addition and subtraction. In programs, addition of high-precision numbers needs to carry, and subtraction of high-precision numbers needs to borrow.

Suppose x and y are two nonnegative high-precision integers, and $n1$ is the number of digits of x, and $n2$ is the number of digits of y. x and y are stored in array a and array b in the above format. The program segment for addition of x and y is as follows:

```
for (i=0; i<( n1>n2 ? n1 : n2 ); i++ ){  Addition of two integers whose
numbers of digits are n1 and n2 respectively
a[i]=a[i]+b[i];              // Bitwise addition
if ( a[i]>9 ) {              //  Carry
a[i]=a[i]-10;
a[i+1]++;
}
}
```

Suppose x and y are two nonnegative high-precision integers $(x > y)$, and n is the number of digits of x. x and y are stored in array a and array b in the above format. If $x < y$, then a and b exchange each other and take a negative after the subtraction. The program segment for subtraction of x and y is as follows:

```
for (i=0; i<n; i++) {
        if (a[i]>=b[i])
           a[i]= a[i]-b[i];
        else                    // Borrow
          { a[i]= a[i]+10-b[i];
            a[i+1]--;
          }
        }
```

   - Multiplication and division of high-precision numbers: Based on addition and subtraction of high-precision numbers, algorithms for multiplication and division of high-precision numbers are given.

For multiplication of high-precision numbers, first the number of digits of the product must be determined. Suppose a and b are two positive high-precision integers. LA is the number of digits for a, and LB is the number of digits for b. The number of digits for the product of a and b is at least $LA + LB - 1$. And the upper limit of the number of digits is $LA + LB$.

The algorithm for multiplication of two positive high-precision integers is as follows: First, calculate the product of each digit of the multiplicand and each digit of the multiplier, where the product of $a[i]$ and $b[j]$ is accumulated in array $c[i + j]$. Then the carry process is done in array c. The program segment is as follows:

```
for (i=0; i<=LA-1, i++) // The product of each digit of multiplicand a
and multiplier b is accumulated to corresponding digits of array c
    for (j=0; j<=LB-1; j++)
        c[i+j] += a[i]*b[j];
for (i=0; i<LA+LB; i++)   // Carry
    if(c[i] >= 10)
    {
        c[i+1] += c[i]/10;
        c[i] %=10;
    }
```

The algorithm calculating the quotient and the remainder for positive high-precision integer $a \div$ positive high-precision integer b is as follows:

If $a < b$, then the quotient is 0 and the remainder is a; else, start division of positive high-precision integers based on subtraction. First, based on the difference between the digit number of a and the digit number of b d_1, determine how many times a_1 that a can subtract $b*10^{d1}$ and get remainder $y_1 = a - a_1*b*10^{d1}$. Then based on the difference between the digit number of y_1 and b d_2, determine how many times a_2 that y_1 can subtract $b*10^{d2}$ and get remainder $y_2 = y_1 - a_2*b*10^{d2}$. Repeat the above steps until times a_k that y_{k-1} can subtract b. Finally, get remainder $y = y_{k-1} - a_k*b$ and the quotient is $a_1( )a_2 ... ( )a_k$. (**Note:** ( ) means if $d_i - d_{i+1} > 1$, then $d_i - d_{i+1} - 1$ zeros should be added before a_{i+1}, $2 \le i \le k - 1$.)

For example, $a = 12345$ and $b = 12$. Then the difference between the digit number of a and the digit number of b is $d_1 = 3$. 12345 can subtract $(12*10^3)$ one time, that is, $a_1 = 1$, and get remainder $y_1 = 12345 - 12*10^3 = 345$. And the difference between the digit number of 345 and the digit number of 12 is $d_2 = 1$. 345 can subtract $(12*10^1)$ two times, that is, $a_2 = 2$, and get remainder $y_2 = 345 - 2*12*10^1 = 105$. y_2 can subtract 12 eight times, $a_3 = 8$, and finally get remainder $y = 105 - 8*12 = 9$ and quotient 1028.

4.2.1 Adding Reversed Numbers

The Antique Comedians of Malidinesia prefer comedies to tragedies. Unfortunately, most of the ancient plays are tragedies. Therefore, the dramatic advisor of the ACM has decided to transfigure some tragedies into comedies. Obviously, this work is very hard because the basic sense of the play must be kept intact, although all the things change to their opposites. For example, with the numbers, if any number appears in the tragedy, it must be converted to its reversed form before being accepted into the comedy play.

A reversed number is a number written in arabic numerals, but the order of digits is reversed. The first digit becomes last and vice versa. For example, if the main hero had 1245 strawberries in the tragedy, he has 5421 of them now. Note that all the leading zeros are omitted. That means if the number ends with a zero, the zero is lost by reversing (e.g., 1200 gives 21). Also note that the reversed number never has any trailing zeros.

ACM needs to calculate with reversed numbers. Your task is to add two reversed numbers and output their reversed sum. Of course, the result is not unique because any particular number is a reversed form of several numbers (e.g., 21 could be 12, 120, or 1200 before reversing). Thus, we must assume that no zeros were lost by reversing (e.g., assume that the original number was 12).

Input

The input consists of N cases. The first line of the input contains only positive integer N. Then follow the cases. Each case consists of exactly one line with two positive integers separated by space. These are the reversed numbers you are to add.

Output

For each case, print exactly one line containing only one integer—the reversed sum of two reversed numbers. Omit any leading zeros in the output.

Sample Input	Sample Output
3	34
24 1	1998
4358 754	1
305 794	

Source: ACM Central Europe 1998.

IDs for online judges: POJ 1504, ZOJ 2001, UVA 713.

Analysis

Suppose *Num*[0][0] stores the length of the first addend, and the first addend is stored in *Num*[0][1 .. *Num*[0][0]]; *Num*[1][0] stores the length of the second addend, and the second addend is stored in *Num*[1][1 .. *Num*[1][0]]; *Num*[2][0] stores the length of the sum, and the sum is stored in *Num*[2][1 .. *Num*[2][0]].

The algorithm is as follows.

First, strings for the first addend and the second addend are input and zeros that the two numbers end with are deleted. The two addends are stored in *Num*[0] and *Num*[1]. Then they are changed into reversed numbers.

Second, two reversed numbers, *Num*[0] and *Num*[1], are added. Then their reversed sums *Num*[2] are output. And any leading zeros should be omitted.

Program

```
#include <iostream>       // Preprocessor Directive
#include <cstdio>
#include <cstring>
#include <string>
using namespace std;       // Using C++ Standard Library
int Num[3][1000];
void Read(int Ord)         // If Ord==0, input the first addend ; If Ord==1,
input the second addend
{
```

```
        int flag=0;
        string Tmp;
        cin>>Tmp;                       // Input the string represent the integer
        for(int i=Tmp.length( )-1;i>=0;i--)  // Analyze each character from
right to left
        {
            if(Tmp[i] > '0')                    // Store the integer into Num[Ord]
                flag = 1;
            if(flag)
                Num[Ord][++Num[Ord][0]] = Tmp[i] - '0';
        }
        for(int i=Num[Ord][0],j=1;i>j;i--,j++)        // Get reversed number
Num[Ord]
        {
            flag = Num[Ord][i];
            Num[Ord][i] = Num[Ord][j];
            Num[Ord][j] = flag;
        }
}
void Add( )
{
    Num[2][0] = max(Num[0][0],Num[1][0]);   // the number of additions is
the maximum length of two addends
    for(int i=1;i<=Num[2][0];i++)                   // Bitwise addition
        Num[2][i] = Num[0][i] + Num[1][i];
    for(int i=1;i<=Num[2][0];i++)                   // Carry
    {
        Num[2][i+1] += Num[2][i]/10;
        Num[2][i] %= 10;
    }
    if(Num[2][Num[2][0]+1] > 0)                // Carry
        Num[2][0] ++;
    int flag = 0;
    for(int i=1;i<=Num[2][0];i++)       // Output the reversed sum of two
reversed numbers
    {
        if(Num[2][i] > 0)
            flag = 1;
        if(flag)
            printf("%d",Num[2][i]);
    }
    printf("\n");
}
int main( )                             // Main function
{
    int N;                              // The number of test cases
    cin>>N;
    for(N;N;N--)                        // Input and process for each test case
    {
        memset(Num,0,sizeof(Num));   // Initialize arrays of high
precision numbers 0
        Read(0);                     // The first addend
        Read(1);                     // The second addend
```

```
        Add( );              // Add two reversed numbers, and output their
reversed sum
    }
    return 0;
}
```

In some cases, calculation of high-precision numbers is reduplicated, such as computing power or polynomial. An object-oriented programming method is suitable for these problems. It makes the program structure more clear.

4.2.2 Very Easy!

Most of the time, the students of Computer Science and Engineering Department of the Bangladesh University of Engineering and Technology (BUET) deal with bogus, tough, and very complex formulas. That is why, sometimes, even for an easy problem, they think very hard and make the problem more complex to solve. But, the team members of the BUET Pessimistic team are the only exceptions. In the opposite manner, they treat every hard problem as easy and so cannot do well in any contest. Today, they try to solve a series but fail for treating it as hard. Let them help.

Input

You are given the values of N and A (integer, $1 \leq N \leq 150$; integer, $0 \leq A \leq 15$), respectively. Just try to determine the answer for the following series:

$$\sum_{i=1}^{N} i * A^i$$

Output

For each line of the input, your correct program should output the *integer* value of the sum in separate lines for each pair of values of N and A.

Sample Input	Sample Output
3 3	102
4 4	1252

Source: The Sams' Contest.

ID for online judge: UVA 10523.

Analysis
Based on upper limits of N (150) and A (15) for

$$\sum_{i=1}^{N} i * A^i$$

calculation of high-precision numbers must be used. Because the calculation process requires repeated high-precision multiplication and addition, the object-oriented programming method is more appropriate.

Define class *bigNumber*, where its private section is array *a* representing a high-precision number whose length is *len*, and its public section includes

> *bigNumber*( ): Initialize array *a* 0.
> int *length*( ): Return the length of array *a*.
> int *at*(int *k*): Return *a*[*k*].
> void *setnum*(char *s*[ ]): Change string *s*[ ] into array *a*.
> bool *isZero*( ): Determine whether array *a* is 0 or not.
> void *add*(*bigNumber* &*x*): Addition of high-precision integers: *a*←*a* + *x*.
> void *multi*(*bigNumber* &*x*): Multiplication of high-precision integers: *a*←*a**x.

Based on the above definitions, the algorithm calculating

$$\sum_{i=1}^{N} i * A^i$$

becomes simple and clear:

1. Define base number *a* and power *ap* as objects of class *bigNumber* (*bigNumber a, ap*); translate string *s* representing a base number into array *a* (*a.setnum*(*s*)) representing a high-precision number; initialize *ap* 1 (*ap.setnum*("1")); and define *sum* as an object of class *bigNumber* (*bigNumber sum*).
2. Loop *n* times. Each loop calculates *i***A^i* and accumulates it into *sum*:
   Define the current item *num* as an object for class *bigNumber* (*bigNumber num*).
   Initialize *num i* (sprintf(*s*, "%d", *i*; *num.setnum*(*s*)).
   Calculate power *ap*←*ap**a and the current item *num*←*num**ap (*ap.multi*(*a*); *num.multi*(*ap*));
   Accumulate the current item *sum*←*sum*+*num* (*sum.add*(*num*)).
3. Output $\sum_{i=1}^{N} i * A^i$.

Program

```
#include <cstdio>
#include <cstring>
const int maxlen = 500;      // The size of array a representing high
precision number
const int maxs = 5;          // The size of string s representing base number
class bigNumber {            //   Class declaration of bigNumber
    private:          //Private: array a representing high precision
number whose length is len.
         int a[maxlen];
         int len;
    public:                   //Public:
        bigNumber ( ) {       // Initialize array a 0
            memset (a, 0, sizeof (a));
```

```
        len = 1;
    }

    int length( ) {          // Return the length of array a
        return len;
    }
    int at(int k) {          // Return a[k]
        if (0 <= k && k < len)
            return a[k];
        return -1;
    }
    void setnum(char s[ ]) {  // Change string s[ ] into array a
        len = 0;
        for (int i = strlen(s) - 1; i >= 0; i--)
            a[len++] = int(s[i] - '0');
    }
    bool isZero( ) {                    // Determine whether array a
is 0 or not
        return len == 1 && a[0] == 0;
    }
    void add(bigNumber &x) {          // Addition of high precision
integers: a←a+x
        for (int i = 0; i < x.len; i++) {    // Bitwise addition
            a[i] += x.a[i];
            a[i + 1] += a[i] / 10;
            a[i] %= 10;
        }
        int k = x.len;                   //Carry
        while (a[k]) {
            a[k + 1] += a[k] / 10;
            a[k++] %= 10;
        }
        len = len > k ? len : k;         // The number of digits
    }
    void multi(bigNumber &x) {    // Multiplication of high precision
integers: a←a*x
        if (x.isZero( ))
            setnum("0");
        int product[maxlen];
        memset(product, 0, sizeof(product));
        for (int i = 0; i < len; i++)    // product of a*x is stored
in array product
            for (int j = 0; j < x.length( ); j++)
                product[i + j] += a[i] * x.at(j);

        int k = 0;                       // Become decimal number
        while (k < len + x.length( ) - 1) {
            product[k + 1] += product[k] / 10;
            product[k++] %= 10;
        }
        while (product[k]) {         // Carry
            product[k + 1] += product[k] / 10;
            product[k++] %= 10;
        }
```

```
            len = k;                        // The length of product
            memcpy(a, product, sizeof(product));    //  product is
assigned to a
          }
};
int main(void)
{
    int n;                     // The number of items
    char s[maxs];              // String for base number
    while (scanf("%d%s", &n, s) != EOF) {    // Input the number of the
items and base number
        bigNumber a, ap;    // Define base number a and power ap as
objects of class bigNumber
        a.setnum(s);           // Change string s[ ] into array a
        ap.setnum("1");        // Initialize ap 1
        bigNumber sum;    // Define sum as an object of class bigNumber
        for (int i = 1; i <= n; i++) {
            bigNumber num;    // Define the current item num as an object
as class bigNumber
            sprintf(s, "%d", i); // Initialize num i
            num.setnum(s);
            ap.multi(a);       // Calculate power ap←ap*a
            num.multi(ap);     // Calculate the current item num←num*ap
            sum.add(num);      // Accumulate the current item sum←sum+num
        }
        for (int i = sum.length( ) - 1; i >= 0; i--)    // Output
            printf("%d", sum.at(i));
        putchar('\n');
    }
    return 0;
}
```

The output comment contains the formula $\sum_{i=1}^{N} i * A^i$

4.3 Application of Arrays 3: Representation and Computation of Polynomials

Representation and computation of polynomials is one of the applications of linear lists accessed directly. A polynomial of one indeterminate is as follows:

$$P_n(x) = a_0 + a_1 x + a_2 x^2 + \ldots a_n x^n = \sum_{i=0}^{n} a_i x^i$$

There are two storage methods for polynomials of one indeterminate:

1. Numeric array a is used to store a polynomial of one indeterminate. All elements' coefficients are stored in a array $a[0 .. n]$ in exponents' ascending order (n is the highest degree). The index for a shows the number of exponents for the current element. For example, if the ith element is empty, that is, in the polynomial the ith element's coefficient $a_i = 0$, then the

corresponding array element $a[i] = 0$. Obviously, the length of array a lies on the highest degree of the polynomial.

2. Structure array a is used to store a polynomial of one indeterminate. Indexes for array a are serial numbers of elements. An array element is a structure containing its coefficient $a[i].coef$ and exponent $a[i].exp$. Obviously, the length of array a is the length of the polynomial.

Based on the above data structures, computations of polynomials are introduced. For example,

$$\sum_{i=0}^{k1} a_i x^i + \sum_{i=0}^{k2} a_i x^i = \sum_{i=0}^{\max\{k1,k2\}} (a_i + b_i) x^i$$

$$\sum_{i=0}^{k1} a_i x^i * \sum_{j=0}^{k2} b_j x^j = \sum_{i=0}^{k1} (a_i x^i * \sum_{j=0}^{k2} (b_j x^j))$$

Similarly, subtraction and division of two polynomials and other polynomials' computations can also be implemented. If storage method 1 is used, the storage of memory will be larger and algorithms will be simple. If storage method 2 is used, the memory consumption will be reduced, but the algorithms will be complex.

4.3.1 Polynomial Showdown

Given the coefficients of a polynomial from degree 8 down to 0, you are to format the polynomial in a readable format with unnecessary characters removed. For instance, given the coefficients 0, 0, 0, 1, 22, –333, 0, 1, and –1, you should generate an output line that displays $x^5 + 22x^4 - 333x^3 + x - 1$.

The formatting rules that must be adhered to are as follows:

1. Terms must appear in decreasing order of degree.
2. Exponents should appear after a caret, ^.
3. The constant term appears as only the constant.
4. Only terms with nonzero coefficients should appear, unless all terms have zero coefficients, in which case the constant term should appear.
5. The only spaces should be a single space on either side of the binary + and – operators.
6. If the leading term is positive, then no sign should precede it; a negative leading term should be preceded by a minus sign, as in $-7x^2 + 30x + 66$.
7. Negated terms should appear as a subtracted unnegated term (with the exception of a negative leading term, which should appear as described above). That is, rather than $x^2 + -3x$, the output should be $x^2 - 3x$.
8. The constants 1 and –1 should appear only as the constant terms. That is, rather than $-1x^3 + 1x^2 + 3x^1 - 1$, the output should appear as $-x^3 + x^2 + 3x - 1$.

Input

The input will contain one or more lines of coefficients delimited by one or more spaces. There are nine coefficients per line, each coefficient being an integer with a magnitude of less than 1000.

Output

The output should contain the formatted polynomials, one per line.

Sample Input	Sample Output
0 0 0 1 22 –333 0 1 –1	x^5 + 22x^4 – 333x^3 + x – 1
0 0 0 0 0 0 –55 5 0	–55x^2 + 5x

Source: ACM Mid-Central United States 1996.

IDs for online judges: POJ 1555, ZOJ 1720, UVA 392.

Analysis

Coefficient a_{n-i-1} whose exponent is $n - i - 1$ is stored in array element $a[i]$. Array element $a[n - 1]$ is the constant term. Initially, based on exponents' order from high to low, coefficients are input into $a[0 .. n - 1]$.

Nonzero term $a[i]$ ($a[i] \neq 0$, $i = 0 .. n - 1$) is analyzed from the exponents' order from high to low. There are two cases: $a[i]$ is the first term or is not the first term:

1. $a[i]$ is the first term of the polynomial:
   Coefficient: If $a[i] == -1$ and it is not a constant term ($i < n - 1$), then output '–' directly; otherwise, if $a[i] \neq 1$ or $a[i]$ is a constant term ($i == n - 1$), then output coefficient $a[i]$.
   Power: If the exponent is 1 ($i == n - 2$), then output 'x' directly; else, if it is not a constant term ($i < n - 1$), output 'x^' ($n - i - 1$).
   Reserve the mark of the first term of the polynomial.
2. $a[i]$ is not the first term of the polynomial:
   Sign: Output ($a[i] < 0$? '–' : '+').
   Coefficient: If $a[i] \neq 1$ or -1, or $a[i]$ is a constant term, output the absolute value of $a[i]$.
   Power: If the exponent is 1 ($i == n - 2$), then output 'x' directly; else, if it is not a constant term ($i < n - 1$), output "x^"($n - i - 1$).

After dealing with the polynomial, if the mark of the first term of the polynomial is not changed, all coefficients are 0. Then output 0.

Program

```
#include <iostream>                    // Preprocessor Directive
using namespace std;                   // Using C++ Standard Library
const int n = 9;                       //The number of terms of a polynomial
inline int fabs(int k)                 // Return the absolute value of k
{
    return k < 0 ? -k : k;
}
int main(void)                         // Main function
{
    int a[n];
    while (cin >> a[0]) {              // Input coefficients
        for (int i = 1; i < n; i++)
            cin >> a[i];
        bool first = true;            // Set the mark for the first term
```

```
        for (int i = 0; i < n; i++)
            if (a[i]) {                      //Output non-zero terms
                if (first) {                 // Deal with the first term
                    if (a[i] == -1 && i < n - 1)    // The current term
is -1
                        cout << '-';
                    else if (a[i] != 1 || i == n - 1)    // The current
is not 1
                        cout << a[i];
                    if (i == n - 2) // If the exponent is 1, don't output
the exponent; else output the exponent.
                        cout << 'x';
                    else if (i < n - 1)
                        cout << "x^" << n - i - 1;
                    first = false;   // Reserve the mark of the first term
of the polynomial
                } else {             //Output the sign and the absolute value
                    cout << ' ' << (a[i] < 0 ? '-' : '+') << ' ';   //
Output the sign
                    if (fabs(a[i]) != 1 || i == n - 1)            //If the
coefficient is 1, don't output it
                        cout << fabs(a[i]);
                    if (i == n - 2)   // If the exponent is 1, don't
output the exponent; else output the exponent.
                        cout << 'x';
                    else if (i < n - 1)
                        cout << "x^" << n - i - 1;
                }
            }
        if (first)                        // If all coefficients are 0, output 0.
            cout << 0;
        cout << endl;
    }
    return 0;
}
```

Making use of arrays to store polynomials can not only normalize polynomials' output, but also be convenient for the computation of polynomials.

4.3.2 Modular Multiplication of Polynomials

Consider polynomials whose coefficients are 0 and 1. Addition of two polynomials is achieved by adding the coefficients for the corresponding powers in the polynomials. The addition of coefficients is performed by addition modulo 2, that is, $(0 + 0) \mod 2 = 0$, $(0 + 1) \mod 2 = 1$, $(1 + 0) \mod 2 = 1$, and $(1 + 1) \mod 2 = 0$. Hence, it is the same as the exclusive-or (xor) operation.

$$(x^6 + x^4 + x^2 + x + 1) + (x^7 + x + 1) = x^7 + x^6 + x^4 + x^2$$

Subtraction of two polynomials is done similarly. Since subtraction of coefficients is performed by subtraction modulo 2, which is also the xor operation, subtraction of polynomials is identical to the addition of polynomials:

$$(x^6 + x^4 + x^2 + x + 1) - (x^7 + x + 1) = x^7 + x^6 + x^4 + x^2$$

Multiplication of two polynomials is done in the usual way (of course, addition of coefficients is performed by addition modulo 2):

$$(x^6 + x^4 + x^2 + x + 1)(x^7 + x + 1) = x^{13} + x^{11} + x^9 + x^8 + x^6 + x^5 + x^4 + x^3 + 1$$

Multiplication of two polynomials $f(x)$ and $g(x)$ modulo a polynomial $h(x)$ is the remainder of $f(x)g(x)$ divided by $h(x)$:

$$(x^6 + x^4 + x^2 + x + 1)(x^7 + x + 1) \text{ modulo } (x^8 + x^4 + x^3 + x + 1) = x^7 + x^6 + 1$$

The largest exponent of a polynomial is called its degree. For example, the degree of $x^7 + x^6 + 1$ is 7.

Given three polynomials $f(x)$, $g(x)$, and $h(x)$, you are to write a program that computes $f(x)g(x)$ modulo $h(x)$.

We assume that the degrees of both $f(x)$ and $g(x)$ are less than the degree of $h(x)$. The degree of a polynomial is less than 1000.

Since coefficients of a polynomial are 0 or 1, a polynomial can be represented by $d + 1$ and a bit string of length $d + 1$, where d is the degree of the polynomial and the bit string represents the coefficients of the polynomial. For example, $x^7 + x^6 + 1$ can be represented by 8 1 1 0 0 0 0 0 1.

Input

The input consists of T test cases. The number of test cases (T) is given in the first line of the input file. Each test case consists of three lines that contain three polynomials $f(x)$, $g(x)$, and $h(x)$, one per line. Each polynomial is represented as described above.

Output

The output should contain the polynomial $f(x)g(x)$ modulo $h(x)$, one per line.

Sample Input	Sample Output
2	8 1 1 0 0 0 0 0 1
7 1 0 1 0 1 1 1	14 1 1 0 1 1 0 0 1 1 1 0 1 0 0
8 1 0 0 0 0 0 1 1	
9 1 0 0 0 1 1 0 1 1	
10 1 1 0 1 0 0 1 0 0 1	
12 1 1 0 1 0 0 1 1 0 0 1 0	
15 1 0 1 0 1 1 0 1 1 1 1 1 0 0 1	

Source: ACM Taejon 2001.

IDs for online judges: POJ 1060, ZOJ 1026, UVA 2323.

Analysis

Suppose the length of the bit string for polynomial $f(x)$ is lf, and all coefficients are stored in $f[lf - 1 .. 0]$; the length of the bit string for polynomial $g(x)$ is lg, and all coefficients are stored in

$g[lg - 1 .. 0]$; the length of the bit string for polynomial $h(x)$ is lh, and all coefficients are stored in $h[lh - 1 .. 0]$.

Array *Sum* is used to store the product of $f(x)*g(x)$ and the result of $(f(x)*g(x))$ modulo $h(x)$, and its bit string length is ls. All coefficients are stored in $f[ls - 1 .. 0]$.

1. **Calculate *Sum*(x) = f(x)*g(x)**
   Because coefficients for $f(x)$ and $g(x)$ are 0 or 1, the length of the bit string for $f(x)*g(x)$ is $ls = lf + lg - 1$. Multiply x^i's coefficient $f[i]$ in f and x^j's coefficient $g[j]$ in g. And add the result $f[i]$ & $g[j]$ to x^{i+j}'s coefficient in the product polynomial, that is, $sum[i + j] = sum[i + j] \wedge (f[i]$ & $g[j])$ $(0 \le i \le lf - 1, 0 \le j \le lg - 1)$.

2. **Calculate *Sum*(x) = *Sum*(x) modulo h(x)**
   $Sum(x) = Sum(x)$ modulo $h(x)$, that is, $Sum(x)$ is repeatedly divided by $h(x)$ until the remainder is less than $h(x)$. The remainder is the result of modulo. The process comparing $Sum(x)$ with $h(x)$ is as follows.

   If $ls > lh$, then $Sum(x)$ is greater. If $ls < lh$, then $h(x)$ is greater. If $ls == lh$, then from the highest power $ls - 1$ compare coefficients termwise until the following cases holds: $sum[i] == 1$ and $h[i] == 0$, or $sum[i] == 0$ and $h[i] == 1$. If $sum[i] == 1$ and $h[i] == 0$, then $Sum(x)$ is greater. If $sum[i] == 0$, $h[i] == 1$, then $h(x)$ is greater.

   If $Sum(x)$ is greater than $h(x)$, $Sum(x)$ is divided by $h(x)$: From the lowest bit of $h(x)$, the division operation is done in the order of powers from low to high. $h[i]$ xor $sum[i + ls - lh]$, that is, $sum[i + d] = sum[i + d] \wedge h[i]$ $(i = 0 .. ls - 1)$. Then the degree of $sum(x)$ is adjusted (while $(ls$ && !$sum[ls - 1])$ $-- ls)$.

Repeat the above process until $Sum(x)$ is not greater than $h(x)$. ls is the number of terms of the remainder of the polynomial. All coefficients are in $Sum[ls - 1 .. 0]$.

Program

```
#include <iostream>                    // Preprocessor Directive
using namespace std;                   // Using C++ Standard Library
const int maxl = 1000 + 5;             // The size of product array sum
int compare(int a[ ], int la, int b[ ], int lb)   //Compare polynomials a
and b
{
    if (la > lb)                       //Compare degrees of a and b
        return 1;
    if (la < lb)
        return -1;
    for (int i = la - 1; i >= 0; i--)      // If degrees are equal,
compare coefficients termwise.
        if (a[i] && !b[i])
            return 1;
        else if (!a[i] && b[i])
            return -1;
    return 0;
}
int main(void)                         // Main function
{
    int loop;
    cin >> loop;                       //Input the number of test cases.
    while (loop--) {
```

```
        int f[maxl], g[maxl], h[maxl];
        int lf, lg, lh;
        cin >> lf;                      //Input the length of bit string
for polynomial f
        for (int i = lf - 1; i >= 0; i--)      // Input coefficients in f
            cin >> f[i];
        cin >> lg;                      // Input the length of bit string
for polynomial g
        for (int i = lg - 1; i >= 0; i--)      // Input coefficients in g
            cin >> g[i];
        cin >> lh;                      // Input the length of bit string
for polynomial h
        for (int i = lh - 1; i >= 0; i--)      // Input coefficients in h
            cin >> h[i];
        int sum[maxl + maxl], ls = lf + lg - 1;   //Initialize product
array sum and its length
        for (int i = 0; i < ls; i++)
            sum[i] = 0;
        for (int i = 0; i < lf; i++)          // Calculate product array sum
            for (int j = 0; j < lg; j++)
                sum[i + j] ^= (f[i] & g[j]);
                // Calculate sun[ ]modulo h[ ]
        while (compare(sum, ls, h, lh) >= 0) { // If the current sum is
not less than h, then continue
            int d = ls - lh;                 // Calculate the remainder that
sum is divided by h
            for (int i = 0; i < lh; i++)
                sum[i + d] ^= h[i];
            while (ls && !sum[ls - 1])       //Determine the degree of sum
                --ls;
        }
        if (ls == 0)                    //Calculate and output the length of sum
            ls = 1;
        cout << ls << ' ';
        for (int i = ls - 1; i > 0; i--)      //Output coefficients in sum
            cout << sum[i] << ' ';
        cout << sum[0] << endl;
    }
    return 0;
}
```

4.4 Application of Arrays 4: Calculation of Numerical Matrices

Matrices are used widely in many fields. Normally two-dimensional arrays are used to represent numerical matrices. Suppose the numbers of the row and column of a matrix are m and n, respectively. A two-dimensional array a is used to represent a matrix, where $a[i-1][j-1]$ represents the element in row i and column j in the matrix. Any element in a matrix can be accessed through the array subscript directly. Therefore, using a two-dimensional array to store a numerical matrix is a kind of linear list accessed directly.

Using two-dimensional arrays can implement many calculations of numerical matrices, such as matrix transpose, addition, subtraction, and multiplication of two numerical matrices.

4.4.1 Error Correction

A Boolean matrix has the *parity property* when each row and each column has an even sum, that is, contains an even number of bits that are set. Here's a 4 × 4 matrix that has the parity property:

```
1 0 1 0
0 0 0 0
1 1 1 1
0 1 0 1
```

The sums of the rows are 2, 0, 4, and 2. The sums of the columns are 2, 2, 2, and 2.

Your job is to write a program that reads in a matrix and checks if it has the parity property. If not, your program should check if the parity property can be established by changing only 1 bit. If this is not possible either, the matrix should be classified as *corrupt*.

Input

The input file will contain one or more test cases. The first line of each test case contains one integer n ($n < 100$), representing the size of the matrix. On the next n lines, there will be n integers per line. No other integers than 0 and 1 will occur in the matrix. Input will be terminated by a value of 0 for n.

Output

For each matrix in the input file, print one line. If the matrix already has the parity property, print "OK." If the parity property can be established by changing one bit, print "Change bit (i, j)," where i is the row and j the column of the bit to be changed. Otherwise, print "Corrupt."

Sample Input	Sample Output
4	OK
1 0 1 0	Change bit (2, 3)
0 0 0 0	Corrupt
1 1 1 1	
0 1 0 1	
4	
1 0 1 0	
0 0 1 0	
1 1 1 1	
0 1 0 1	
4	
1 0 1 0	

(Continued)

Sample Input	Sample Output
0 1 1 0	
1 1 1 1	
0 1 0 1	
0	

Source: Ulm Local Contest 1998.

IDs for online judges: POJ 2260, ZOJ 1949.

Analysis

Suppose there is matrix a, where the sum of numbers in the ith row is $row[i]$ and the sum of numbers in the jth column is $col[j]$.

First, after matrix a is input, the sums of all numbers in every row and all numbers in every column are calculated and stored in arrays row and col, respectively.

Second, the number of rows cr and the number of columns cc whose sum is odd are calculated. The last row number i and the last column number j whose sum is odd are stored. (if ($row[k]$ & 1), then {cr++; $i = k$;}; if ($col[k]$ & 1), then {cc++; $j = k$;}; ($0 \le k \le n - 1$)).

Finally, determine the parity property of matrix a:

1. If sums of all elements in every row and all elements in every column are even ($cc ==$ 0 and $cr ==$ 0), then matrix a has the parity property, and "OK" is output.
2. If there is only one row and one column that has an odd sum ($cc ==$ 1 and $cr ==$ 1), then the bit in (i, j) makes sums odd, and the bit is changed to make matrix a have the parity property. Because rows and columns in array a are numbered from 0, $(i + 1, j + 1)$ is output.
3. Otherwise, "corrupt" is output.

Program

```
#include <stdio.h>                      // Preprocessor Directive
#include <assert.h>
#define MAXN 512                        //The max size of matrix
int n;                                  // The size of matrix
int a[MAXN][MAXN], row[MAXN], col[MAXN];//* matrix a, the sum of numbers
in the ith row is row[i] and the sum of numbers in the jth column is
col[j]
FILE *input;                            //Pointer variable for input
int read_case( )                        //Input matrix
{
  int i,j;
  fscanf(input,"%d",&n);                //Input the size of matrix
  if (n==0) return 0;                   //If input ends, return 0
  for (i=0; i<n; i++)                   //Input the matrix and return 1
    for (j=0; j<n; j++)
      fscanf(input,"%d",&a[i][j]);
  return 1;
}
void solve_case( )     //Determine whether the boolean matrix has the
parity property or not
{
```

```
int cc,cr,i,j,k;
  for (i=0; i<n; i++)                 //Initialization
    row[i] = col[i] = 0;
  for (i=0; i<n; i++)          // Calculate the sums of the rows and the
sums of the columns
    for (j=0; j<n; j++)
    {
  row[i] += a[i][j];
  col[j] += a[i][j];
    }
  cr = cc = 0;
  for (k=0; k<n; k++)          // Calculate the number of rows cr and the
number of columns cc whose sum is odd
    {
      if (row[k]&1) { cr++; i=k; }
      if (col[k]&1) { cc++; j=k; }
    }
  if (cc==0 && cr==0) printf("OK\n");   // determine the parity property
of matrix a
  else if (cc==1 && cr==1) printf("Change bit (%d,%d)\n",i+1,j+1);
  else                     printf("Corrupt\n");
}
int main( )
{
  input = fopen("error.in","r");
  assert(input!=NULL);
  while(read_case( )) solve_case( );
  fclose(input);
  return 0;
}
```

4.4.2 Matrix Chain Multiplication

Suppose you have to evaluate an expression like $A*B*C*D*E$, where A, B, C, D, and E are matrices. Since matrix multiplication is associative, the order in which multiplications are performed is arbitrary. However, the number of elementary multiplications needed strongly depends on the evaluation order you choose.

For example, let A be a 50*10 matrix, B a 10*20 matrix, and C a 20*5 matrix. There are two different strategies to compute $A*B*C$, namely, $(A*B)*C$ and $A*(B*C)$. The first one takes 15,000 elementary multiplications, but the second one only 3,500.

Your job is to write a program that determines the number of elementary multiplications needed for a given evaluation strategy.

Input

Input consists of two parts: a list of matrices and a list of expressions.

The first line of the input file contains one integer n ($1 \le n \le 26$), representing the number of matrices in the first part. The next n lines each contain one capital letter, specifying the name of the matrix, and two integers, specifying the number of rows and columns of the matrix.

The second part of the input file strictly adheres to the following syntax (given in EBNF):

SecondPart = Line {Line}
Line = Expression

Expression = Matrix | "(" Expression Expression ")"
Matrix = "A" | "B" | "C" | .. | "X" | "Y" | "Z"

Output

For each expression found in the second part of the input file, print one line containing the word "error" if evaluation of the expression leads to an error due to nonmatching matrices. Otherwise, print one line containing the number of elementary multiplications needed to evaluate the expression in the way specified by the parentheses.

Sample Input	Sample Output
9	0
A 50 10	0
B 10 20	0
C 20 5	error
D 30 35	10000
E 35 15	error
F 15 5	3500
G 5 10	15000
H 10 20	40500
I 20 25	47500
A	15125
B	
C	
(AA)	
(AB)	
(AC)	
(A(BC))	
((AB)C)	
(((((DE)F)G)H)I)	
(D(E(F(G(HI)))))	
((D(EF))((GH)I))	

Source: Ulm Local Contest 1996.

IDs for online judges: POJ 2246, ZOJ 1094.

Analysis

Suppose matrix *A*, whose size is $m*n$, and matrix *B*, whose size is $n*l$, multiply. The product is matrix *C*, whose size is $m*l$, where

$$C[i][j] = \sum_{k=0}^{n-1} A[i][k] * B[k][j] (0 \le i \le m-1,\ 0 \le j \le l-1)$$

It takes $m*n*l$ elementary multiplications.

Suppose e is the string array for an expression and p is the character pointer for e.

The number of elementary multiplications *mults* and the number of rows and columns in the product matrix *rows* and *cols* constitute a structure, defined by *triple*. Function *expression( )* is used to calculate the number of elementary multiplications. And the current product matrix t and matrix $t1$ and matrix $t2$, which are needed to multiply, are all instances of *triple*.

When the first part, a list of matrices, is input, the number of rows *rows[c]* and columns *cols[c]* ('A' $\le c \le$ 'Z') are recorded, and a capital letter represents a matrix. Then expressions are input repeatedly. Before expression e is input, character pointer p and error mark for e are initialized 0. Function *expression( )* is called to calculate product matrix t:

1. If the current character is '(' and then character pointer p++, recursively calculate expressions $t1$ and $t2$ in brackets ($t1$ = expression( ); $t2$ = expression( )) and character pointer p++. If the number of columns for $t1$ is not equal to the number of rows for $t2$ ($t1.cols! = t2.rows$), then set error mark ($error = 1$); else, calculate product matrix t, which $t1$ and $t2$ multiply ($t.rows = t1.rows$; $t.cols = t2.cols$; $t.mults = t1.mults + t2.mults + t1.rows*t1.cols*t2.cols$).
2. If the current character is a letter, then record the number of rows and columns and initialize the number of multiplications 0 ($t.rows = rows[e[p]]$; $t.cols = cols[e[p]]$; $t.mults = 0$) and character pointer p++.

Finally, return product matrix t. After calling function *expression( )*, if $error == 1$, then expression calculation is wrong; else, the number of elementary multiplications is $t.mults$.

Program

```
#include <stdio.h>            // Preprocessor Directive
typedef struct {int mults; int rows; int cols;} triple;  // The number of
elementary multiplication mults, the number of rows and columns in the
product matrix rows and cols constitute struct triple.
int rows[256],cols[256];   // Matrix c with the number of rows rows[c] and
columns cols[c]
char e[100];               // Character array e stores an expression
int p;                     // character point e
char error;                // error mark
triple expression( )       //Calculate the product matrix for expression e
{
   triple t;               //the product matrix
   if (e[p]=='(')   // If the current character is '(', then character
pointer p++, and take out expressions t1 and t2 in the bracket
   {
       triple t1,t2;
       p++;
       t1 = expression( );
       t2 = expression( );
       p++;        // then character pointer p++
```

```
      if (t1.cols!=t2.rows) error = 1;// If the number of columns for t1 is
not equal to the number of rows for t2, then the set error mark (error=1)
      t.rows  = t1.rows;            // calculate the number of
columns and rows for product matrix t and the number of elementary
multiplications
      t.cols  = t2.cols;
      t.mults = t1.mults+t2.mults+t1.rows*t1.cols*t2.cols;
    }
  else                     // If the current character is a letter, then
record the number of rows and columns and initialize the number of
multiplications 0
    {
      t.rows = rows[e[p]];
      t.cols = cols[e[p]];
      t.mults = 0;
      p++;                 // character pointer p++
    }
  return t;                // Return product matrix t
}
main( )
{
  FILE* input = fopen("matrix.in","r");
  char c;
  int i,n,ro,co;
  triple t;                           //Product matrix t is an instance
of class triple
    fscanf(input,"%d%c",&n,&c);          // the number of matrices
  for (i=0; i<n; i++)                   //Input n matrices' information
    {
      fgets(e,99,input);
      sscanf(e,"%c %d %d",&c,&ro,&co);     // one capital letter,
specifying the name of the matrix, and two integers, specifying the
number of rows and columns of the matrix
      rows[c] = ro;
      cols[c] = co;
    }
  while (1)                          //Input and deal with each expression
    {
      fgets(e,99,input);            //Input the expression
      if (feof(input)) break;       //If input ends, end the loop
      p = error = 0;                //Initialize character pointer and
error mark
      t = expression( );            //Calculate the product matrix for
expression e
      if (error)                    // if evaluation of the expression
leads to an error, then output "error"; else output he number of
elementary multiplications
    puts("error");
        else
          printf("%dn",t.mults);
    }
  fclose(input);
  return 0;
}
```

4.5 Character Strings 1: Storage Structure of Character Strings

A character string is a sequence of characters. The length of a string is the number of characters in the sequence. If the length is zero, the string is called an empty string. Character string $s = \text{``} a_0 a_1 \ldots a_{n-1}\text{,''}$ where s is the name of the string, $a_0 a_1 \ldots a_{n-1}$ is the value of the string, and a_i $(0 \le i \le n-1)$ is a character in the string. The length of the string is n. '\0' is the end mark for a string and isn't regarded as a character in the string. Obviously, a character string is a linear list whose element is a character.

Normally, a character string is stored in an array. There are many library functions for character string in the C++ standard library.

4.5.1 TEX Quotes

TEX is a typesetting language developed by Donald Knuth. It takes source text together with a few typesetting instructions and produces, one hopes, a beautiful document. Beautiful documents use double-left-quote and double-right-quote to delimit quotations, rather than the mundane ", which is what is provided by most keyboards. Keyboards typically do not have an oriented double-quote, but they do have a left-single-quote, `, and a right-single-quote, ´. Check your keyboard now to locate the left-single-quote key ` (sometimes called the backquote key) and the right-single-quote key ´ (sometimes called the apostrophe or just quote). Be careful not to confuse the left-single-quote ` with the backslash key, \. TEX lets the user type two left-single-quotes `` to create a left-double-quote and two right-single-quotes ´´ to create a right-double-quote. Most typists, however, are accustomed to delimiting their quotations with the unoriented double-quote, ".

If the source contained

"To be or not to be," quoth the bard, "that is the question."

then the typeset document produced by TEX would not contain the desired form: "To be or not to be," quoth the bard, "that is the question." In order to produce the desired form, the source file must contain the sequence:

``To be or not to be,´´ quoth the bard, ``that is the question.´´

You are to write a program that converts text containing double-quote (") characters into text that is identical except that double-quotes have been replaced by the two-character sequences required by TEX for delimiting quotations with oriented double-quotes. The double-quote (") characters should be replaced appropriately by either `` if the " opens a quotation or ´´ if the " closes a quotation. Notice that the question of nested quotations does not arise: The first " must be replaced by ``, the next by ", the next by ``, the next by ", the next by ``, the next by ", and so on.

Input

Input will consist of several lines of text containing an even number of double-quote (") characters. Input is ended with an end-of-file character.

Output

The text must be output exactly as it was input except that

- The first " in each pair is replaced by two ` characters: ``
- The second " in each pair is replaced by two ' characters: "

Sample Input	Sample Output
"To be or not to be," quoth the Bard, "that is the question."	``To be or not to be," quoth the Bard, ``that is the question."
The programming contestant replied: "I must disagree.	The programming contestant replied: ``I must disagree.
To `C' or not to `C', that is The Question!"	To `C' or not to `C', that is The Question!"

Source: ACM East Central North America 1994.

IDs for online judges: POJ 1488, UVA 272.

Analysis
Substitution forms for each pair of double-quotes appear alternately. That is, the first " is replaced by two ` characters, and the second " is replaced by two ' characters. Suppose $p[0]$ is the first substitution form """ for the first ", and $p[1]$ is the second substitution form """ for the second ".
    Initially $k = 0$. Then the string is scanned. If the current character is not a double-quote, then output it; else, replace it by $p[k]$, and change the substitution form ($k = !k$).

Program
```
#include <cstdio>                      // Preprocessor Directive
#include <cstring>
const char p[ ][5] = { "``", "''" };       //p[0] is two ` characters,
and p[1] is two ' characters
int main(void)
{
int k = 0;                       // The first " is replaced by two `
characters
    char c;
    while ((c = getchar( )) != EOF) {  // Input character c repeatedly
        if (c == '"') {                // If the current character is not
double-quote, then output it; else replace it by p[k], and change the
substitution form (k = !k) .
            printf("%s", p[k]);
            k = !k;
        } else
            putchar(c);
    }
    return 0;
}
```

4.6 Character Strings 2: Pattern Matching of Character Strings

Pattern matching is an important operation for character strings. Suppose T and P are two character strings, where the length of T is n and the length of P is m, $1 \leq m \leq n$. T is the target. P is the pattern. Pattern matching is to search whether there is a substring in T that equals P and return the substring's position in T, if it exists. There are two kinds of algorithms for pattern matching:

1. Brute force algorithm: A naive pattern matching algorithm. That is, characters in P are compared with characters in T successively. Its time complexity is $O((n - m + 1)m))$.
2. KMP algorithm: Shown by D.E. Knuth, J.H. Morris, and V.R. Pratt.

4.6.1 Blue Jeans

The Genographic Project is a research partnership between IBM and the National Geographic Society that is analyzing DNA from hundreds of thousands of contributors to map how the earth was populated.

As an IBM researcher, you have been tasked with writing a program that will find commonalities among given snippets of DNA that can be correlated with individual survey information to identify new genetic markers.

A DNA base sequence is noted by listing the nitrogen bases in the order in which they are found in the molecule. There are four bases: adenine (A), thymine (T), guanine (G), and cytosine (C). A six-base DNA sequence could be represented as TAGACC.

Given a set of DNA base sequences, determine the longest series of bases that occurs in all of the sequences.

Input

Input to this problem will begin with a line containing a single integer n indicating the number of data sets. Each data set consists of the following components:

■ A single positive integer m ($2 \le m \le 10$) indicating the number of base sequences in this data set
■ m lines each containing a single base sequence consisting of 60 bases

Output

For each data set in the input, output the longest base subsequence common to all of the given base sequences. If the longest common subsequence is less than three bases in length, display the string "no significant commonalities" instead. If multiple subsequences of the same longest length exist, output only the subsequence that comes first in alphabetical order.

Sample Input	Sample Output
3	no significant commonalities
2	AGATAC
GATACCAGATACCAGATACCAGATACCAG ATACCAGATACCAGATACCAGATACCAGATA	CATCATCAT
AAAAAAAAAAAAAAAAAAAAAAAAAAAAAAAAAAA AAAAAAAAAAAAAAAAAAAAAAAAAAAAAA	
3	
GATACCAGATACCAGATACCAGATACCAG ATACCAGATACCAGATACCAGATACCAGATA	

(Continued)

Sample Input	Sample Output
GATACTAGATACTAGATACTAGATACTAAAGG AAAGGGAAAAGGGGAAAAAGGGGGAAAA	
GATACCAGATACCAGATACCAGATACCAAAGG AAAGGGAAAAGGGGAAAAAGGGGGAAAA	
3	
CATCATCATCCCCCCCCCCCCCCCCCCCCCCCC CCCCCCCCCCCCCCCCCCCCCCCCCCCCCCCC	
ACATCATCATAAAAAAAAAAAAAAAAAAAAAAAA AAAAAAAAAAAAAAAAAAAAAAAAAAAAA	
AACATCATCATTTTTTTTTTTTTTTTTTTTTTTTTTTTT TTTTTTTTTTTTTTTTTTTTTTTTTTT	

Source: ACM South Central United States 2006.

IDs for online judges: POJ 3080, ZOJ 2784, UVA 3628.

Analysis

Suppose the longest common subsequence is *ans* and its length is *len*.

m base sequences are stored in $p[0] .. p[m-1]$. Because the common subsequence is a substring for all base sequences, all subsequences for $p[0]$ are enumerated: a subsequence for $p[0]$ *s* is a pattern, and $p[1] .. p[m-1]$ are targets. The process of pattern matching is as follows.

If *s* is a common subsequence for $p[1] .. p[m-1]$ ($strstr(p[k], s)! = NULL, 1 <= k <= m-1$), and (the length of $s > len$) or (the length of $s == len$) and *s* is the first for the current longest common subsequence *ans* in alphabetical order ($strcmp(ans, s) > 0$), then *s* is adjusted as the current longest common subsequence ($len =$ the length of s; $strcpy(ans, s)$).

After all subsequences for $p[0]$ are enumerated, the longest common subsequence *ans* is the solution to the problem.

Because the length of the base sequence is only 60, the brute force algorithm can be used. Some library functions for character strings, such as *strlen*( ), *strcmp*( ), and *strcpy*( ), are used to make the program concise and clear.

Program

```
#include <cstdio>               // Preprocessor Directive
#include <cstring>
const int maxm = 10 + 5;        //The upper limit of the number of base
sequences
const int maxs = 60 + 5;        // The upper limit of the length of a
base sequence
int main(void)
{
    int loop;
    scanf("%d", &loop);         //The number of test cases
    while (loop--) {
        int m;
        char p[maxm][maxs];
        scanf("%d", &m);        // The number of base sequences
```

```
    for (int i = 0; i < m; i++)    //The i-th base sequence
        scanf("%s", p[i]);
    int len;         // The length of longest common base subsequence
    char ans[maxs];          // The longest common base subsequence
    len = 0;
    // All subsequences for p[0] are enumerated and determined
    for (int i = 0; i < strlen(p[0]); i++)   //  The starting position
of the enumerated subsequence i
        for (int j = i + 2; j < strlen(p[0]); j++) {  // The end
position of the enumerated subsequence j
            char s[maxs];                     //The enumerated
subsequence s
            strncpy(s, p[0] + i, j - i + 1);
            s[j - i + 1] = '0';
            bool ok = true; // Determine whether s is a common
subsequence for p[1]...p[m-1]
            for (int k = 1; ok && k < m; k++)
                if (strstr(p[k], s) == NULL)
                    ok = false;
// If s is a common subsequence for p[1]..p[m-1]; and (the length of s >
len), or (the length of s == len) and s is the first for current longest
common subsequence ans in alphabetical order ; then s is adjusted as the
current longest common subsequence.
            if (ok && (j - i + 1 > len || j - i + 1 == len &&
strcmp(ans, s) > 0)) {
                len = j - i + 1;
                strcpy(ans, s);
            }
        }
    if (len < 3) // If the longest common subsequence is less than
three bases in length, display the string "no significant commonalities";
else output the longest common base subsequence
        printf("%s\n", "no significant commonalities");
    else {
        printf("%s\n", ans);
    }
    }
    return 0;
}
```

In fact, the brute force algorithm is not perfect. There are a large number of repeated operations. For example, in Figure 4.1, the sixth character in pattern $P =$ "ATATACG" can't be matched. The brute force algorithm makes s increase 1 and begins the matching again from the first character in P. We can also make s increase 2 and begin the matching again from the fourth character in P (Figure 4.2), for the previous matching will lead to failure.

In order to avoid repeated operation and improve time complexity, the KMP algorithm is introduced as follows.

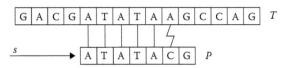

Figure 4.1 The sixth character in pattern *P* = ATATACG can't be matched.

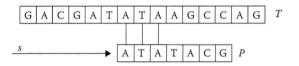

Figure 4.2   The matching from the fourth character in P.

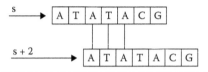

Figure 4.3   s is increased by 2.

For Figure 4.2, why can *s* be increased by 2, and why can the matching begin from the fourth character in *P*? It lies on pattern *P*.

In Figure 4.3, a prefix for *P* "ATA" is just a suffix for a prefix for *P* "ATATA". Therefore, if "ATATA" can be matched and "ATATAC" can't be matched, the substring $T[s + 2 .. s + 4]$ must be "ATA" in *T*. (Because in position *s* "ATATA" can be matched, $T[s .. s + 4] ==$ "ATATA".) Therefore, the prefix for *P* "ATA" must be matched in position $s + 2$. The KMP algorithm improves the time complexity to $O(n + m)$ by making use of the property.

The key to the KMP algorithm is prefix function *suffix*[ ] for *P*, where suffix[q] = max{k | $(k < q) \wedge (P[0 .. k - 1] ==$ the suffix of $P[0 .. q - 1])$}. That is, suffix[q] represents the length of the longest matching substring for the suffix of $P[0 ... (q - 1)]$ and the prefix of *P*. Calculating *suffix*[ ] is a process matching *P* with *P*, and it can be implemented as follows:

```
suffix[0] = -1;              //Set the boundary value for prefix function
suffix[ ] for P
        suffix[1] = 0;
        int k = 0;                        //Initialize the pointer of the
prefix function
        for (int i = 2; i <= m; i++) {
            while (k>=0 && P[k]!= P[i-1])     //Through the pointer of the
prefix function to find position k that P is equal to P[i-1], that is, if
the target current can't match P[i], it must be compared with P[k+1]
                k = suffix[k];
            suffix[i] = ++k;
        }
```

```
With prefix function suffix[ ], the process that P matches T can be
implemented by a loop, matching T[2],..., T[n-1] with P[0] one by one.
If P[j] can match T[i] successfully, then P[j+1] and T[i+1] are matched;
If P[j] can't match T[i], let T[i] match P[suffix[j]],
P[P[suffix[j]]],........, until it matches successfully (T[i]=P[...
P[suffix[j]]]) or the match fails (P[...P[suffix[j]]]=-1). If it matches
successfully, then compare P[P[...P[suffix[j]]]+1] with T[i+1]; and if it
the match fails, then compare P[0] with T[i+1] (0<=i<=n-1, 0<=j<=m-1).
i=0;                            // Initialize matching pointers of T and P
j=0;
while (i<=n-1 && j<=m-1)
if (j==-1||T[i]==P[j])          // If prefix function can't match T[i], then
compare P[0] with T[i+1]; If the matching is successful, then compare
P[j+1] with T[i+1]
```

```
{ i++;   j++;
}
else j= suffix[j];    // If T[i] and P[j] can't match, backtrack along the
pointer of prefix function
if (j>m-1)       //If all characters in P successfully match, return the
matching position in T; else return -1
    return(i-(m-1));
else
return(-1);
```

4.6.2 Oulipo

The French author Georges Perec (1936–1982) once wrote a book, *La disparition*, without the letter *e*. He was a member of the Oulipo group. A quote from the book:

> Tout avait Pair normal, mais tout s'affirmait faux. Tout avait Fair normal, d'abord, puis surgissait l'inhumain, l'affolant. Il aurait voulu savoir où s'articulait l'association qui l'unissait au roman : stir son tapis, assaillant à tout instant son imagination, l'intuition d'un tabou, la vision d'un mal obscur, d'un quoi vacant, d'un non-dit : la vision, l'avision d'un oubli commandant tout, où s'abolissait la raison : tout avait l'air normal mais.

Perec would probably have scored high (or rather, low) in the following contest. People are asked to write a perhaps even meaningful text on some subject with as few occurrences of a given "word" as possible. Our task is to provide the jury with a program that counts these occurrences, in order to obtain a ranking of the competitors. These competitors often write very long texts with nonsense meaning; a sequence of 500,000 consecutive 'T's' is not unusual. And they never use spaces.

So we want to quickly find out how often a word, that is, a given string, occurs in a text. More formally, given the alphabet {'A', 'B', 'C,' ..., 'Z'} and two finite strings over that alphabet, a word W and a text T, count the number of occurrences of W in T. All the consecutive characters of W must exactly match consecutive characters of T. Occurrences may overlap.

Input

The first line of the input file contains a single number: the number of test cases to follow. Each test case has the following format:

- One line with the word W, a string over {'A', 'B', 'C', ..., 'Z'}, with $1 \le |W| \le 10{,}000$ (here $|W|$ denotes the length of the string W)
- One line with the text T, a string over {'A', 'B', 'C', ..., 'Z'}, with $|W| \le |T| \le 1{,}000{,}000$

Output

For every test case in the input file, the output should contain a single number, on a single line: the number of occurrences of the word W in the text T.

Sample Input	Sample Output
3	1
BAPC	3

(Continued)

Sample Input	Sample Output
BAPC	0
AZA	
AZAZAZA	
VERDI	
AVERDXIVYERDIAN	

Source: BAPC 2006, Qualification.

ID for online judge: POJ 3461.

Analysis

Because the problem requires calculating the number of occurrences of the word w in the text t, t is the target and w is the pattern. First, the KMP algorithm is used to calculate the prefix function *next* for w, and then *next* is used to calculate the number of occurrences *cnt* for the word w in the text t. The process is as follows.

Suppose p is the matching pointer for w, and *scur* is the matching pointer for t. Initially $p = 0$, *cur* = 0. Then t is scanned one by one:

1. If current characters for t and w are the same $(t[cur] == w[p])$, then pointers for t and w are moved forward $(++cur, ++p)$.
2. If current characters for t and w are different $(t[cur] \neq w[p])$, then
   a. If analysis of characters for w is not finished $(p >= 0)$, then pointer for w is moved based on array *next* $(p = next[p])$.
   b. If analysis of characters for w is finished $(p < 0)$, then try to match the next character of t with the first character of w $(++cur, p = 0)$.
3. If the match is successful $(p == $ the length of $w)$, then the number of occurrences of the word w in the text t is *cnt*++; the pointer for w is moved based on array *next* $(p = next[p])$. Try to match the next character for t.

Repeat the match process until $cur \geq t$. And *cnt* is the number of occurrences of the word w in the text t.

Program

```
#include <cstdio>
#include <cstring>
const int maxw = 10000 + 10;          //upper limit of the length of word w
const int maxt = 1000000 + 10;                 // upper limit of the length
of text t
int match(char w[ ], char s[ ], int next[ ])
{                         // calculate the number of occurrences of the word
w[ ] in the text s[ ]
    int cnt = 0;
    int slen = strlen(s);
    int wlen = strlen(w);
    int p = 0, cur = 0;
    while (cur < slen) {               // while the scan isn't finished
        if (s[cur] == w[p]) { // If current characters for t and w are
the same (t[cur]==w[p]), then pointers for t and w are moved forward
```

```
                ++cur;
                ++p;
        } else if (p >= 0) {                   // If analysis of characters
for w is not finished (p>=0), then pointer for w is moved based on array
next (p = next[p]); else try to match the next character of t with the
first character of w (++cur, p=0)
                p = next[p];
        } else {
                ++cur;
                p = 0;
        }
                if (p == wlen) {                    // If the match is
successful ( p==the length of w ), then the number of occurrences of the
word w in the text t cnt++; pointer for w is moved based on array next
(p = next[p]); try to match the next character for t.
                ++cnt;
                p = next[p];
        }
    }
    return cnt;
}
int main(void)
{
    int loop;
    scanf("%d", &loop);              // Number of test cases
    while (loop--) {
        char w[maxw], t[maxt];
        scanf("%s%s", w, t);          // word w and text t
        int suffix[maxw];             // prefix function for word w
        suffix[0] = -1;
        suffix[1] = 0;
        int p = 0;
        for (int cur = 2; cur <= strlen(w); cur++) {
            while (p >= 0 && w[p] != w[cur - 1])
                p = suffix[p];
            suffix[cur] = ++p;
        }
        printf("%d\n", match(w, t, suffix));   // the number of
occurrences of the word w in the text t
    }
    return 0;
}
```

4.7 Problems

4.7.1 Moscow Time

In email, the following format for date and time setting is used:

  EDATE::=Day_of_week, Day_of_month Month Year Time Time_zone

  Here *EDATE* is the name of the date and time format; the text to the right from ":=" defines how date and time are written in this format. Below the descriptions of *EDATE* fields are presented:

Day_of_week	The name of a day of the week. Possible values are MON, TUE, WED, THU, FRI, SAT, SUN. The name is followed by the "," character (a comma).
Day_of_month	A day of the month. Set by two decimal digits.
Month	The name of the month. Possible values are JAN, FEB, MAR, APR, MAY, JUN, JUL, AUG, SEP, OCT, NOV, and DEC.
Year	Set by two or four decimal digits. If a year is set by two decimals, it is assumed that this is a number of the year of the twentieth century. For instance, 74 and 1974 set a year of 1974.
Time	Local time in the format hours:minutes:seconds, where hours, minutes, and seconds are made up of two decimal digits. The time keeps within the limits from 00:00:00 to 23:59:59.
Time_zone	Offset of local time from Greenwich Mean Time. It is set by the difference sign "+" or "–" and by a sequence of four digits. The first two digits set the hours and the last two the minutes of the offset value. The absolute value of the difference does not exceed 24 hours. The time zone can also be presented by one of the following names:

Name	Digital Value
UT	–0000
GMT	–0000
EDT	–0400
CDT	–0500
MDT	–0600
PDT	–0700

Both adjacent fields of *EDATA* are separated with exactly one space. Names of day of the week, month, and time zone are written in capitals. For instance, 10 a.m. of the contest day in St. Petersburg can be presented as

TUE, 03 DEC 96 10:00:00 +0300

Write a program that transforms the given date and time in *EDATE* format to the corresponding date and time in Moscow time zone. So-called "summer time" is not taken into consideration. Your program should rely on the predefined correctness of the given *Day_of_week* and *Time_zone*.

Note

■ Moscow time is 3 hours later than Greenwich Mean Time (time zone +0300).
■ January, March, May, July, August, October, and December have 31 days. April, June, September, and November have 30 days. February, as a rule, has 28 days, save for the case of the leap year (29 days).
■ A year is a leap year if one of the two following conditions is valid:

1. Its number is divisible by 4 and is not divisible by 100.
2. Its number is divisible by 400.

Input

Input contains date and time in *EDATE* format in the first line. The minimum permissible year in the input data is 0001, maximum 9998. The input *EDATA* string does not contain leading and trailing spaces.

Output

Output must contain a single line with date and time of the Moscow time zone in *EDATE* format. In the output *EDATE* string a year can be presented in any of the two allowed ways. The output string should not include leading and trailing spaces.

Sample Input	Sample Output
SUN, 03 DEC 1996 09:10:35 GMT	SUN, 03 DEC 1996 12:10:35 +0300

Source: ACM Northeastern Europe 1996.

IDs for online judges: POJ 1446, ZOJ 1323, UVA 505.

Hint

First, date and time in *EDATE* format are represented as numbers. Suppose the following:

> *week* is the variable for the day of the week, where *week* == 0 corresponds to "SUN," *week* == 1 corresponds to "MON," ..., *week* == 6 corresponds to "SAT."
> *year* is the variable for the number of the year; if a year is given by two decimals, 1900 must be added.
> *month* is the variable for the month, where *month* == 1 corresponds to "JAN," ..., *month* == 12 corresponds to "DEC."
> *day* is the variable for a day of the month.
> *hour*, *minute*, and *second* are variables for hour, minute, and second. *hour* and *minute* need to be adjusted based on the current time zone, and *second* need not be changed.
> *zt* is the variable for time zone, where *zt* == 0 corresponds to UT, ..., *zt* == 5 corresponds to PDT. Based on Greenwich Time, the adjustment for the *zt*th time zone is $d[zt]$. Accordi g to the problem description, $d[0] = 3$, $d[1] = 3$, $d[2] = 7$, $d[3] = 8$, $d[4] = 9$, and $d[5] = 10$. If the input time is in time zone *zt*, then hour is adjusted *hour* += $d[zt]$; otherwise, it is Greenwich Time. Sign *c* should be input to calculate the adjustment to hour *dh* and minute *dm*. If sign *c* is negative, then *hour* −= *dh*, *minute* += *dm*; if sign *c* is positive, then *hour* += *dh*, *minute* −= *dm*.

Then time and date are normalized based on 60 minutes an hour and 24 hours a day. When minutes are normalized, hour and date (*day*, *month*, *week*, and *year*) may be changed.

1. Minute: If *minute* < 0, then *minute* += 60, and −−*hour*; if *minute* ≥ 60, then *minute* −=60, and ++*hour*.
2. Hour: If *hour* < 0, then *hour* += 24, *week* = (*week* −1 + 7)% 7, and −−*day*. If *day* ≤ 0, then −−*month*. If *month* < 1, then *month* = 12, −−*year*, and *day* = *day* + the number of days in the month of the year. If *hour* ≥ 24, then *hour* − = 24, *week* = (*week* + 1)% 7, and ++*day*. If *day* > the number of days in the month of the year, then *day* = 1, and ++*month*. If *month* > 12, then *month* = 1, ++*year*.

Then date and time of the Moscow time zone in *EDATE* format is obtained: *week* is the string for the day of the week; *day* and *month* is the string for the month; and *year*, *hour*, *minute*, *second*.

4.7.2 Double Time

In 45 BC, a standard calendar was adopted by Julius Caesar—each year would have 365 days, and every fourth year have an extra day—the 29th of February. However, this calendar was not quite accurate enough to track the true solar year, and it became noticeable that the onset of the seasons was shifting steadily through the year. In 1582 Pope Gregory XIII ruled that a new style calendar should take effect. From then on, century years would only be leap years if they were divisible by 400. Furthermore, the current year needed an adjustment to realign the calendar with the seasons. This new calendar and the correction required were adopted immediately by Roman Catholic countries, where the day following Thursday, October 4, 1582, was Friday, October 15, 1582. The British and Americans (among others) did not follow suit until 1752, when Wednesday, September 2 was followed by Thursday, September 14. (Russia did not change until 1918, and Greece waited until 1923.) Thus, there was a long period of time when history was recorded in two different styles.

Write a program that will read in a date, determine which style it is in, and then convert it to the other style.

Input

Input will consist of a series of lines, each line containing a day and date (such as Friday, December 25, 1992). Dates will be in the range January 1, 1600 to December 31, 2099, although converted dates may lie outside this range. Note that all names of days and months will be in the style shown; that is, the first letter will be capitalized with the rest lowercase. The file will be terminated by a line containing a single '#'.

Output

Output will consist of a series of lines, one for each line of the input. Each line will consist of a date in the other style. Use the format and spacing shown in the example and described above. Note that there must be exactly one space between each pair of fields. To distinguish between the styles, dates in the old style must have an asterisk ('*') immediately after the day of the month (with no intervening space). Note that this will not apply to the input.

Sample Input	Sample Output
Saturday 29 August 1992	Saturday 16* August 1992
Saturday 16 August 1992	Saturday 29 August 1992
Wednesday 19 December 1991	Wednesday 1 January 1992
Monday 1 January 1900	Monday 20* December 1899
#	

Source: New Zealand Contest 1992.

ID for online judge: UVA 150.

Hint

Julius Caesar's calendar is called old calendar, and Pope Gregory XIII's calendar is called new calendar. Suppose *week* represents day of week, and *day*, *month*, and *year* represent input date (day, month, and year). Based on the problem description, in the old calendar, a *year* that is a multiple of 4 is a leap year, and in the new calendar, a *year* that is a multiple of 4 is a leap year unless it is a multiple of 100 that is not a multiple of 400.

According to the old calendar, the number of days that elapsed from January 1, 0000 AD to *month day* – 1, *year* is calculated as follows:

$$d1 = ((year - 1) * 365 + \frac{year - 1}{4} + \sum_{i=1}^{month-1} (Number\_of\_days\_for\_the\_ith\_month) - 2)$$

$$+ day - 1$$

According to the new calendar, the number of days that elapsed from January 1, 0000 AD to *month day* – 1, *year* is calculated as follows:

$$d2 = ((year - 1) * 365 + \frac{year - 1}{4} - \frac{year - 1}{100} + \frac{year - 1}{400}$$

$$+ \sum_{i=1}^{month-1} (Number\_of\_days\_for\_the\_ith\_month)) + day - 1$$

1. If (1 + *d*1) % 7 == *week*, then the current calendar is the old calendar, and the date of the new calendar should be output. The process is as follows:
   – Adjustment is based on the rule of leap year:

$$day = day + \frac{year - 1}{100} - \frac{year - 1}{400} - 2$$

   – Normalize the date: If *day* is larger than the number of days in *month, year* for the old calendar, then decrease the number of days in the month, and *month++*. If *month* is larger than 12, then *month* = 1 and *year++*.
2. If (1 + *d*1) % 7 ≠ *week*, then the current calendar is the new calendar, and the date of the old calendar should be output. The process is as follows:
   – Adjust *day* based on leap years:

$$day = day - \frac{year - 1}{100} + \frac{year - 1}{400} + 2$$

   – Normalize the date: If *day* is less than 1, then *month*––. If *month* is less than 1, then *month* = 12 and *year*––. Then calculate the number of days *d* in *month, year* under the old calendar, *day* = *day* + *d*.

4.7.3 Maya Calendar

During his last sabbatical, professor M.A. Ya made a surprising discovery about the old Maya calendar. From an old knotted message, the professor discovered that the Maya civilization used a 365-day-long year, called Haab, which had 19 months. Each of the first 18 months was 20 days long, and the names of the months were pop, no, zip, zotz, tzec, xul, yoxkin, mol, chen, yax, zac, ceh, mac, kankin, muan, pax, koyab, and cumhu. Instead of having names, the days of the months were denoted by numbers starting from 0 to 19. The last month of Haab was called uayet and had 5 days denoted by numbers 0, 1, 2, 3, and 4. The Maya believed that this month was unlucky, the court of justice was not in session, the trade stopped, and people did not even sweep the floor.

For religious purposes, the Maya used another calendar in which the year was called Tzolkin (holly year). The year was divided into 13 periods, each 20 days long. Each day was denoted by a pair consisting of a number and the name of the day. They used 20 names, imix, ik, akbal, kan, chicchan, cimi, manik, lamat, muluk, ok, chuen, eb, ben, ix, mem, cib, caban, eznab, canac, and ahau, and 13 numbers, both in cycles.

Notice that each day has an unambiguous description. For example, at the beginning of the year the days were described as follows: 1 imix, 2 ik, 3 akbal, 4 kan, 5 chicchan, 6 cimi, 7 manik, 8 lamat, 9 muluk, 10 ok, 11 chuen, 12 eb, 13 ben, 1 ix, 2 mem, 3 cib, 4 caban, 5 eznab, 6 canac, 7 ahau, and again in the next period, 8 imix, 9 ik, 10 akbal, and so forth.

Years (both Haab and Tzolkin) were denoted by numbers 0, 1, : : :, where the number 0 was the beginning of the world. Thus, the first day was

Haab: 0. pop 0
Tzolkin: 1 imix 0

Help Professor M.A. Ya and write a program to convert the dates from the Haab calendar to the Tzolkin calendar.

Input

The date in Haab is given in the following format:

NumberOfTheDay. Month Year

The first line of the input file contains the number of the input dates in the file. The next n lines contain n dates in the Haab calendar format, each in separate lines. The year is smaller than 5000.

Output

The date in Tzolkin should be in the following format:

Number NameOfTheDay Year

The first line of the output file contains the number of the output dates. In the next n lines, there are dates in the Tzolkin calendar format, in the order corresponding to the input dates.

Sample Input	Sample Output
3	3
10. zac 0	3 chuen 0
0. pop 0	1 imix 0
10. zac 1995	9 cimi 2801

Source: ACM Central Europe 1995.

IDs for online judges: POJ 1008, UVA 300.

Hint

Suppose the date of the Haab calendar is *month date, year*. The number of days from the beginning of the world is *day* = 365*year + (*month* – 1)*20 + *date* + 1.

Suppose the date of the Tzolkin calendar is *num* (period) *word* (day), *year*. There are 260 days every year for the Tzolkin calendar. (There are 13 periods every year, and there are 20 days for each period.) If *day* % 260 = 0, the day must be the last day of a year in Tzolkin calendar, that is, *year* = *day*/260 – 1; *num* = 13; *word* = 20. If *day* % 260 ≠ 0, then *year* = *day*/260; *num* = (*day* % 13 == 0 ? 13 : *day* % 13); *word* = (*day* – 1) % 20 + 1.

4.7.4 Time Zones

Prior to the late nineteenth century, timekeeping was a purely local phenomenon. Each town would set their clocks to noon when the sun reached its zenith each day. A clockmaker or town clock would be the official time, and the citizens would set their pocket watches and clocks to the time of the town—enterprising citizens would offer their services as mobile clock setters, carrying a watch with the accurate time to adjust the clocks in customers' homes on a weekly basis. Travel between cities meant having to change one's pocket watch upon arrival.

However, once railroads began to operate and move people rapidly across great distances, time became much more critical. In the early years of the railroads, the schedules were very confusing because each stop was based on a different local time. The standardization of time was essential to efficient operation of railroads.

In 1878, Canadian Sir Sanford Fleming proposed the system of worldwide time zones that we use today. He recommended that the world be divided into 24 time zones, each spaced 15° of longitude apart. Since the earth rotates once every 24 hours and there are 360° of longitude, each hour the earth rotates 1/24 of a circle or 15° of longitude. Sir Fleming's time zones were heralded as a brilliant solution to a chaotic problem worldwide.

U.S. railroad companies began utilizing Fleming's standard time zones on November 18, 1883. In 1884 an International Prime Meridian Conference was held in Washington, DC, to standardize time and select the Prime Meridian. The conference selected the longitude of Greenwich, England, as 0° longitude and established the 24 time zones based on the Prime Meridian. Although the time zones had been established, not all countries switched immediately. Though most U.S. states began to adhere to the Pacific, Mountain, Central, and Eastern time zones by 1895, Congress didn't make the use of these time zones mandatory until the Standard Time Act of 1918.

Today, many countries operate on variations of the time zones proposed by Sir Fleming. All of China (which should span five time zones) uses a single time zone—8 hours ahead of Coordinated Universal Time (known by the abbreviation UTC, based on the time zone running through Greenwich at 0° longitude). Russia adheres to its designated time zones, although the entire country is on permanent Daylight Saving Time and is an hour ahead of their actual zones. Australia uses three time zones—its central time zone is a half hour ahead of its designated time zone. Several countries in the Middle East and South Asia also utilize half-hour time zones.

Since time zones are based on segments of longitude and lines of longitude narrow at the poles, scientists working at the North and South Poles simply use UTC. Otherwise, Antarctica would be divided into 24 very thin time zones.

Time zones have recently been given standard capital-letter abbreviations as follows:

UTC	Coordinated Universal Time
GMT	Greenwich Mean Time, defined as UTC
BST	British Summer Time, defined as UTC + 1 hour
IST	Irish Summer Time, defined as UTC + 1 hour
WET	Western Europe Time, defined as UTC
WEST	Western Europe Summer Time, defined as UTC + 1 hour
CET	Central Europe Time, defined as UTC + 1
CEST	Central Europe Summer Time, defined as UTC + 2
EET	Eastern Europe Time, defined as UTC + 2
EEST	Eastern Europe Summer Time, defined as UTC + 3
MSK	Moscow Time, defined as UTC + 3
MSD	Moscow Summer Time, defined as UTC + 4
AST	Atlantic Standard Time, defined as UTC − 4 hours
ADT	Atlantic Daylight Time, defined as UTC − 3 hours
NST	Newfoundland Standard Time, defined as UTC − 3.5 hours
NDT	Newfoundland Daylight Time, defined as UTC − 2.5 hours
EST	Eastern Standard Time, defined as UTC − 5 hours
EDT	Eastern Daylight Saving Time, defined as UTC − 4 hours
CST	Central Standard Time, defined as UTC − 6 hours
CDT	Central Daylight Saving Time, defined as UTC − 5 hours
MST	Mountain Standard Time, defined as UTC − 7 hours
MDT	Mountain Daylight Saving Time, defined as UTC − 6 hours
PST	Pacific Standard Time, defined as UTC − 8 hours
PDT	Pacific Daylight Saving Time, defined as UTC − 7 hours
HST	Hawaiian Standard Time, defined as UTC − 10 hours
AKST	Alaska Standard Time, defined as UTC − 9 hours
AKDT	Alaska Standard Daylight Saving Time, defined as UTC − 8 hours
AEST	Australian Eastern Standard Time, defined as UTC + 10 hours
AEDT	Australian Eastern Daylight Time, defined as UTC + 11 hours
ACST	Australian Central Standard Time, defined as UTC + 9.5 hours
ACDT	Australian Central Daylight Time, defined as UTC + 10.5 hours
AWST	Australian Western Standard Time, defined as UTC + 8 hours

Given the current time in one time zone, you are to compute what time it is in another time zone.

Input

The first line of input contains N, the number of test cases. For each case a line is given with a time and two time zone abbreviations. Time is given in standard a.m./p.m. format, with midnight denoted "midnight" and noon denoted "noon" (12:00 a.m. and 12:00 p.m. are oxymorons).

Output

For each case, assuming the given time is the current time in the first time zone, give the current time in the second time zone.

Sample Input	Sample Output
4	midnight
noon HST CEST	4:29 p.m.
11:29 a.m. EST GMT	12:01 a.m.
6:01 p.m. CST UTC	6:40 p.m.
12:40 p.m. ADT MSK	

Source: Waterloo Local Contest, September 28, 2008.

IDs for online judges: POJ 2351, ZOJ 1916, UVA 10371.

Hint

Suppose $td[i]$ is the hour increment for the ith time zone with respect to UTC ($1 \leq i \leq 24$). For a test case, the first time zone is time zone i, and the second time zone is time zone j. The time of a time zone is represented with *hour* and *minute*.

1. For convenience, the time of time zone i is represented with the 24-hour system.
   a. If "noon" is input, then *hour* = 12 and *minute* = 0.
   b. If "midnight" is input, then *hour* = 0 and *minute* = 0.
   c. If *hour*, *minute*, and "p.m." (or "a.m.") are input, then
      If *hour* == 12 and "a.m." is input, then *hour* is adjusted (*hour* = 0).
      If *hour* ≠ 12 and "p.m." is input, then *hour* = *hour* + 12.
2. The time in time zone i is transferred into the time in time zone j.
   Because the time difference between the two time zones is $td[j] - td[j]$, *hour* = *hour* + *int*(*td*[j] − *td*[j]), *minute* = *minute* + *int*(60.0*(delta − *int*(*td*[j] − *td*[j]))), then *hour* and *minute* are adjusted as follows:
   a. If the time difference between two time zones is one day, then
      If *minute* < 0, then *hour* = *hour* − 1, *minute* = *minute* + 60.
      If *minute* ≥ 60, then *hour* = *hour* + 1, *minute* = *minute* − 60.
   b. *hour* must be adjusted in [0 .. 23]:
      If *hour* < 0, then *hour* = *hour* + 24.
      If *hour* ≥ 24, then *hour* = *hour* − 24.
3. The time in time zone j is transferred into the 12-hour system:
   a. If *hour* == 12 and *minute* == 0, then output "noon."
   b. If *hour* == 0 and *minute* == 0, then output "midnight."
   c. If *hour* == 0, then output 12; else, output (*hour* > 12 ? *hour* − 12 : *hour*).
   d. Output *minute*.
   e. If *hour* < 12, then output "a.m."; else. output "p.m."

4.7.5 Polynomial Remains

Given the polynomial $a(x) = a_n x^n + \ldots + a_1 x + a_0$, compute the remainder $r(x)$ when $a(x)$ is divided by $x^k + 1$ (Figure 4.4).

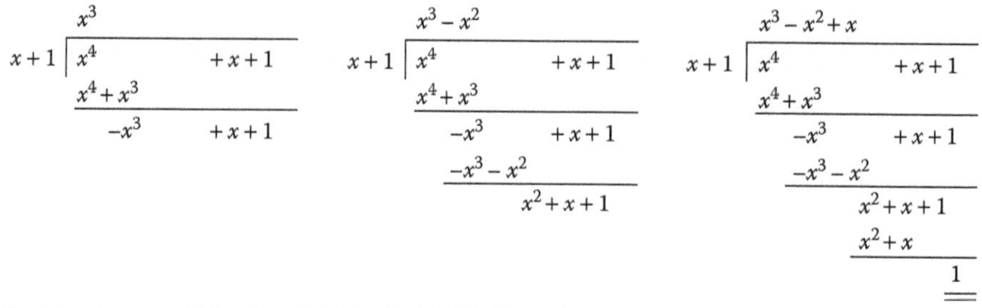

$$
\begin{array}{r}
x^3 \qquad\qquad\qquad \\
x+1\,\overline{)\,x^4 \qquad\qquad +x+1} \\
\underline{x^4+x^3\qquad\qquad\qquad} \\
-x^3 \qquad +x+1
\end{array}
\qquad
\begin{array}{r}
x^3-x^2 \qquad\qquad \\
x+1\,\overline{)\,x^4 \qquad\qquad +x+1} \\
\underline{x^4+x^3\qquad\qquad\qquad} \\
-x^3 \qquad +x+1 \\
\underline{-x^3-x^2\qquad\qquad} \\
x^2+x+1
\end{array}
\qquad
\begin{array}{r}
x^3-x^2+x \qquad\quad \\
x+1\,\overline{)\,x^4 \qquad\qquad +x+1} \\
\underline{x^4+x^3\qquad\qquad\qquad} \\
-x^3 \qquad +x+1 \\
\underline{-x^3-x^2\qquad\qquad} \\
x^2+x+1 \\
\underline{x^2+x\qquad} \\
1
\end{array}
$$

Figure 4.4 Polynomial remains.

Input

The input consists of a number of cases. The first line of each case specifies the two integers n and k $(0 \le n, k \le 10{,}000)$. The next $n + 1$ integers give the coefficients of $a(x)$, starting from a_0 and ending with a_n. The input is terminated if $n = k = -1$.

Output

For each case, output the coefficients of the remainder on one line, starting from the constant coefficient r_0. If the remainder is 0, print only the constant coefficient. Otherwise, print only the first $d + 1$ coefficients for a remainder of degree d. Separate the coefficients by a single space.

You may assume that the coefficients of the remainder can be represented by 32-bit integers.

Sample Input	Sample Output
5 2	3 2
6 3 3 2 0 1	−3 −1
5 2	−2
0 0 3 2 0 1	−1 2 −3
4 1	0
1 4 1 1 1	0
6 3	1 2 3 4
2 3 −3 4 1 0 1	
1 0	
5 1	
0 0	
7	
3 5	

Sample Input	Sample Output
1 2 3 4	
−1 −1	

Source: Alberta Collegiate Programming Contest, October 18, 2003.

ID for online judge: POJ 2527.

Hint

Suppose array *a* is used to store the remainder, the length of the array is *n* + 1, and coefficients of the remainder are stored in *a*[0 .. *n*]. Initially, an array is used to store polynomial *a*(*x*):

```
for (i=0; i<=n; i++)
        scanf ("%d", &a[i]);
The algorithm that a(x) is repeatedly divided by x^k+1 is as follows:
Initially i=n. If i>=k, then a(x) can be divided by x^k+1 repeatedly until
i<k:
for (i=n; i>=k; i--)         // for each non-zero coefficient in a[n] ..
a[k],division is done
            if (a[i]!=0) {         // coefficients for the remainder must
be adjusted
                a[i-k]  += (-a[i]);
a[i]=0;
                }
Then the length of array a is adjusted:
while (n >= 0 && ! a[n]) n--;
Finally a[0..n-1] is the coefficients of the remainder and outputted:
for (i=0; i<n; i++)
        printf ("%d ", a[i]);
```

4.7.6 *Factoring a Polynomial*

Recently Georgie has learned about polynomials. A polynomial in one variable can be viewed as a formal sum $a_n x^n + a_{n-1} x^{n-1} + \ldots + a_1 x + a_0$, where x is the formal variable and a_i are the coefficients of the polynomial. The greatest i such that $a_i \neq 0$ is called the degree of the polynomial. If $a_i = 0$ for all i, the degree of the polynomial is considered to be $-\infty$. If the degree of the polynomial is zero or $-\infty$, it is called trivial; otherwise, it is called nontrivial.

What really impressed Georgie while studying polynomials was the fact that in some cases one can apply different algorithms and techniques developed for integer numbers to polynomials. For example, given two polynomials, one may sum them up, multiply them, or even divide one of them by the other.

The most interesting property of polynomials, from Georgie's point of view, is the fact that a polynomial, just like an integer number, can be factorized. We say that the polynomial is irreducible if it cannot be represented as the product of two or more nontrivial polynomials with real coefficients. Otherwise, the polynomial is called reducible. For example, the polynomial $x^2 - 2x + 1$ is reducible because it can be represented as $(x - 1)(x - 1)$, while the polynomial $x^2 + 1$ is not. It is well known that any polynomial can be represented as the product of one or more irreducible polynomials.

Given a polynomial with integer coefficients, Georgie would like to know whether it is irreducible. Of course, he would also like to know its factorization, but such a problem seems to be too difficult for him now, so he just wants to know about reducibility.

Input

The first line of the input contains n—the degree of the polynomial ($0 \leq n \leq 20$). The next line contains $n + 1$ integer numbers, $a_n, a_{n-1}, \ldots, a_1, a_0$—polynomial coefficients ($-1000 \leq a_i \leq 1000$, $a_n! = 0$).

Output

Output YES if the polynomial given in the input file is irreducible and NO in the other case.

Sample Input	Sample Output
2	NO
1 –2 1	

Source: ACM Northeastern Europe 2003, Northern Subregion.

ID for online judge: POJ 2126.

Hint

If the degree of the polynomial $n < 2$, then the polynomial is irreducible; if $n > 2$, then we can prove that the polynomial is reducible; and if $n == 2$, we can determine whether $ax^2 + bx + c$ is reducible or not based on Vieta's theorem: if $b^2 - 4ac \geq 0$, the polynomial is reducible; otherwise, the polynomial is irreducible.

4.7.7 What's Cryptanalysis?

Cryptanalysis is the process of breaking someone else's cryptographic writing. This sometimes involves some kind of statistical analysis of a passage of (encrypted) text. Your task is to write a program that performs a simple analysis of a given text.

Input

The first line of input contains a single positive decimal integer n. This is the number of lines that follow in the input. The next n lines will contain zero or more characters (possibly including white space). This is the text that must be analyzed.

Output

Each line of output contains a single uppercase letter, followed by a single space, then followed by a positive decimal integer. The integer indicates how many times the corresponding letter appears in the input text. Upper- and lowercase letters in the input are to be considered the same. No other characters must be counted. The output must be sorted in descending count order; that is, the most frequent letter is on the first output line, and the last line of output indicates the least frequent letter. If two letters have the same frequency, then the letter that comes first in the alphabet must appear first in the output. If a letter does not appear in the text, then that letter must not appear in the output.

Sample Input	Sample Output
3	S 7
This is a test.	T 6
Count me 1 2 3 4 5.	I 5
Wow!!!! Is this question easy?	E 4
	O 3
	A 2
	H 2
	N 2
	U 2
	W 2
	C 1
	M 1
	Q 1
	Y 1

Source: University of Valladolid Contest, September 2000.

ID for online judge: UVA 10008.

Hint
In the set of letters, the serial number for "A"("a") is 0, the serial number for "B"("b") is 1, …, and the serial number for "Z"("z") is 25. That is, for letter c, the serial number is $tolower(c) - $ "a". Suppose $cnt[i]$ indicates how many times the corresponding letter appears in the input text $(0 \leq i \leq 25)$.

First, characters are repeatedly input, and how many times every letter appears is accumulated, until EOF is input.

Then, for array cnt, repeat the process until elements in array cnt all become 0: every time, find the serial number k that appears most, and output its corresponding letter (its ASCII is $k + $ "A") and $cnt[k]$; then $cnt[k]$ is set to 0.

4.7.8 Run-Length Encoding

Your task is to write a program that performs a simple form of run-length encoding, as described by the rules below.

Any sequence of between two and nine identical characters is encoded by two characters. The first character is the length of the sequence, represented by one of the characters 2–9. The second character is the value of the repeated character. A sequence of more than nine identical characters is dealt with by first encoding nine characters, then the remaining ones.

Any sequence of characters that does not contain consecutive repetitions of any characters is represented by a 1 character followed by the sequence of characters, terminated with another 1. If a 1 appears as part of the sequence, it is escaped with a 1; thus, two 1 characters are output.

Input

The input consists of letters (both upper- and lowercase), digits, spaces, and punctuation. Every line is terminated with a new-line character, and no other characters appear in the input.

Output

Each line in the input is encoded separately, as described above. The new line at the end of each line is not encoded, but is passed directly to the output.

Sample Input	Sample Output
AAAAAABCCCC	6A1B14C
12344	11123124

Source: Ulm Local Contest 2004.

IDs for online judges: POJ 1782, ZOJ 2240.

Hint by the Problemsetter (http://www.informatik.uni-ulm.de/acm/Locals/2004/)

The input is processed line by line. Every line is processed by a loop that checks if a sequence of consecutive repetitions starts at the current position. If this is the case, the length of the repetition is calculated up to a maximal length of 9 characters, and its encoding is output. Otherwise, the next sequence of consecutive repetitions, if any, is located and the intermediate characters are encoded and output. This procedure is continued until the end of the line is reached.

4.7.9 Zipper

Given three strings, you are to determine whether the third string can be formed by combining the characters in the first two strings. The first two strings can be mixed arbitrarily, but each must stay in its original order.

For example, consider forming "tcraete" from "cat" and "tree":

String A: cat
String B: tree
String C: tcraete

As you can see, we can form the third string by alternating characters from the two strings. As a second example, consider forming "catrtee" from "cat" and "tree":

String A: cat
String B: tree
String C: catrtee

Finally, notice that it is impossible to form "cttaree" from "cat" and "tree".

Input

The first line of input contains a single positive integer from 1 through 1000. It represents the number of data sets to follow. The processing for each data set is identical. The data sets appear on the following lines, one data set per line.

For each data set, the line of input consists of three strings, separated by a single space. All strings are composed of upper- and lowercase letters only. The length of the third string is always the sum of the lengths of the first two strings. The first two strings will have lengths between 1 and 200 characters, inclusive.

Output

For each data set, print

Data set n: yes

if the third string can be formed from the first two or

Data set n: no

if it cannot. Of course, *n* should be replaced by the data set number. See the sample output below for an example.

Sample Input	Sample Output
3	Data set 1: yes
cat tree tcraete	Data set 2: yes
cat tree catrtee	Data set 3: no
cat tree cttaree	

Source: ACM Pacific Northwest 2004.

IDs for online judges: POJ 2192, ZOJ 2401, UVA 3195.

Hint

Suppose $A = a_0a_1 \dots a_{n-1}$, where its prefix $A_i = a_0a_1 \dots a_i$ $(0 \le i \le n - 1)$; $B = b_0b_1 \dots b_{m-1}$, where its prefix $B_j = b_0b_1 \dots b_j$ $(0 \le j \le m - 1)$; $C = c_0c_1 \dots c_{n+m-1}$, where its prefix $C_k = c_0c_1 \dots c_k$ $(0 \le k \le n + m - 1)$; and $can[i][j]$ is the sign that A_{i-1} (the prefix of A whose length is i) and B_{j-1} (the prefix of B whose length is j) can successfully constitute C_{i+j-1} (the prefix of C whose length is $i + j$) or not. Obviously, $can[0][0] = \text{true}$.

If $i \ge 1$ and $c_{i+j-1} == a_{i-1}$, $can[i][j] = can[i][j] \,||\, can[i - 1][j]$; that is, $can[i][j]$ is determined whether A_{i-2} and B_{j-1} can constitute C_{i+j-2} or not. If $j \ge 1$ and $c_{i+j-1} == b_{j-1}$, $can[i][j] = can[i][j] \,||\, can[i][j - 1]$; that is, $can[i][j]$ is determined whether A_{i-1} and B_{j-2} can constitute C_{i+j-2} or not $(0 \le i \le n, 0 \le j \le m)$.

Finally, $can[n][m]$ is the sign whether A and B can constitute C or not.

4.7.10 *Anagram Groups*

World-renowned Prof. A.N. Agram's current research deals with large anagram groups. He has just found a new application for his theory on the distribution of characters in English language texts. Given such a text, you are to find the largest anagram groups.

A text is a sequence of words. A word *w* is an anagram of a word *v* if and only if there is some permutation *p* of character positions that takes *w* to *v*. Then, *w* and *v* are in the same anagram group. The size of an anagram group is the number of words in that group. Find the five largest anagram groups.

Input

The input contains words composed of lowercase alphabetic characters, separated by white space. It is terminated by EOF.

Output

Output the five largest anagram groups. If there are less than five groups, output them all. Sort the groups by decreasing size. Break ties lexicographically by the lexicographical smallest element. For each group output, print its size and its member words. Sort the member words lexicographically and print equal words only once.

Sample Input	Sample Output
undisplayed	Group of size 5: caret carte cater crate trace.
trace	Group of size 4: abet bate beat beta.
tea	Group of size 4: ate eat eta tea.
singleton	Group of size 1: displayed.
eta	Group of size 1: singleton.
eat	
displayed	
crate	
cater	
carte	
caret	
beta	
beat	
bate	
ate	
abet	

Source: Ulm Local Contest 2000.

IDs for online judges: POJ 2408, ZOJ 1960.

Hint by the Problemsetter (http://www.informatik.uni-ulm.de/acm/Locals/2000/)

Since anagram groups are classes of an equivalence relation, we pick as a representative element of each class the sorted version of a member. Then, it takes logarithmic time to find the representative and update the class, as we read the words. We sort the classes by their size and their smallest member. We take the first five and output them in sorted order.

Although the problem is not too difficult, many mistakes can be made when writing a program. Some of them are

- Too small-dimensioned arrays (assuming too few words or groups)
- Sorting equal-sized groups not by their smallest elements but by their (sorted) representatives
- Words that occur multiple times in the input must be counted multiple times, but output only once
- Too many groups in the output
- Inefficient handling of the sorted data structures (leads to Time Limit Exceeded (TLE))

4.7.11 English Number Translator

In this problem, you will be given one or more integers in English. Your task is to translate these numbers into their integer representation. The numbers can range from –999,999,999 to 999,999,999. The following is an exhaustive list of English words that your program must account for: *negative, zero, one, two, three, four, five, six, seven, eight, nine, ten, eleven, twelve, thirteen, fourteen, fifteen, sixteen, seventeen, eighteen, nineteen, twenty, thirty, forty, fifty, sixty, seventy, eighty, ninety, hundred, thousand, million.*

Input

The input consists of several instances. Notes on input follow:

1. Negative numbers will be preceded by the word *negative*.
2. The word "hundred" is not used when "thousand" could be. For example, 1500 is written "one thousand five hundred," not "fifteen hundred."

The input is terminated by an empty line.

Output

The answers are expected to be on separate lines with a new line after each.

Sample Input	Sample Output
six	6
negative seven hundred twenty-nine	–729
one million one hundred one	1000101
eight hundred fourteen thousand twenty-two	814022

Source: CTU Open 2004.

IDs for online judges: UVA 486, POJ 2121, ZOJ 2311.

Hint
  Suppose *word* is a list used to store 32 words for 0, 1, 2, ..., 20, 30, 40, 50, 60, 70, 80, 90, hundred, thousand, million, negative.
  Indexes for *word*[0] ... *word*[20] correspond to their strings; that is, *word*[i] corresponds to integer i ($0 \le i \le 20$); from *word*[21] to *word*[27], *word*[i] corresponds to integer (i – 18)*10 ($21 \le i \le 27$). *word*[28], *word*[29], and *word*[30] correspond to the strings "hundred," "thousand," and "million," respectively. *isNeg* is the negative sign.
  Input words repeatedly. For each word *s*,

1. Initially *isNeg* is false. If *s* is *word*[31], then *isNeg*=true; input the next word *s*;
2. Calculate:
   num=0;
   Enter the loop:
   Calculate the index *r* for *s* in *word*:
   If *r*∈[0, 27], then

   $$num=num+\begin{cases} r & r \leq 20 \\ (r-18)*10 & 21 \leq r \leq 27 \end{cases};$$

   If *r*∈[28, 31], then

   $$num=num \text{ \% } b*b + \left\lfloor \frac{num}{b} \right\rfloor * b$$

   (*b* is the number that *word*[r] corresponds to);
   Get character *c*; if *c* is '\n' or EOF, then exit the loop; else, input word *s*;
3. If *isNeg* is true, then *num* is negative; output *num*.

4.7.12 Message Decowding

The cows are thrilled because they've just learned about encrypting messages. They think they will be able to use secret messages to plot meetings with cows on other farms.

Cows are not known for their intelligence. Their encryption method is nothing like Data Encryption Standard (DES) or Blowfish or any of those really good secret coding methods. No, they are using a simple substitution cipher.

The cows have a decryption key and a secret message. Help them decode it. The key looks like this:

yrwhsoujgcxqbativndfezmlpk

An 'a' in the secret message really means 'y', a 'b' in the secret message really means 'r', a 'c' decrypts to 'w', and so on. Blanks are not encrypted; they are simply kept in place.

Input text is in upper- or lowercase; both decrypt using the same decryption key, keeping the appropriate case, of course.

Input

Line 1: 26 lowercase characters representing the decryption key
Line 2: As many as 80 characters that are the message to be decoded

Output

Line 1: A single line that is the decoded message. It should have the same length as the second line of input.

Sample Input	Sample Output
eydbkmiqugjxlvtzpnwohracsf	Jump the fence when you seeing me coming
Kifq oua zarxa suar bti yaagrj fa xtfgrj	

Source: USACO, March 2003, Orange.

ID for online judge: POJ 2141.

Hint

Suppose *key* is the decryption key. In a secret message, 'a' is represented by *key*[0], ..., and 'z' is represented by *key*[25].

Character *c* is input repeatedly until *c* is EOF. For each character *c*, we deal with it as follows:

1. If *c* isn't a letter, then *c* is output directly.
2. If *c* is a letter, then
   If *c* is a lowercase letter, then *key*[*c* – 'a'] is output.
   If *c* is an uppercase letter, then key[*c* – 'A'] – 'a' + 'A' is output.

4.7.13 *Common Permutation*

Given two strings of lowercase letters, *a* and *b*, print the longest string *x* of lowercase letters such that there is a permutation of *x* that is a subsequence of *a* and there is a permutation of *x* that is a subsequence of *b*.

Input

Input file contains several lines of input. Consecutive two lines make a set of input. That means in the input file, lines **1** and **2** are a set of input, lines **3** and **4** are a set of input, and so on. The first line of a pair contains *a* and the second contains *b*. Each string is on a separate line and consists of at most **1000** lowercase letters.

Output

For each set of input, output a line containing *x*. If several *x* satisfy the criteria above, choose the first one in alphabetical order.

Sample Input	Sample Input
pretty	e
women	nw
walking	et
down	
the	
street	

Source: World Finals Warm-Up Contest, Problem Source: University of Alberta Local Contest.

ID for online judge: UVA 10252.

Hint

Because a permutation of *x* is a common subsequence of *a* and *b*, the number of a letter in *x* can't exceed the number of the letter in *a* and *b*; that is, the number of letter *c* in *x* = min {the number of letter *c* in *a*, the number of letter *c* in *b*}.

Suppose *a*[*i*] is the number of the letter whose serial number is *i* in *a*, and *b*[*i*] is the number of the letter whose serial number is *i* in *b*; that is, the serial number of 'a' is 0, the serial number of 'b' is 1, ..., and the serial number of 'z' is 25; $0 \leq i \leq 25$.

When two strings are input, arrays *a* and *b* are calculated. Then every letter's serial number *i* is enumerated ($0 \leq i \leq 25$), and the number of the corresponding letter (its ASCII code is *i* + 'a') in *x* is min{$a[i]$, $b[i]$}.

4.7.14 Human Gene Functions

It is well known that a human gene can be considered a sequence, consisting of four nucleotides, which are simply denoted by four letters, A, C, G, and T. Biologists have been interested in identifying human genes and determining their functions, because these can be used to diagnose human diseases and design new drugs for them.

A human gene can be identified through a series of time-consuming biological experiments, often with the help of computer programs. Once a sequence of a gene is obtained, the next job is to determine its function.

One of the methods for biologists to use in determining the function of a new gene sequence that they have just identified is to search a database with the new gene as a query. The database to be searched stores many gene sequences and their functions—many researchers have been submitting their genes and functions to the database and the database is freely accessible through the Internet.

A database search will return a list of gene sequences from the database that are similar to the query gene.

Biologists assume that sequence similarity often implies functional similarity. So, the function of the new gene might be one of the functions that the genes from the list have. To exactly determine which one is right, another series of biological experiments will be needed.

Your job is to make a program that compares two genes and determines their similarity as explained below. Your program may be used as a part of the database search if you can provide an efficient one.

Given two genes *AGTGATG* and *GTTAG*, how similar are they? One of the methods to measure the similarity of two genes is called alignment. In an alignment, spaces are inserted, if necessary, in appropriate positions of the genes to make them equally long and score the resulting genes according to a scoring matrix.

For example, one space is inserted into *AGTGATG* to result in *AGTGAT–G*, and three spaces are inserted into *GTTAG* to result in *–GT––TAG*. A space is denoted by a minus sign (–). The two genes are now of equal length. These two strings are aligned:

$$A \quad G \quad T \quad G \quad A \quad T \quad - \quad G$$
$$- \quad G \quad T \quad - \quad - \quad T \quad A \quad G$$

In this alignment, there are four matches, namely, *G* in the second position, *T* in the third, *T* in the sixth, and *G* in the eighth. Each pair of aligned characters is assigned a score according to the scoring matrix shown in Figure 4.5. It denotes that a space–space match is not allowed. The score of the alignment above is $(-3) + 5 + 5 + (-2) + (-3) + 5 + (-3) + 5 = 9$.

Of course, many other alignments are possible. One is shown below (a different number of spaces are inserted into different positions):

$$A \quad G \quad T \quad G \quad A \quad T \quad G$$
$$- \quad G \quad T \quad T \quad A \quad - \quad G$$

	A	C	G	T	—
A	5	−1	−2	−1	−3
C	−1	5	−3	−2	−4
G	−2	−3	5	−2	−2
T	−1	−2	−2	5	−1
—	−3	−4	−2	−1	*

Figure 4.5   Scoring matrix.

This alignment gives a score of (−3) + 5 + 5 + (−2) + 5 + (−1) + 5 = 14. So, this one is better than the previous one. In fact, this one is optimal since no other alignment can have a higher score. So, it is said that the similarity of the two genes is 14.

Input

The input consists of T test cases. The number of test cases (T) is given in the first line of the input file. Each test case consists of two lines: each line contains an integer, the length of a gene, followed by a gene sequence. The length of each gene sequence is at least 1 and does not exceed 100.

Output

The output should print the similarity of each test case, one per line.

Sample Input	Sample Output
2	14
7 AGTGATG	21
5 GTTAG	
7 AGCTATT	
9 AGCTTTAAA	

Source: ACM Taejon 2001.

IDs for online judges: POJ 1080, ZOJ 1027, UVA 2324.

Hint
The space and four nucleotides are labeled as integers: [0(space), 1(A), 2(C), 3(G), 4(T)]. Gene sequence a and gene sequence b are transferred into integer sequence $s1$ and $s2$, respectively. Because the last digit in a gene sequence may be matched with a space, a 0 is added to the end of $s1$ and $s2$. That is, the lengths of $s1$ and $s2$ are $len1 + 1$ and $len2 + 1$, respectively.

Suppose $score[\ ][\ ]$ is used to represent the scoring matrix. Based on the problem description,

$$score[\ ][\ ] = \begin{vmatrix} 0 & -3 & -4 & -2 & -1 \\ -3 & 5 & -1 & -2 & -1 \\ -4 & -1 & 5 & -3 & -2 \\ -2 & -2 & -3 & 5 & -2 \\ -1 & -1 & -2 & -2 & 5 \end{vmatrix}$$

$f[i, j]$ is the maximal score for alignment of the prefix whose length is i in a and the prefix whose length is j in b.

Of course, i and j can't be 0 simultaneously.

If $i > 0$, the maximal score for alignment of a_{i-1} and space is $f[i-1][j] + score[0][s1[i-1]]$.

If $j > 0$, the maximal score for alignment of b_{j-1} and space is $f[i][j-1] + score[0][s2[j-1]]$.

When i and j are both larger than 0, the maximal score for alignment of a_{i-1} and b_{j-1} is $f[i-1][j-1] + score[s1[i-1]][s2[j-1]]$.

Therefore, $f[i][j] = max\{f[i-1][j] + score[0][s1[i-1]], f[i][j-1] + score[0][s2[j-1]],$ and $f[i-1][j-1]+ score[s1[i-1]][s2[j-1]]\}$; $0 \leq i \leq len1 + 1$, $0 \leq j \leq len2 + 1$.

Obviously, the maximal score of two gene sequences when the last two characters (a_{len1-1} and b_{len2-1}) match is $f[len1][len2]$, but the last character in a gene sequence may match with a space.

The maximal score that a_{len1-1} and space match is $f[len1][len2+1]$, and that space and b_{len2-1} match is $f[len1 + 1][len2]$.

The similarity of gene sequence a and gene sequence b is $max\{f[len1][len2], f[len1][len2 + 1], f[len1 + 1][len2]\}$.

4.7.15 Palindrome

A palindrome is a symmetrical string, that is, a string read identically from left to right as well as from right to left. You are to write a program that, given a string, determines the minimal number of characters to be inserted into the string in order to obtain a palindrome.

As an example, by inserting two characters, the string "Ab3bd" can be transformed into a palindrome ("dAb3bAd" or "Adb3bdA"). However, inserting fewer than two characters does not produce a palindrome.

Input

Your program is to read from standard input. The first line contains one integer: the length of the input string N, $3 <= N <= 5000$. The second line contains one string with length N. The string is formed from uppercase letters from A to Z, lowercase letters from a to z, and digits from 0 to 9. Uppercase and lowercase letters are to be considered distinct.

Output

Your program is to write to standard output. The first line contains one integer, which is the desired minimal number.

Sample Input	Sample Output
5	2
Ab3bd	

Source: IOI 2000.

ID for online judge: POJ 1159.

Hint

Suppose $C(i, j)$ is the minimal number of characters to be inserted into the string $s_i \dots s_j$ in order to obtain a palindrome. Therefore, the problem requires calculating $C(1, n)$.

The following formula holds:

$$C(i, j) = \begin{cases} 0 & i \geq j \\ C(i+1, j-1) & s_i = s_j \\ \min(C(i+1, j), C(i, j-1)) + 1 & s_i \neq s_j \end{cases}$$

4.7.16 Power Strings

Given two strings a and b, we define $a*b$ to be their concatenation. For example, if $a = $ "abc" and $b = $ "def", then $a*b = $ "abcdef." If we think of concatenation as multiplication, exponentiation by a nonnegative integer is defined in the normal way: $a^\wedge 0 = $ "" (the empty string) and $a^\wedge(n+1) = a*(a^\wedge n)$.

Input

Each test case is a line of input representing s, a string of printable characters. The length of s will be at least 1 and will not exceed 1 million characters. A line containing a period follows the last test case.

Output

For each s you should print the largest n such that $s = a^\wedge n$ for some string a.

Sample Input	Sample Output
abcd	1
aaaa	4
ababab	3
.	

Source: Waterloo Local Contest, July 1, 2002.

IDs for online judges: POJ 2406, ZOJ 1905.

Hint

Based on the definition of $s = a^\wedge n$, the length of a must be the shortest if n is the largest. Suppose *len* is the length of s.

The KMP algorithm is used to produce the prefix function *suffix*[ ] for s. If *suffix*[*cur*] == k, $s[0 ... (k-1)] == s[(cur-k) .. (cur-1)]$, and k is the length of the longest matching substrings for the prefix of s and the suffix of $s[0 ... (cur-1)]$.

Because $s[0 ... suffix[len] - 1] == s[(len - suffix[len]) ... (len - 1)]$, if $(len - suffix[len])$ is the divisor of *len*, then $s[0 ... (len - suffix[len] - 1)]$ is the shortest repeated substring, its length is $len - suffix[len]$, and

$$n = \frac{len}{len - suffix[len]}$$

4.7.17 Period

For each prefix of a given string S with N characters (each character has an ASCII code between 97 and 126, inclusive), we want to know whether the prefix is a periodic string. That is, for each i

$(2 \leq i \leq N)$ we want to know the largest $K > 1$ (if there is one) such that the prefix of S with length i can be written as A^K, that is, A concatenated K times, for some string A. Of course, we also want to know the period K.

Input

The input consists of several test cases. Each test case consists of two lines. The first one contains N ($2 \leq N \leq 1000000$)—he size of the string S. The second line contains the string S. The input file ends with a line, having the number zero in it.

Output

For each test case, output "Test case #" and the consecutive test case number on a single line; then, for each prefix with length i that has a period $K > 1$, output the prefix size i and the period K separated by a single space; the prefix sizes must be in increasing order. Print a blank line after each test case.

Sample Input	Sample Output
3	Test case #1
aaa	2 2
12	3 3
aabaabaabaab	
0	Test case #2
	2 2
	6 2
	9 3
	12 4

Source: ACM Southeastern Europe 2004.

IDs for online judges: POJ 1961, ZOJ 2177, UVA 3026.

Hint
First, the KMP algorithm is used to produce the prefix function *suffix*[ ] for s. If *suffix*[*cur*] == k, then $s[0 \dots (k-1)]$ == $s[(cur-k) \dots (cur-1)]$, and k is the length of the longest matching substring for the prefix of s and the suffix of $s[0 \dots (cur-1)]$.

Second, for all prefixes of s, $s[0] \dots s[m-1]$ ($2 \leq m \leq n$) are enumerated. Because $s[0 \dots suffix[m] - 1]$ == $s[(m - suffix[m]) \dots (m-1)]$, if $(m - suffix[m])$ is the divisor of m, then $s[0 \dots (m - suffix[m] - 1)]$ must be the shortest repeated substring for $s[\ ]$.

4.7.18 Seek the Name, Seek the Fame

The little cat is so famous that many couples tramp over hill and dale to Byteland and ask the little cat to give names to their newly born babies. They seek the name, and at the same time seek the fame. In order to escape from such a boring job, the innovative little cat works out an easy but fantastic algorithm:

Step 1: Connect the father's name and the mother's name to a new string *S*.

Step 2: Find a proper prefix–suffix string of *S* (which is not only the prefix, but also the suffix of *S*).

Example: Father = ala, Mother = la; we have *S* = ala + la = alala. Potential prefix–suffix strings of *S* are {a, ala, alala}. Given the string *S*, could you help the little cat to write a program to calculate the length of possible prefix–suffix strings of *S*? (He might thank you by giving your baby a name.)

Input

The input contains a number of test cases. Each test case occupies a single line that contains the string *S* described above.

Restrictions: Only lowercase letters may appear in the input: 1 <= length of *S* <= 400,000.

Output

For each test case, output a single line with integer numbers in increasing order, denoting the possible length of the new baby's name.

Sample Input	Sample Output
ababcabababcabab	2 4 9 18
aaaaa	1 2 3 4 5

Source: POJ Monthly, January 22, 2006, Zeyuan Zhu.

ID for online judge: POJ 2752.

Hint

First, the KMP algorithm is used to produce the prefix array *suffix*[ ]. If *suffix*[*cur*] == *k*, then *S*[0... (*k* – 1)] == *S*[(*cur* – *k*) .. (*cur* – 1)], and *k* is the length of the longest matching substring for the suffix of *S*[0 ... (*cur* – 1)] and the prefix of *S*.

Based on the KMP algorithm, through *suffix*[*len*], *suffix*[*suffix*[*len*]], *suffix*[*suffix*[*suffix*[*len*]]], ..., the length of all possible prefix–suffix strings of *S* can be calculated.

4.7.19 Excuses, Excuses!

Judge Ito is having a problem with people subpoenaed for jury duty giving rather lame excuses in order to avoid serving. In order to reduce the amount of time required listening to goofy excuses, Judge Ito has asked that you write a program that will search for a list of keywords in a list of excuses identifying them as lame. Keywords can be matched in an excuse regardless of case.

Input

Input to your program will consist of multiple sets of data. Line 1 of each set will contain exactly two integers. The first number (1 ≤ *K* ≤ 20) defines the number of keywords to be used in the search. The second number (1 ≤ *E* ≤ 20) defines the number of excuses in the set to be searched. Lines 2 through *K* + 1 each contain exactly one keyword. Lines *K* + 2 through *K* + 1 + *E* each contain exactly one excuse. All keywords in the keyword list will contain only contiguous lowercase alphabetic characters of length *L* (1 ≤ *L* ≤ 20) and will occupy columns 1 through *L* in the input line. All excuses can contain any upper- or lowercase alphanumeric character, a space, or any of the following punctuation marks [".,!?], not including the square brackets, and will not exceed 70 characters in length. Excuses will contain at least one nonspace character.

Output

For each input set, you are to print the worst excuse(s) from the list. The worst excuse(s) is defined as the excuse(s) that contains the largest number of incidences of keywords. If a keyword "occurs" more than once in an excuse, each occurrence is considered a separate incidence. A keyword occurs in an excuse if and only if it exists in the string in contiguous form and is delimited by the beginning or end of the line or any nonalphabetic character or a space.

For each set of input, you are to print a single line with the number of the set immediately after the string "Excuse Set #" (see the sample output). The following line(s) is to contain the worst excuse(s), one per line, exactly as read in. If there is more than one worst excuse, you may print them in any order. After each set of output, you should print a blank line.

Sample Input	Sample Output
5 3	Excuse Set #1
dog	Can you believe my dog died after eating my canary … AND MY HOMEWORK?
ate	
homework	
canary	
died	
My dog ate my homework.	
Can you believe my dog died after eating my canary … AND MY HOMEWORK?	
This excuse is so good that it contains 0 keywords.	
6 5	Excuse Set #2
superhighway	I am having a superhighway built in my bedroom.
crazy	There was a thermonuclear war!
thermonuclear	
bedroom	
war	
building	
I am having a superhighway built in my bedroom.	

Sample Input	Sample Output
I am actually crazy. 1234567890.....,,,,,0987654321?????!!!!!! There was a thermonuclear war! I ate my dog, my canary, and my homework ... note outdated keywords?	

Source: ACM South Central United States 1996.

IDs for online judges: POJ 1598, UVA 409.

Hint
Suppose *key* is the set of keywords, where *key*[*i*] is the *i*th keyword; *next* is the set of prefix functions, where *next*[*i*] is the prefix function for the *i*th keyword; *keycnt* is the number of keywords occurring in the current excuse, where *keycnt*[*i*] is the number of *i*th keywords occurring in the current excuse; and *sentence* is the set of excuses, where *sentence*[*j*] is the *j*th excuse; $(0 \leq i \leq e-1, 0 \leq j \leq k-1)$.

The problem requires outputting the worst excuses; that is, the excuses contain the largest number of incidences of keywords. Therefore, the key to the problem is to calculate the number of incidences *cnt* of the *j*th keyword *key*[*j*] in an excuse *sentence*[*i*]. Suppose the number of characters in excuse *sentence*[*i*] is *n*, and the number of characters in the *j*th key *key*[*j*] is *m*; *cur* is the matching pointer for *sentence*[*i*], and *p* is the matching pointer for *key*[*j*].

The method calculating *cnt* is as follows.

Initially, *cnt* is 0, from the first characters in *sentence*[*i*] and *key*[*j*] ($p = 0, cur = 0$); the KMP algorithm is used as follows:

1. If *sentence*[*i*][*cur*] and *key*[*j*][*p*] are the same, then the next characters in the two strings are compared (++*cur*; ++*p*;).
2. If *sentence*[*i*][*cur*] and *key*[*j*][*p*] aren't the same, if there are matching characters, that is, $p \geq 0$, then *sentence*[*i*][*cur*] is compared with the *next*[*j*][*p*]th character in *key*[*j*] ($p = next[j][p]$); otherwise, *sentence*[*i*][*cur*+1] is compared with *key*[*j*][0] (++*cur*; *p*=0).
3. If the matching is successful ($p == m$), and if *sentence*[*i*][*cur*] and *sentence*[*i*][*cur*−*p*−1] aren't letters, then ++*cnt*. Then $p = next[r][p]$.

Repeat the above process until $cur \geq n$. *cnt* is the number of incidences of keyword *key*[*j*] in excuse *sentence*[*i*].

The main algorithm is as follows:

1. Input *key*[*i*], and calculate its prefix function *next*[*i*] $(0 \leq i \leq k-1)$.
2. Input excuse *sentence*[*i*] $(0 \leq i \leq e-1)$ one by one, and calculate the number of incidences of the *k* keywords

$$keycnt_i = \sum_{j=0}^{k-1} keycnt[j]$$

in *sentence*[*i*].
3. The sentence with the largest number of incidences of keywords $\max_{0 \leq i \leq e-1} \{keycnt_i\}$ is the solution.

4.7.20 Product

The problem is to multiply two integers X, Y ($0 <= X, Y < 10^{250}$).

Input

The input will consist of a set of pairs of lines. Each line in a pair contains one multiplier.

Output

For each input pair of lines the output line should consist of one integer, the product.

Sample Input	Sample Output
12	144
12	4444444444444444444444444
2	
2222222222222222222222222	

Source: Sergant Pepper's Lonely Programmers Club Junior Contest 2001.

ID for online judge: UVA 10106.

Hint
The problem is solved by multiplication of high-precision numbers. Suppose X is the string representing the multiplicand, $L1$ is the length for X; and Y is the string representing the multiplier, $L2$ is the length for Y. *Ans* is the array representing the product, where *Ans*[0] is used as the length of the array, and its length's upper limit is $L1 + L2$, and *Ans*[*Ans*[0] .. 1] is the result.
    The algorithm has been shown in Section 4.2.

4.7.21 Expression Evaluator

This problem is about evaluating some C-style expressions. The expressions to be evaluated will contain only simple integer variables and a limited set of operators; there will be no constants in the expressions. There are 26 variables in the program, named by lowercase letters a through z. Before evaluation, the initial values of these variables are $a = 1$, $b = 2$, ..., $z = 26$.
    The operators allowed are addition and subtraction (binary + and –), with their known meanings. So, the expression $a + c - d + b$ has the value 2 ($1 + 3 - 4 + 2$). Additionally, ++ and –– operators are allowed in the input expression too, which are unary operators, and may come before or after variables. If the ++ operator comes before a variable, then that variable's value is increased (by 1) before the variable's value is used in calculating the value of the whole expression. Thus, the value of ++ $c - b$ is 2. When ++ comes after a variable, that variable is increased (by 1) after its value is used to calculate the value of the whole expression. So, the value of the c ++ – b is 1, though c is incremented after the value for the entire expression is computed; its value will be 4 too. The –– operator behaves the same way, except that it decreases the value of its operand.
    More formally, an expression is evaluated in the following manner:

- Identify every variable that is preceded by ++. Write an assignment statement for incrementing the value of each of them, and omit the ++ from before that variable in the expression.
- Do similarly for the variables with ++ after them.

- At this point, there is no ++ operator in the expression. Write a statement evaluating the remaining expression after the statements determined in step 1 and before those determined in step 2.
- Execute the statements determined in step 1, then those written in step 3, and finally the one written in step 2.

This way, evaluating ++ *a* + *b* ++ is the same as computing *a* = *a* + 1, *result* = *a* + *b*, and *b* = *b* + 1.

Input

The first line of the input contains a single integer *T* that is the number of test cases, followed by *T* lines each containing the input expression for a test case. Ignore blanks in the input expression. Be sure that no ambiguity is in the input expressions (like *a*+++*b*). Similarly, ++ or −− operators do not appear both before and after one single variable (like ++*a*++). You may safely assume each variable appears only once in an expression.

Output

For each test case, write each expression as it appears in the input (exactly), and then write the value of the complete expression. After this, on separate lines, write the value of each variable after evaluating the expression (write them in sorted order of the variable names). Write only the values of the variables that are used in the expressions. To find out about the output format, follow the style used in the sample output below.

Sample Input	Sample Output
2	Expression: a + b
a + b	value = 3
c + f —+— a	a = 1
	b = 2
	Expression: c + f —+— a
	value = 9
	a = 0
	c = 3
	f = 5

Source: ACM Tehran 2006, Preliminary.

ID for online judge: POJ 3337.

Hint
The problem is a simulation problem. You are asked to solve the problem following rules in the problem description. From left to right, every character in the expression is evaluated.

4.7.22 Integer Inquiry

One of the first users of BIT's new supercomputer was Chip Diller. He extended his exploration of powers of 3 to go from 0 to 333, and he explored taking various sums of those numbers.

"This supercomputer is great,'" remarked Chip. "I only wish Timothy were here to see these results.'" (Chip moved to a new apartment, once one became available on the third floor of the Lemon Sky apartments on Third Street.)

Input

The input will consist of at most 100 lines of text, each of which contains a single VeryLongInteger. Each VeryLongInteger will be 100 or fewer characters in length and will only contain digits (no VeryLongInteger will be negative).

The final input line will contain a single zero on a line by itself.

Output

Your program should output the sum of the VeryLongIntegers given in the input.

Sample Input	Sample Output
1234567890123456789012345678901234567890	37037036703703703670370370367037036703670
1234567890123456789012345678901234567890	
1234567890123456789012345678901234567890	
0	

Source: ACM East Central North America 1996.

IDs for online judges: POJ 1503, ZOJ 1292, UVA 424.

Hint

Because the length of a single VeryLongInteger is 100, arrays are used to store a high-precision numbers. Additions of high-precision numbers are used to get the result.

4.7.23 Super-Long Sums

The creators of a new programming language D++ have found out that whatever limit for SuperLongInt type they make, sometimes programmers need to operate even larger numbers. A limit of 1000 digits is so small. You have to find the sum of two numbers with a maximal size of 1 million digits.

Input

The first line of an input file is an integer N, and then a blank line followed by N input blocks. The first line of each input block contains a single number M ($1 \leq M \leq 1,000,000$) — the length of the integers (in order to make their lengths equal, some leading zeros can be added). It is followed by these integers written in columns. That is, the next M lines contain two digits each, divided by a space. Each of the two given integers is not less than 1, and the length of their sum does not exceed M.

There is a blank line between input blocks.

Output

Each output block should contain exactly M digits in a single line representing the sum of these two integers.

There is a blank line between output blocks.

Sample Input	Sample Output
2	4750
4	470
0 4	
4 2	
6 8	
3 7	
3	
3 0	
7 9	
2 8	

Source: Ural State University Collegiate Programming Contest, March 25, 2000, Problem Authors: Stanislav Vasilyev and Alexander Klepinin.

IDs for online judges: UVA 10013, Ural 1048.

Hint
The problem is for addition of high-precision numbers. Based on the input format, a for statement *for* (int $i = m - 1$; $i >= 0$; i––) is used to calculate the result.

4.7.24 Exponentiation

Problems involving the computation of exact values of very large magnitude and precision are common. For example, the computation of the national debt is a taxing experience for many computer systems.

This problem requires that you write a program to compute the exact value of R^n, where R is a real number ($0.0 < R < 99.999$) and n is an integer such that $0 < n <= 25$.

Input

The input will consist of a set of pairs of values for R and n. The R value will occupy columns 1–6, and the n value will be in columns 8 and 9.

Output

The output will consist of one line for each line of input giving the exact value of $R^\wedge n$. Leading zeros should be suppressed in the output. Insignificant trailing zeros must not be printed. Don't print the decimal point if the result is an integer.

Sample Input	Sample Output
95.123 12	548815620517731830194541.8990253434157159735359672218698527211
0.4321 20	.0000000051485546410769561219945112767671548384817602007263512038 35429763013462401
5.1234 15	43992025569.928573701266488041146654993318703707511666295476720497 3953024
6.7592 9	29448126.7641210216181644302069090371732766721
98.999 10	90429072743629540498.107596019456651774561044010001
1.0100 12	1.126825030131969720661201

Source: ACM East Central North America 1988.

IDs for online judges: POJ 1001, UVA 748.

Hint

Power is based on multiplication of high-precision numbers. The problem requires computing the exact value of R^n, where R is a real number. The process is as follows:

1. When R is stored in an array as a high-precision number, the position of decimal point *dec* must be noted.
2. When real a and real b are multiplied (the lengths of arrays storing a and b are l_a and l_b, respectively, and the positions of decimal points are k_a and k_b),
   a. Multiply high-precision numbers $c = a*b$ and note the position of decimal point $k_a + k_b + 1$.
   b. Carry c and calculate its length l_c ($l_a + l_b - 1$ or $l_a + l_b$).
   c. Delete redundant 0 after the decimal point.

4.7.25 Number Base Conversion

Write a program to convert numbers in one base to numbers in a second base. There are 62 different digits: {0–9, A–Z, a–z}.

Hint: If you make a sequence of base conversions using the output of one conversion as the input to the next, when you get back to the original base, you should get the original number.

Input

The first line of input contains a single positive integer. This is the number of lines that follow. Each of the following lines will have a (decimal) input base followed by a (decimal) output base followed by a number expressed in the input base. Both the input base and the output base will be in the range from 2 to 62. That is (in decimal), $A = 10$, $B = 11$, ..., $Z = 35$, $a = 36$, $b = 37$, ..., $z = 61$ (0–9 have their usual meanings).

Output

The output of the program should consist of three lines of output for each base conversion performed. The first line should be the input base in decimal followed by a space and then the input number (as given expressed in the input base). The second output line should be the output base followed by a space and then the input number (as expressed in the output base). The third output line is blank.

Sample Input
8
62 2 abcdefghiz
10 16 1234567890123456789012345678901234567890
16 35 3A0C92075C0DBF3B8ACBC5F96CE3 F0AD2
35 23 333YMHOUE8JPLT7OX6K9FYCQ8A
23 49 946B9AA02MI37E3D3MMJ4G7B L2F05
49 61 1VbDkSIMJL3JjRgAdIUfcaWj
61 5 dl9MDSWqwHjDnToKcsWE1S
510 42104444441001414401221302402201233340311104212022213303

Sample Output
62 abcdefghiz
2 1101110000010001011111001001011001111100100110001101010001
10 1234567890123456789012345678901234567890
16 3A0C92075C0DBF3B8ACBC5F96CE3F0AD2
16 3A0C92075C0DBF3B8ACBC5F96CE3F0AD2
35 333YMHOUE8JPLT7OX6K9FYCQ8A
35 333YMHOUE8JPLT7OX6K9FYCQ8A
23 946B9AA02MI37E3D3MMJ4G7BL2F05
23 946B9AA02MI37E3D3MMJ4G7BL2F05
49 1VbDkSIMJL3JjRgAdIUfcaWj
61 dl9MDSWqwHjDnToKcsWE1S

Source: ACM Greater New York 2002.

IDs for online judges: POJ 1220, ZOJ 1325, UVA 2559.

Hint
Suppose the first base is *ibase*, the string representing the number in *ibase* is *s*, and the second base is *obase*.

First, every digit for *s* is transferred into its corresponding number and stored in a high-precision array *a*. Second, the number in *ibase* is transferred into the number in decimal base. Third, it is transferred into the number in *obase*. Finally, the number in *obase* is transferred into a string.

4.7.26 Super-Long Sums

"Oh! If I could do the easy mathematics like my school days! I can guarantee that I'd not make any mistake this time!" says a smart university student.

But his even smarter teacher said, "OK! I'll assign you such projects in your software lab. Don't be so sad."

"Really!" the students happily exclaims. He is so happy that he cannot see the smile on his teacher's face.

The first project for the poor student was to make a calculator that can just perform the basic arithmetic operations.

But like many other university students, he doesn't like to do any project by himself. He just wants to collect programs from here and there. As you are a friend of him, he asks you to write the program. But, you are also intelligent enough to tackle this kind of problem. You agreed to write only the (integer) division and mod (% in C/C++) operations for him.

Input

The input is a sequence of lines. Each line will contain an input number, one or more spaces, a sign (division or mod), again spaces, and another input number. Both input numbers are non-negative integers. The first one may be arbitrarily long. The second number *n* will be in the range $0 < n < 2^{31}$.

Output

The output is a line for each input, each containing an integer. See the sample input and output. The output should not contain any extra spaces.

Sample Input	Sample Output
110 / 100	1
99 % 10	9
2147483647/ 2147483647	1
2147483646% 2147483647	2147483646

Source: Monthly Contest, May 2003.

ID for online judge: UVA 10494.

Hint
The problem is for the division of high-precision numbers. The algorithm is shown in Section 4.2.

4.7.27 Simple Arithmetics

One part of the new WAP portal is also a calculator computing expressions with very long numbers. To make the output look better, the result is formatted in the same way as it is usually used with manual calculations.

Your task is to write the core part of this calculator. Given two numbers and the requested operation, you are to compute the result and print it in the form specified below. With addition and subtraction, the numbers are written below each other. Multiplication is a little bit more complex: first, we make a partial result for every digit of one of the numbers, and then sum the results together.

Input

There is a single positive integer T on the first line of input. It stands for the number of expressions to follow. Each expression consists of a single line containing a positive integer number, an operator (one of +, −, and *), and the second positive integer number. Every number has at most 500 digits. There are no spaces on the line. If the operation is subtraction, the second number is always lower than the first one. No number will begin with zero.

Output

For each expression, print two lines with two given numbers, the second number below the first one, and last digits (representing unities) must be aligned in the same column. Put the operator right in front of the first digit of the second number. After the second number, there must be a horizontal line made of dashes (–).

For each addition or subtraction, put the result right below the horizontal line, with the last digit aligned to the last digit of both operands.

For each multiplication, multiply the first number by each digit of the second number. Put the partial results one below the other, starting with the product of the last digit of the second number. Each partial result should be aligned with the corresponding digit. That means the last digit of the partial product must be in the same column as the digit of the second number. No product may begin with any additional zeros. If a particular digit is zero, the product has exactly one digit—zero. If the second number has more than one digit, print another horizontal line under the partial results, and then print the sum of them.

There must be a minimal number of spaces on the beginning of lines, with respect to other constraints. The horizontal line is always as long as necessary to reach the left and right end of both numbers (and operators) right below and above it. That means it begins in the same column where the leftmost digit or operator of those two lines (one below and one above) is. It ends in the column where the rightmost digit of those two numbers is. The line can be neither longer nor shorter than specified.

Print one blank line after each test case, including the last one.

Sample Input	Sample Output
4	12345
12345+67890	+67890
324-111	------
325*4405	80235
1234*4	
	324
	−111

Sample Input	Sample Output

	213
	325
	*4405

	1625
	0
	1300
	1300

	1431625
	1234
	*4

	4936

Source: ACM Central Europe 2000.

IDs for online judges: POJ 1396, ZOJ 2017, UVA 2153.

Hint

The problem is for addition, subtraction, and multiplication of high-precision numbers.

4.7.28 $a^b - b^a$

You are given natural numbers a and b. Find $a^b - b^a$.

Input

The input contains numbers a and b ($1 \le a, b \le 100$).

Output

Write the answer to the output.

Sample Input	Sample Output
2 3	−1

ID for online judge: SGU 112.

Hint

Because of the reduplicated addition of high-precision numbers, the object-oriented programming method is suitable to solve this problem. Class *bigNumber* is defined, where its private section is a high-precision array *a* whose length is *len*, and its public section includes *bigNumber*( ): High-precision array *a* is initialized 0.

int *length*( ): Return the length of high-precision array *a*.

int *at*(int *k*): Return *a*[*k*].

void *setnum*(char *s*[ ]): A string *s*[ ] is transferred into a high-precision array *a* whose length is *len*.

isNeg( ): Determine whether high-precision array *a* is negative or not.

void *add*(*bigNumber* &*x*): Addition of high-precision integers: *a*←*a* + *x*.

void *multi*(*bigNumber* &*x*): Multiplication of high-precision integers: *a*←*a***x*.

int *compare*(*bigNumber* &*x*): Compare *a* with *x*, and return

$$\begin{cases} 1 & a > x \\ -1 & a < x \\ 0 & a = x \end{cases}$$

void *minus*(*bigNumber* &*x*): Subtraction of high-precision integers: *a*←*a* − *x*.

void *power*(int *k*): Exponentiation of high-precision integers: *num*←*num*^{k}.

Based on class *bigNumber*, the main algorithm is as follows:

1. Define *bna* and *bnb* as objects of class *bigNumber* (*bigNumber bna*, *bnb*) and *a* and *b* as arrays of *bna* and *bnb* (*bna.setnum*(*a*); *bnb.setnum*(*b*)).
2. Compute *bna*←*bna*^{b}, *bnb*←*bnb*^{a} (*bna.power*(*b*); *bnb.power*(*a*)); *bna*←*bna*-*bnb* (*bna. minus*(*bnb*)).
3. If *bna* is negative (*bna.isNeg*( ) == true), then output minus sign; output *bna.at*(*bna. length*( )–1..*bna.at*(0).

4.7.29 Fibonacci Number

A Fibonacci sequence is calculated by adding the previous two members of the sequence, with the first two members both being 1.

$$f(1) = 1, f(2) = 1, f(n > 2) = f(n - 1) + f(n - 2)$$

Input and Output

Your task is to take numbers as input (one per line) and print the corresponding Fibonacci number.

Sample Input	Sample Output
3	2
100	354224848179261915075

Source: UVA Local Qualification Contest 2003.

Note: No generated fibonacci number in excess of 1000 digits will be in the test data; that is, *f* (20) = 6765 has four digits.

ID for online judge: UVA 10579.

Hint

Because the upper limit of Fibonacci numbers is 1000 digits, a high-precision array is used to store these numbers. Because of the reduplicated addition of high-precision numbers, the object-oriented programming method can be used.

4.7.30 How Many Fibs

Recall the definition of the Fibonacci numbers:

$$F_1 = 1$$
$$F_2 = 2$$
$$F_n = F_{n-1} + F_{n-2} \ (n \geq 3)$$

Given two numbers a and b, calculate how many Fibonacci numbers are in the range $[a, b]$.

Input

The input contains several test cases. Each test case consists of two nonnegative integer numbers a and b. Input is terminated by $a = b = 0$. Otherwise, $a <= b <= 10^{100}$. The numbers a and b are given with no superfluous leading zeros.

Output

For each test case, output on a single line the number of Fibonacci numbers F_i with $a <= F_i <= b$.

Sample Input	Sample Output
10 100	5
1234567890 9876543210	4
0 0	

Source: Ulm Local Contest 2000.

IDs for online judges: POJ 2413, ZOJ 1962.

Hint

Because the 500th number in the Fibonacci sequence exceeds 10^{100}, the offline method is used and the first 500 Fibonacci numbers $fib[1] \ldots fib[500]$ are calculated by the high-precision method. For each test case, a and b, in array $fib[\ ]$ the largest $fib[left]$, which is no less than a ($fib[left] \geq a$), and the smallest $fib[right]$, which is no more than b ($fib[right] \leq b$), are found. Obviously, the number of Fibonacci numbers F_i with $a \leq F_i \leq b$ is *right–left*.

4.7.31 Heritage

Your rich uncle died recently, and the heritage needs to be divided among your relatives and the church (your uncle insisted in his will that the church must get something). There are N relatives ($N \leq 18$) that were mentioned in the will. They are sorted in descending order according to their importance (the first one is the most important). Since you are the computer scientist in the family, your relatives asked you to help them. They need help because there are some blanks in the will left to be filled. Here is how the will looks:

Relative 1 will get 1/... of the whole heritage.
Relative 2 will get 1/... of the whole heritage.
...
Relative *n* will get 1/... of the whole heritage.

The logical desire of the relatives is to fill the blanks in such way that the uncle's will is preserved (i.e., the fractions are nonascending and the church gets something) and the amount of heritage left for the church is minimized.

Input

The only line of input contains the single integer N ($1 \leq N \leq 18$).

Output

Output the numbers that the blanks need to fill (on separate lines), so that the heritage left for the church is minimized.

Sample Input	Sample Output
2	2
	3

Source: Bulgarian Online Contest, September 2001.

IDs for online judges: POJ 1405, Ural 1108.

Hint

Suppose $a[i]$ is the number of the $i + 1$th blank to be filled. That is, the $i + 1$th relative gets $1/a[i]$ of the whole heritage ($0 <= i <= n - 1$).

$$a[i] = \begin{cases} 2 & i = 0 \\ a[i-1]*a[i-1]-a[i-1]+1 & 1 <= i <= n-1 \end{cases}$$

can be proved.
    The heritage left for the church is

$$l = 1 - \frac{1}{a[0]} - \frac{1}{a[1]} - \cdots \frac{1}{a[n-1]} = \frac{1}{a[0]} - \frac{1}{a[1]} - \cdots \frac{1}{a[n-1]}$$

Because $a[0] \ldots a[n-1]$ are all positive integers, in order to minimize the heritage left for the church l, we should prove

$$\frac{1}{a[0]} - \frac{1}{a[1]} - \cdots - \frac{1}{a[n-1]} = \frac{1}{a[0]a[1]\cdots a[n-1]}$$

Proof.

There are recursive formulas as follows:

$$a[1] = a[0]*a[0] - a[0] + 1 = a[0]*(a[0] - 1) + 1 = a[0] + 1$$
$$a[2] = a[1]*(a[1] - 1) + 1 = a[0]*a[1] + 1$$

$$\cdots$$

Based on it, we can imply $a[i] = a[0]*a[1]* \ldots a[i-1] + 1$ ($1 <= i <= n - 1$). The following formula can be implied:

$$\left(\frac{1}{a[0]} - \frac{1}{a[1]} \right) - \cdots - \frac{1}{a[n-1]}$$

$$= \left(\frac{1}{a[0]a[1]} - \frac{1}{a[2]} \right) - \cdots - \frac{1}{a[n-1]}$$

$$= \left(\frac{1}{a[0]a[1]a[2]} - \frac{1}{a[3]} \right) - \cdots - \frac{1}{a[n-1]}$$

$$\cdots \cdots$$

$$= \frac{1}{a[0]a[1]a[2]a[3]\cdots a[n-2]} - \frac{1}{a[n-1]}$$

$$= \frac{1}{a[0]a[1]a[2]a[3]\cdots a[n-2]a[n-1]}$$

■

Chapter 5

Applications of Linear Lists for Sequential Access

Linear lists for sequential access are linear lists in which all elements are stored and accessed in order. The first data element in a linear list is at the front position, and the last one is at the rear position.

There are two features for linear lists for sequential access:

1. There is no size restriction for the length of the list after it is created. That is, the list can be dynamically changed.
2. Elements in the list can only be accessed in sequential order and can't be accessed directly. In order to access an element, we should visit elements in the list one by one from the first element (or from the last element).

A simple example for linear lists for sequential access is a shopping list. Write down all items that you want to buy in a shopping list. When you find an item, you cross it out from the list. In this kind of linear list, elements can be listed in order, but they can also be listed in disorder. In disordered linear lists, the order of elements is random, and in ordered linear lists, all elements are listed in some order. For example, in the following list, names are in alphabetical order.

Beata Bernica David Frank Jennifer Mike Raymond Terrill

The search efficiency for an ordered list is higher. For example, the efficiency of binary search for an ordered linear list is much higher than the efficiency of the sequential search. Based on the access mode, there are two kinds of linear lists for sequential access:

1. Linear lists for sequential access based on positions of data elements, such as arrays and linked lists
2. Stacks and queues

5.1 Application of Sequence Lists

A sequence list is a linear list storing n data elements, where n is the length of the list, and if $n == 0$, the list is null. The data type for each element in the list is the same. The length of the list can be changed by inserting or deleting elements.

The relationship of data elements in the sequence list is linear. Elements can be accessed through their position in the list. There are two kinds of storage structures:

1. Array: The type of array elements can not only be simple, but also be a structure type. An array element can be accessed through its index. Inserting or deleting an element not only increases or decreases the length of the list, but also needs to move elements in the list.
2. Linked list: If a linear list is a linked list, inserting or deleting an element doesn't need to move elements in the list, and only needs to change related pointers.

There are several kinds of linked lists. Singly linked lists are the simplest and classical linked lists. And based on them, there are doubly linked lists and circular linked lists. Linked lists are widely used.

5.1.1 Children

N children constitute a circle. Children are numbered from 1 to N. Then children begin to circularly count off from the Wth child. Every time, the Sth child gets out from the circle and the next child begins to count off. The process repeats until all children get out from the circle. Output the sequence of children getting out from the circle.

Input

The first line is the number of children N ($N \leq 64$).

Then children's names are shown, one name per line. The length of a child's name doesn't exceed 15.

W, S ($W < N$) are shown in the last line and separated by a comma.

Output

Output the sequence of children getting out from the circle, one name per line.

Sample Input	Sample Output
5	Zhangsan
Xiaoming	Xiaohua
Xiaohua	Xiaoming
Xiaowang	Xiaowang
Zhangsan	Lisi
Lisi	
2,3	

Source: Preliminary Contest for C Programming Language.

ID for online judge: POJ 3750.

Analysis

It is a Josephus problem. *N* children constitute a circle. *N* children can be represented as an array and can also be represented as a circular linked list. Each child corresponds to an element. The element type is a string storing a child's name.

The problem is a simulation problem. The process in the problem description is simulated to solve the problem.

Program

```
#include <stdio.h>
#include <stdlib.h>
int main( )
{
    int child[65];
    char name[65][16];      // children's linear list storing names
    int n, s, w, ss, i;
    scanf("%d", &n);     // the number of children
    for (i = 0; i < n; i++)   // Input children's names
    {
        child[i] = 1;
        scanf("%s", name[i]);
    }
    scanf("%d,%d", &w, &s);   // w and s specified in the problem
description
    w--;
    for (i = n; i >= 1; i--)
// Every time the sth child get out from the circle and the next child
begin to count off.
    {
        ss = s % n;
        while (1)
        {
            if (child[w] == 1)
            {
                ss--;
                if (ss == 0) break;
            }
            w = (w + 1) % n;
        }
        printf("%sn", name[w]);
        child[w] = 0;
    }
    return 0;
}
```

There are some modifications for the Josephus problem. The following problem is an example.

5.1.2 *The Dole Queue*

In a serious attempt to downsize (reduce) the dole queue, the New National Green Labor Rhinoceros Party has decided on the following strategy. Every day, all dole applicants will be placed in a large circle, facing inward. Someone is arbitrarily chosen as number 1, and the rest are numbered counterclockwise up to *N* (who will be standing on 1's left). Starting from 1 and moving counterclockwise, one labor official counts off k applicants, while another official starts from N and moves clockwise, counting m applicants. The two who are chosen are then sent off for retraining; if both officials pick the same person, he or she is sent off to become a politician. Each official then starts counting again at the next available person, and the process continues until no one is left. Note that the two victims (sorry, trainees) leave the ring simultaneously, so it is possible for one official to count a person already selected by the other official.

Input

Write a program that will successively read in (in that order) the three numbers (N, k, and m; k, $m > 0$, $0 < N < 20$) and determine the order in which the applicants are sent off for retraining. Each set of three numbers will be on a separate line, and the end of data will be signaled by three zeros (0 0 0).

Output

For each triplet, output a single line of numbers specifying the order in which people are chosen. Each number should be in a field of three characters. For pairs of numbers, list the person chosen by the counterclockwise official first. Separate successive pairs (or singletons) by commas (but there should not be a trailing comma).

Sample Input	Sample Output
10 4 3	ΔΔ4ΔΔ8,ΔΔ9ΔΔ5,ΔΔ3ΔΔ1,ΔΔ2ΔΔ6,Δ10,ΔΔ7
0 0 0	

Source: New Zealand contest 1990.

Δ Represents a space.

ID for online judge: UVA 133.

Analysis

The problem is a modified Josephus problem. There are two directions: counterclockwise and clockwise, counting k and m applicants.

Suppose *left* is the number of persons in the current circle. Marks for persons in the circle are *exist*, where *exist*[i] == true represents the ith person is in the circle. Suppose p_i is the ith person whom the first official selects, and q_i is the ith person whom the second official selects. Obviously, $p_0 = 0$, $q_0 = n + 1$, *left* = n, and *exist*[1 .. n] all are true.

The program should simulate the process until *left* == 0.

Program

```
#include <cstdio>
#include <cstring>
const int maxn = 20;        //The upper limit of the number of persons in
    the circle
int main(void)
{
    int n, k, m;
    scanf("%d%d%d", &n, &k, &m);  // a test case
    while (n || k || m) {
```

```
      bool exist[maxn];    //exist[i] represents the i-th person is in the
circle or not
      memset(exist, true, sizeof(exist));
      int p=0, q=n+1;
      int left = n;              //Initialization
      while (left) {                 //Simulate the process until left is 0
      int cnt = (k % left ? k % left : left);
       while (cnt--)
          do {
             p = ((p + 1) % n ? (p + 1) % n : n);
          } while (!exist[p]);
      cnt = (m % left ? m % left : left);
      while (cnt--)
          do {
             q = ((q - 1 + n) % n ? (q - 1 + n) % n : n);
          } while (!exist[q]);
      if (left < n)        // Output
          putchar(',');
      printf("%3d", p);
      if (p != q)
          printf("%3d", q);
      exist[p] = exist[q] = false;    // Out the circle
      left -= (p == q ? 1 : 2);     //Calculate the number of persons in
the circle
      }
      putchar('n');
      scanf("%d%d%d", &n, &k, &m);          // The next test case
   }
   return 0;
}
```

5.2 Application of Stacks

A stack is a linear list in which insertions and deletions take place at the same end. Therefore, a stack is a last-in, first-out (LIFO) structure. Several operations are defined on stacks. Two of the most important operations are PUSH and POP. PUSH adds an element at the top of the stack. POP, in contrast, reduces the stack size by one by removing the last element at the top of the stack. Suppose *top* is the stack pointer (SP) pointing to the top of the stack. If a new element is added into the stack, the stack pointer *top* ++ and the new element are stored in the address. If an element is deleted from the stack, *top* –– (Figure 5.1).

5.2.1 Rails

There is a famous railway station in PopPush City. The country there is incredibly hilly. The station was built in the last century. Unfortunately, funds were extremely limited at that time. It was possible to establish only a surface track. Moreover, it turned out that the station could be only a dead-end one (Figure 5.2), and due to lack of available space, it could have only one track.

The local tradition is that every train arriving from direction A continues in direction B with coaches reorganized in the some way. Assume that the train arriving from direction A has $N \leq 1000$ coaches numbered in increasing order, 1, 2, ..., N. The chief for train reorganizations must know

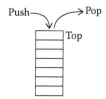

Figure 5.1 Normally arrays are used as storage structures for stacks, to avoid using pointers, to save time.

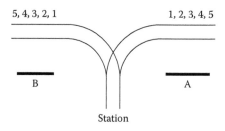

Figure 5.2 Railway station.

whether it is possible to marshal coaches continuing in direction B so that their order will be a_1, a_2, ..., a_N. Help him and write a program that decides whether it is possible to get the required order of coaches. You can assume that single coaches can be disconnected from the train before they enter the station and that they can move themselves until they are on the track in direction B. You can also suppose that at any time there can be located as many coaches as necessary in the station. But once a coach has entered the station, it cannot return to the track in direction A, and also, once it has left the station in direction B, it cannot return back to the station.

Input

The input consists of blocks of lines. Each block except the last describes one train and possibly more requirements for its reorganization. In the first line of the block there is the integer N described above. In each of the next lines of the block there is a permutation of 1, 2, ..., N. The last line of the block contains just 0.

The last block consists of just one line containing 0.

Output

The output contains the lines corresponding to the lines with permutations in the input. A line of the output contains "Yes" if it is possible to marshal the coaches in the order required on the corresponding line of the input. Otherwise, it contains "No." In addition, there is one empty line after the lines corresponding to one block of the input. There is no line in the output corresponding to the last null block of the input.

Sample Input	Sample Output
5	Yes
1 2 3 4 5	No

Sample Input	Sample Output
5 4 1 2 3	
0	Yes
6	
6 5 4 3 2 1	
0	
0	

Source: ACM Central Europe 1997.

IDs for online judges: POJ 1363, ZOJ 1259.

Analysis
The railway station is a stack. The train arriving from direction *A* has $n \leq 1000$ coaches numbered in increasing order, 1, 2, …, *n*. And the coaches leaving in direction *B* are numbered as a permutation for 1, 2, …, *n*. The permutation is implemented by stack operations.

There are two methods to solve the problem.

Method 1

Based on last-in-first-out, for an element *x*, elements that are larger than *x* are pushed into the stack after *x* is pushed into the stack, and elements that are less than *x* are pushed into the stack before *x* is pushed into the stack. Therefore, when an element *x* is popped from the stack, elements that are larger than *x* in the stack must be popped before, and elements that are less than *x* must be in the stack. Suppose *valid* is the legal flag for the permutation, *max* is the maximal value of elements in the stack or popped from the stack, and *p* is the flag of an element's status, where

$$p[x] = \begin{cases} 0 & \text{element } x \text{ hasn't been pushed into the stack} \\ 1 & \text{element } x \text{ is in the stack} \\ 2 & \text{element } x \text{ has been popped from the stack} \end{cases}$$

The algorithm is as follows:

Initially all elements aren't in the stack, that is, all elements in *p* are set 0, *max*=0;
Input elements in the current permutation one by one. For the current element *x*, determine the permutation is valid or not as follows.
If *valid*==true, then

- If there exists an element *t* which is larger than *x* in the stack ($p[t]==1$, and $x+1 \leq t \leq max$), then *valid*=false; for *x* can't be popped from the stack. Based on "last-in-first-out," *p* should be popped before *x*;
- Adjust the maximal value of elements in the stack or popped from the stack ($max = (max > x ? max : x)$);
- Any element $p[j]$ which is less than *x* should be in the stack ($p[j] = 1$, $1 \leq j \leq x - 1$).

After the above process, output the result based on *valid* (*valid*? "Yes" : "No").
Its time complexity is $O(n^3)$.

Program for Method 1

```
#include <iostream>
#include <cstring>
using namespace std;
const int maxn = 1000 + 10;          // The upper limit for the number of
coaches
int main(void)
{
    int n, p[maxn];
    cin >> n;                        //number of elements
    while (n) {
        int x, max = 0;
        cin >> x;                    // the first permutated element
        while (x) {
            memset(p, 0, sizeof(p)); // Initialization (0: element isn't
pushed into the stack; 1: element is in the stack; 2: element is popped
from the stack;)
            bool valid = true;
            for (int i = 1; i <= n; i++) {
                if (valid) {         // check whether there exists an
element t which is larger than x in the stack
                    bool ok = true;
                    for (int i = x + 1; i <= max; i++)
                        if (p[i] == 1) {
                            ok = false;
                            break;
                        }
                    if (!ok)         // if exist
                        valid = false;
                    else {           // adjust the maximal value of
elements in the stack or popped from the stack
                        max = (max > x ? max : x);
                        p[x] = 2; //x is popped from the stack, element
p[j] which is less than x should be in the stack
                        for (int i = x - 1; i > 0 && !p[i]; i--)
                            p[i] = 1;
                    }
                }
                if (i < n)
                    cin >> x;                    // the next permutated element
            }
            cout << (valid ? "Yes" : "No") << endl;    // the permutation
is valid or not
            cin >> x;
        }
        cout << endl;
        cin >> n;                        // the number of elements in the next case
    }
    return 0;
}
```

Method 2

For each test case, the permutation is simulated. That is, coaches 1, 2, ..., n in direction A are pushed into the stack and compared with the elements in the permutation one by one, to determine whether the permutation is valid or not.

1. If the current element in direction A (i.e., the element that will be pushed into the stack) is the same as the current element in the permutation, then the current element in direction A will be pushed into the stack and popped from the stack directly. The next elements in direction A and in the permutation become the current elements.
2. If the element at the top of the stack is the same as the current element in the permutation, then the element will be popped from the stack. The next element in the permutation becomes the current element in the permutation.
3. Otherwise, the current element in direction A is pushed into the stack. The next element in direction A becomes the current element in direction A.

Repeat the above steps. If n elements in the permutation can be popped from the stack, then the permutation is valid; otherwise, the permutation isn't valid. Its time complexity is $O(n)$.

Program for Method 2

```
#include<stdio.h>
int main()
{
    int a[1005], b[1005], i, j, k, n;      // a[0..n-1] stores elements
pushed into the stack , and k is the pointer pointing to the top of the
stack; b[0..n-1] stores elements popped from the stack, that is, the
permutation, and j is the pointer pointing to the top of the permutation.
    while (scanf("%d", &n), n)          //n: the number of coaches
    {
        while (scanf("%d", &b[0]), b[0])
        {
            for (j=1; j<n; j++)  scanf("%d",&b[j]);      // Input the
permutation
    // Determine whether the permutation is valid or not
            for (i=1, j=0, k=0; i<=n&&j<n; i++, k++)       // coaches 1,
2, ..., n in the direction A are pushed into the stack and compared with
the current element in the permutation one by one
            {
                a[k]=i;                    // i is pushed into stack a[ ]
                while (a[k]==b[j])             // If the current element in
the direction A (that is, the element will be pushed in the stack) is
same as the current element in the permutation, then the current element
in the direction A will be pushed in the stack and popped from the stack
directly.
                {
                    if (k>0) k--;
                      else    {  a[k]=0, k--;    }
                    j++;                   //b[j] is popped from the stack
                    if (k==-1) break;
```

```
                }
            }
            if (j==n) printf("Yesn");          //all elements are popped
from the stack
                else printf("Non");
        }
        printf("n");
    }
}
```

A stack is also used for expression evaluation. An expression constitutes

1. Operands: Valid variable names or constants
2. Operators, including
   - Arithmetic operators, including +, −, *, /, %, and unary operator (−)
   - Relation operators, including <, <=, ==, !=, >, >=
   - Logical operators, including &&, ||, !
   - Brackets

In order to correctly evaluate expressions, the priority for operators should be defined. In C++, the priority for operators is defined as follows:

Priority	Operators
7	−, ! (unary)
6	*, /, %
5	+, −
4	<, <=, >, >=
3	==, !=
2	&&
1	‖

For example, for an expression $A + B*(C - D) - E/F$, the order of evaluation is as shown in Figure 5.3, where R_1, R_2, R_3, R_4, and R_5 are intermediate results.

In the process of the expression evaluation, two stacks should be used:

1. Operator stack *op*: Used to store operators
2. Value stack *val*: Used to store operands and intermediate results

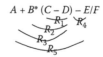

Figure 5.3   Calculation of $A + B*(C - D) - E/F$.

The expression is a string. The algorithm is as follows:

```
While there are still tokens to be read in, get the next token:
        { If the token is an operand and there is no unary operator at the
top of operator stack, the current operand is pushed into the value stack;
else unary operators are popped and are operated on the current operand;
        If the token is an operator (Op), while the operator stack is not
empty, and the top element's priority on the operator stack is the same
or greater precedence as Op;
        { Pop the operator from the operator stack;
          Pop the value stack twice, getting two operands;
          Apply the operator to the operands, in the correct order;
          Push the result into the value stack;
        };
        Push Op into the operator stack;
        }
        While there is no token to be read in, and Operator stack is not
empty
        { Pop the operator from the operator stack;
          Pop the value stack twice, getting two operands;
          Apply the operator to the operands, in the correct order;
          Push the result into the value stack;
        }
```

Finally, the value in the value stack is the result of the expression.

5.2.2 Boolean Expressions

The objective of the program you are going to produce is to evaluate Boolean expressions such as the one shown next:

$$\text{Expression: } (V \mid V) \, \& \, F \, \& \, (F \mid V)$$

where V is for true and F is for false. The expressions may include the following operators: ! for *not*, & for *and*, and | for *or*; the use of parentheses for operations grouping is also allowed.

To perform the evaluation of an expression, it will be considered the priority of the operators, the *not* having the highest and the *or* the lowest. The program must yield V or F as the result for each expression in the input file.

Input

The expressions are of a variable length, although they will never exceed 100 symbols. Symbols may be separated by any number of spaces or no spaces at all; therefore, the total length of an expression, as a number of characters, is unknown.

The number of expressions in the input file is variable and will never be greater than 20. Each expression is presented in a new line, as shown below.

Output

For each test expression, print "Expression" followed by its sequence number, ":," and the resulting value of the corresponding test expression. Separate the output for consecutive test expressions with a new line.

Use the same format as that shown in the sample output below.

Sample Input	Sample Output
(V | V) & F & (F | V)	Expression 1: F
!V | V & V & !F & (F | V) & (!F | F | !V & V)	Expression 2: V
(F&F|V|!V&!F&!(F|F&V))	Expression 3: V

Source: ACM Mexico and Central America 2004.

ID for online judge: POJ 2106.

Analysis

The priority for operators is defined in the following list, where '!' is a unary operator and '|' and '&' are operators.

Operators	Priority
(	0
|	1
&	2
!	3
)	4

Two stacks are used to solve the problem: stack *op* is used to store operators, *otop* is its stack pointer, and stack *val* is used to store values.

Because of the priority, if there are some '!' at the top of stack *op*, before a value is pushed into stack *val*, these '!' should be popped and the final result is pushed into stack *val*.

The algorithm is as follows:

1. Initially stack *val* and stack *op* are empty ($vtop = otop = 0$);
2. Analyze every character *c* in the expression one by one:
   - If $c ==$ '(', then 0 is popped into stack *op*;
   - If $c ==$ ')', then the expression between '(' and ')' is calculated, the result is pushed into stack *val*, and the '(' at the top of stack *op* is popped;
   - If $c ==$ '!', then 3 is pushed into stack op;
   - If $c ==$ '&', then '&' and '!' at the top of stack *op* are popped and calculated, and 2 is pushed into stack *op*;
   - If $c ==$ '|', then '|', '&' and '!' at the top of stack *op* are popped and calculated, and 1 is pushed into stack *op*;
   - If *c* is 'V' or 'F', then transfer it into a number ('V' is transferred into 1, and 'F' is transferred into 0); and is pushed into stack *val*;
3. Operators in stack *op* are popped one by one, and are calculated. Finally the element at the bottom of stack *val* is the result ($val[0]$? 'V' : 'F').

Program

```cpp
#include <cstdio>
const int maxn = 100 + 10;              //the upper limit of the length of
the expression
int val[maxn], vtop;                    // stack val and its stack pointer
int op[maxn], otop;                     // stack op and its stack pointer
void insert(int b)                      //b is pushed into stack val
{
    while (otop && op[otop - 1] == 3) { // while '!' at the top of stack
op
        b = !b;
        --otop;
    }
    val[vtop++] = b;                    // b is pushed into stack val
}
void calc(void)                         // calculation
{
    int b = val[--vtop];                // a and b are popped from stack val
    int a = val[--vtop];
    int opr = op[--otop];               // operator opr is popped from stack op
    int c = (a & b);
    if (opr == 1)
        c = (a | b);
    insert(c);                          // the result is pushed into stack val
}
int main(void)
{
    int loop = 0;
    char c;
    while ((c = getchar( )) != EOF) {   // Character of the expression
        vtop = otop = 0;
        do {                            // the expression is scanned
            if (c == '(') {             // if c = '(', then 0 is pushed into
stack op
                op[otop++] = 0;
            } else if (c == ')') { // if c = ')', the subexpression in
the brackets is calculated, and the result is pushed into stack val
                while (otop && op[otop - 1] != 0)
                    calc( );
                --otop;                 //'(' is popped from stack op
                insert(val[--vtop]);
            } else if (c == '!') {      //if c = '!', then 3 is pushed into
stack op
                op[otop++] = 3;
            } else if (c == '&') {      // if c = '&', then '&' or '!' at the
top of stack op are popped and calculated, and 2 is pushed into stack op
                while (otop && op[otop-1] >= 2)
                    calc( );
                op[otop++] = 2;
            } else if (c == '|') {      //if c = '|', then '|','&' or '!' at
the top of stack op are popped and calculate, and 1 is pushed in stack op
                while (otop && op[otop - 1] >= 1)
                    calc( );
```

```
                op[otop++] = 1;
           } else if (c == 'V' || c == 'F') {// if c is a value, c is
pushed into stack val
                insert(c == 'V' ? 1 : 0);
           }
      } while ((c = getchar( )) != '\n' && c != EOF);
      while (otop)                          // elements are popped from stack
op and calculate
           calc( );
      printf("Expression %d: %c\n", ++loop, (val[0] ? 'V' : 'F')); //
the result of the expression
     }
    return 0;
}
```

5.3 Application of Queues

A queue is also a kind of linear list. Unlike stacks, additions of entities are at the rear terminal position of the linear list, called enqueue, and removals of entities are from the front terminal position, called dequeue. Therefore, queues are first-in, first-out (FIFO) data structures.

5.3.1 A Stack or a Queue?

Do you know stack and queue? They're both important data structures. A stack is a first-in, last-out (FILO) data structure, and a queue is a FIFO one.

Here comes the problem: Given the order of some integers (it is assumed that the stack and queue are both for integers) going into the structure and coming out of it, what kind of data structure could it be—stack or queue?

Notice that here we assume that none of the integers are popped out before all the integers are pushed into the structure.

Input

There are multiple test cases. The first line of input contains an integer T ($T \leq 100$), indicating the number of test cases. Then T test cases follow.

Each test case contains three lines: The first line of each test case contains only one integer N, indicating the number of integers ($1 \leq N \leq 100$). The second line of each test case contains N integers separated by a space, which are given in the order of going into the structure (i.e., the first one is the earliest going in). The third line of each test case also contains N integers separated by a space, which are given in the order of coming out of the structure (the first one is the earliest coming out).

Output

For each test case, output your guess in a single line. If the structure can only be a stack, output "stack," or if the structure can only be a queue, output "queue"; otherwise, if the structure can be either a stack or a queue, output "both," or else output "neither."

Sample Input	Sample Output
4	stack

Sample Input	Sample Output
3	queue
1 2 3	both
3 2 1	neither
3	
1 2 3	
1 2 3	
3	
1 2 1	
1 2 1	
3	
1 2 3	
2 3 1	

Source: 6th Zhejiang Provincial Collegiate Programming Contest.

ID for online judge: ZOJ 3210.

Analysis

Based on definitions of stack and queue, for each test case, if the ith integer in the first sequence is equal to the ith integer in the second sequence, the data structure is a queue, and if the ith integer in the first sequence is equal to the reciprocal ith integer in the second sequence, the data structure is a stack; $0 \le i \le n - 1$.

Suppose $a[i]$ is the ith integer in the first sequence and $b[i]$ is the ith integer in the second sequence, $0 \le i \le n - 1$, and *issta* and *isque* are flags for stack and queue, respectively. Initially, *issta* and *isque* are true.

The algorithm is as follows:

For each element in a and b, if $b[i] \ne a[i]$, the data structure isn't a FIFO structure, and isn't a queue (*isque*=false); and if $b[i] \ne a[n-i-1]$, the data structure isn't a LIFO structure, and isn't a stack (*issta*=false); $0 \le i \le n - 1$.

Finally, output the result based on *issta* and *isque*:

issta	isque	Output
false	false	"neither"
false	true	"queue"
true	false	"stack"
true	true	"both"

Program

```cpp
#include <iostream>
using namespace std;
const int maxn = 100 + 10;              //the upper limit of the length the
structure
int main(void)
{
    int loop;
    cin >> loop;                        //the number of test cases
    while (loop--) {
        int n, a[maxn];                 // the number of integers and the
structure
        cin >> n;
        for (int i = 0; i < n; i++)     // Input integers into the
structure
            cin >> a[i];
        bool isque = true, issta = true; // marks for queue and stack
        for (int i = 0; i < n; i++) {
            int x;
            cin >> x;                   // the i-th integer leaving the
structure
            if (x != a[i])              // the structure isn't a queue
                isque = false;
            if (x != a[n - i - 1]) // the structure isn't a stack
                issta = false;
        }
                if (issta && isque)                     // the structure is both
a stack and a queue
            cout << "both" << endl;
        else if (issta) //                      stack
            cout << "stack" << endl;
        else if (isque) // queue
            cout << "queue" << endl;
        else                            // neither a queue nor a stack
            cout << "neither" << endl;
    }
    return 0;
}
```

5.3.2 Team Queue

Queues and priority queues are data structures that are known to most computer scientists. The team queue, however, is not so well known, though it occurs often in everyday life. At lunchtime the queue in front of the Mensa is a team queue, for example.

In a team queue each element belongs to a team. If an element enters the queue, it first searches the queue from head to tail to check if some of its teammates (elements of the same team) are already in the queue. If yes, it enters the queue right behind them. If not, it enters the queue at the tail and becomes the new last element (bad luck). Dequeuing is done like in normal queues: elements are processed from head to tail in the order they appear in the team queue.

Your task is to write a program that simulates such a team queue.

Input

The input will contain one or more test cases. Each test case begins with the number of teams t $(1 \le t \le 1000)$. Then t team descriptions follow, each one consisting of the number of elements belonging to the team and the elements themselves. Elements are integers in the range 0–999,999. A team may consist of up to 1000 elements.

Finally, a list of commands follows. There are three different kinds of commands:

ENQUEUE x: Enter element x into the team queue.

DEQUEUE: Process the first element and remove it from the queue.

STOP: End of test case.

The input will be terminated by a value of 0 for t.

Warning: A test case may contain up to 200,000 commands, so the implementation of the team queue should be efficient: both enqueing and dequeuing of an element should only take constant time.

Output

For each test case, first print a line saying "Scenario #k," where k is the number of the test case. Then, for each DEQUEUE command, print the element that is dequeued on a single line. Print a blank line after each test case, even after the last one.

Sample Input	Sample Output
2	Scenario #1
3 101 102 103	101
3 201 202 203	102
ENQUEUE 101	103
ENQUEUE 201	201
ENQUEUE 102	202
ENQUEUE 202	203
ENQUEUE 103	
ENQUEUE 203	Scenario #2
DEQUEUE	259001
DEQUEUE	259002
DEQUEUE	259003
DEQUEUE	259004
DEQUEUE	259005
DEQUEUE	260001
STOP	

(Continued)

Sample Input	Sample Output
2	
5 259001 259002 259003 259004 259005	
6 260001 260002 260003 260004 260005 260006	
ENQUEUE 259001	
ENQUEUE 260001	
ENQUEUE 259002	
ENQUEUE 259003	
ENQUEUE 259004	
ENQUEUE 259005	
DEQUEUE	
DEQUEUE	
ENQUEUE 260002	
ENQUEUE 260003	
DEQUEUE	
DEQUEUE	
DEQUEUE	
DEQUEUE	
STOP	
0	

Source: Ulm Local 1998.

IDs for online judges: POJ 2259, ZOJ 1948, UVA 540.

Analysis

In a team queue, teams constitute a queue. And in a team, elements constitute a queue. Therefore, a team queue is a queue of queues. Queues consisting of elements are nested in a team queue. The algorithm is to simulate the problem description and implement a queue of queues.

Program

```
#include <stdio.h>
#include <assert.h>

#define MAXTEAMS 1024
#define MAXTEAMSIZE 1024
#define MAXELEMENTS 1048576
#define DBG(x)
FILE *input;
```

```
int kase=0;
int numteams;
int team[MAXELEMENTS];    /* team[i] = the team element #i belongs to */
int teampos[MAXTEAMS];    /* teampos[i] = position of team #i in the queue
*/
int teamsize[MAXTEAMS];   /* teamsize[i] = number of elements of team #i
currently in the queue */
int queue[MAXTEAMS][MAXTEAMSIZE]; /* the queue of queues */
int queuehead[MAXTEAMS];              /* the heads of the single queues */
int queuetail[MAXTEAMS];              /* the tails of the single queues */
int head,tail;                     /* head and tail of the queue of queues */
int read_case( )
{
  int i,j,n,elmt;

  /* read team descriptions */
  fscanf(input,"%d",&numteams);
  if (numteams==0) return 0;
  for (i=0; i<numteams; i++)
    {
      fscanf(input,"%d",&n);
      for (j=0; j<n; j++)
    {
      fscanf(input,"%d",&elmt);
      DBG(printf("%d ",elmt));
      team[elmt] = i;
    }
      DBG(printf("OK\n"));
    }
  return 1;
}
void enqueue (int element)
{
  int t,pos;
  t = team[element];
  if (teamsize[t]==0)  /* create a new team at the tail */
    {
      queue[tail][0] = element;
      queuehead[tail] = 0;
      queuetail[tail] = 1;
      teampos[t] = tail;
      teamsize[t] = 1;
      tail = (tail+1)%MAXTEAMS;
    }
  else                  /* add element to the team */
    {
      pos = teampos[t];
      queue[pos][queuetail[pos]] = element;
      queuetail[pos] = (queuetail[pos]+1)%MAXTEAMSIZE;
      teamsize[t]++;
    }
}
int dequeue( )
{
```

```
    int element = queue[head][queuehead[head]];
    int t = team[element];
    queuehead[head] = (queuehead[head]+1)%MAXTEAMSIZE;
    teamsize[t]--;
    if (teamsize[t]==0)   /* team is empty => remove it */
      head = (head+1)%MAXTEAMS;
    return element;
}
void solve_case( )
{
  char cmd[30];
  int element,t;
  printf("Scenario #%d\n",++kase);
  /* initialize queue */
  head = tail = 0;
  for (t=0; t<numteams; t++)
    teamsize[t] = 0;
  /* simulation */
  while (1)
    {
      fscanf(input,"%s",cmd);
      if (strcmp(cmd,"ENQUEUE")==0)
    {
      fscanf(input,"%d",&element);
      enqueue(element);
    }
      else if (strcmp(cmd,"DEQUEUE")==0)
    {
      printf("%d\n",dequeue( ));
    }
      else if (strcmp(cmd,"STOP")==0)
    {
      printf("\n");
      return;
    }
      else
    {
      assert(0);
    }
    }
}
int main( )
{
  input = fopen("team.in","r");
  assert(input!=NULL);
  while (read_case( )) solve_case( );
  fclose(input);
  return 0;
}
```

A priority queue is a queue that each element has a priority associated with it. In a priority queue, an element with high priority is served before an element with low priority. If two elements have the same priority, they are served according to their order in the queue.

5.3.3 Printer Queue

The only printer in the computer science students' union is experiencing an extremely heavy workload. Sometimes there are a hundred jobs in the printer queue and you may have to wait for hours to get a single page of output.

Because some jobs are more important than others, the hacker general has invented and implemented a simple priority system for the print job queue. Now, each job is assigned a priority between 1 and 9 (with 9 being the highest priority and 1 being the lowest), and the printer operates as follows:

- The first job *J* in queue is taken from the queue.
- If there is some job in the queue with a higher priority than job *J*, then move *J* to the end of the queue without printing it.
- Otherwise, print job *J* (and do not put it back in the queue).

In this way, all those important muffin recipes that the hacker general is printing get printed very quickly. Of course, those annoying term papers that others are printing may have to wait for quite some time to get printed, but that's life.

Your problem with the new policy is that it has become quite tricky to determine when your print job will actually be completed. You decide to write a program to figure this out. The program will be given the current queue (as a list of priorities) as well as the position of your job in the queue, and it must then calculate how long it will take until your job is printed, assuming that no additional jobs will be added to the queue. To simplify matters, we assume that printing a job always takes exactly 1 minute, and that adding and removing jobs from the queue is instantaneous.

Input

The input is one line with a positive integer: the number of test cases (at most 100). Then, for each test case,

- One line with two integers n and m, where n is the number of jobs in the queue ($1 \leq n \leq 100$) and m is the position of your job ($0 \leq m \leq n - 1$). The first position in the queue is number 0, the second is number 1, and so on.
- One line with n integers in the range 1–9, giving the priorities of the jobs in the queue. The first integer gives the priority of the first job, the second integer the priority of the second job, and so on.

Output

For each test case, print one line with a single integer: the number of minutes until your job is completely printed, assuming that no additional print jobs will arrive.

Sample Input	Sample Output
3	1
1 0	2

(Continued)

Sample Input	Sample Output
5	5
4 2	
1 2 3 4	
6 0	
1 1 9 1 1 1	

Source: ACM Northwestern Europe 2006.

IDs for online judges: POJ 3125, UVA 3638.

Analysis

The problem is a priority queue's problem.

Suppose *a* is the priority queue storing jobs; the size of *a maxn* = 100 + 5. Initially, *n* jobs are stored in *a*, and *a*[*m*] is your job. In order to mark your job, *a*[*m*] = −*a*[*m*], and *cnt* is the number of minutes until your job is completely printed, initially *cnt* = 0.

The algorithm is to simulate the printer operation. The process is shown in the program.

Program

```
#include <iostream>
using namespace std;
const int maxn = 100 + 5;              // the upper limit of the
length of priority queue a
inline int fabs(int k)                  // return absolute value
{
    return k < 0 ? -k : k;
}
int main(void)
{
    int loop;
    cin >> loop;                        // the number of test cases
    while (loop--) {
      int n, m;
      cin >> n >> m;           // the number of jobs in the queue and the
position of your job
      int st, ed, a[maxn];              // priority queue a, points for
the front and the rear
      for (int i = 0; i < n; i++)         // the priorities of n jobs
            cin >> a[i];
      a[m] = -a[m];       // initialization
      st = 0;
      ed = n;
      int cnt = 0;
      while ((ed + 1) % maxn != st) { // a circular queue used to
simulate the printer queue
            int k = a[st];
            st = (st + 1) % maxn;
            bool print = true;
            for (int i = st; i != ed; i = (i + 1) % maxn)
```

```
                    if (fabs(k) < fabs(a[i])) {
                        print = false;
                        a[ed] = k;
                        ed = (ed + 1) % maxn;
                        break;
                    }
                if (print) {
                    ++cnt;
                    if (k < 0) {
                        cout << cnt << endl;
                        break;
                    }
                }
            }
        }
    }
    return 0;
}
```

The priority queue is represented with an array. Its time complexity is $O(n)$ (n is the number of elements in the priority queue.) A binary tree can also be used to store a priority queue to improve time complexity. It will be introduced in Section III.

5.4 Problems

5.4.1 Roman Roulette

The historian Flavius Josephus relates how, in the Romano-Jewish conflict of 67 AD, the Romans took the town of Jotapata, which he was commanding. Escaping, Josephus found himself trapped in a cave with 40 companions. The Romans discovered his whereabouts and invited him to surrender, but his companions refused to allow him to do so. He therefore suggested that they kill each other, one by one, the order to be decided by lot. Tradition has it that the means for effecting the lot was to stand in a circle and, beginning at some point, count round, every third person being killed in turn. The sole survivor of this process was Josephus, who then surrendered to the Romans. This begs the question: Had Josephus previously practiced quietly with 41 stones in a dark corner, or had he calculated mathematically that he should adopt the 31st position in order to survive?

Having read an account of this gruesome event, you become obsessed with the fear that you will find yourself in a similar situation at some time in the future. In order to prepare yourself for such an eventuality, you decide to write a program to run on your handheld PC that will determine the position from which the counting process should start in order to ensure that you will be the sole survivor.

In particular, your program should be able to handle the following variation of the processes described by Josephus: $n > 0$ people are initially arranged in a circle, facing inward, and numbered from 1 to n. The numbering from 1 to n proceeds consecutively in a clockwise direction. Your allocated number is 1. Starting with person number i, counting starts in a clockwise direction, until we get to person number k ($k > 0$), who is promptly killed. We then proceed to count a further k people in a clockwise direction, starting with the person immediately to the left of the victim. The person number k so selected has the job of burying the victim and then returning to the position in the circle that the victim had previously occupied. Counting then proceeeds from the person to his immediate left, with the kth person being killed, and so on, until only one person remains.

For example, when $n = 5$, $k = 2$, and $i = 1$, the order of execution is 2, 5, 3, and 1. The survivor is 4.

Input and Output

Your program must read input lines containing values for n and k (in that order), and for each input line output the number of the person with which the counting should begin in order to ensure that you are the sole survivor. For example, in the above case the safe starting position is 3. Input will be terminated by a line containing values of 0 for n and k.

Your program may assume a maximum of 100 people taking part in this event.

Sample Input	Sample Output
1 1	1
1 5	1
0 0	

Source: New Zealand Contest 1989.

ID for online judge: UVA 130.

Hint

The problem is a simulation problem. The program is to simulate the process in the problem description. Suppose the number for the ith person is $who[i]$, the number of persons in the circle is cnt, and the position of the pth killed person is i_p, $i_p = (i_{p-1} + k)\%cnt$. Initially $who[i] = i + 1$ $(0 \le i \le n - 1)$, $cnt = n$, and $i_0 = -1$.

Simulate the process until $cnt == 1$.

5.4.2 M*A*S*H

Corporal Klinger is a member of the 4077th Mobile Army Surgical Hospital in the Korean War, and he will do just about anything to get out. The U.S. Army has made an offer for a lottery that will choose some number of lucky people (X) to return to the states for a recruiting tour. Klinger needs your help getting out.

The lottery is run by lining up all the members of the unit at attention and eliminating members by counting off the members from 1 to N, where N is a number chosen by pulling cards off of the top of a deck. Every time N is reached, that person falls out of the line, and counting begins again at 1 with the next person in line. When the end of the line has been reached (with whatever number that may be), the next card on the top of the deck will be taken, and counting starts again at 1 with the first person in the remaining line. The last X people in line get to go home.

Klinger has found a way to trade a stacked deck with the real deck just before the selection process begins. However, he will not know how many people will show up for the selection until the last minute. Your job is to write a program that will use the deck Klinger supplies and the number of people in line that he counts just before the selection process begins and tell him what position(s) in the line to get in to assure himself of a trip home. You are assured that Klinger's deck will get the job done by the time the 20th card is used.

A simple example with 10 people, 2 lucky spots, and the numbers from cards 3, 5, 4, 3, and 2 would show that Klinger should get in position 1 or 8 to go home.

Input

For each selection, you will be given a line of 22 integers. The first integer $(1 \le N \le 50)$ tells how many people will participate in the lottery. The second integer $(1 \le X \le N)$ is how many lucky "home" positions will be selected. The next 20 integers are the values of the first 20 cards in the deck. Card values are interpreted to integer values between 1 and 11 inclusive.

Output

For each input line, you are to print the message "Selection #*A*" on a line by itself, where *A* is the number of the selection starting with 1 at the top of the input file. The next line will contain a list of "lucky" positions that Klinger should attempt to get into. The list of "lucky" positions is then followed by a blank line.

Sample Input	Sample Output
10 2 3 5 4 3 2 9 6 10 10 6 2 6 7 3 4 7 4 5 3 2	Selection #1
47 6 11 2 7 3 4 8 5 10 7 8 3 7 4 2 3 9 10 2 5 3	1 8
	Selection #2
	1 3 16 23 31 47

Source: ACM South Central United States 1995.

IDs for online judges: POJ 1591, ZOJ 1326, UVA 402.

Hint

The problem is also a modified Josephus problem. A list is used to represent members. The program simulates the process in the problem description.

5.4.3 Joseph

Joseph's problem is notoriously known. For those who are not familiar with the original problem: from among *n* people, numbered 1, 2, ..., *n*, standing in a circle every *m*th is going to be executed and only the life of the last remaining person will be saved. Joseph was smart enough to choose the position of the last remaining person, thus saving his life to give us the message about the incident. For example, when $n = 6$ and $m = 5$, then the people will be executed in the order 5, 4, 6, 2, and 3, and 1 will be saved.

Suppose that there are *k* good guys and *k* bad guys. In the circle the first *k* are good guys and the last *k* bad guys. You have to determine such minimal *m* that all the bad guys will be executed before the first good guy.

Input

The input file consists of separate lines containing *k*. The last line in the input file contains 0. You can suppose that $0 < k < 14$.

Output

The output file will consist of separate lines containing *m* corresponding to *k* in the input file.

Sample Input	Sample Output
3	5

(Continued)

Sample Input	Sample Output
4	30
0	

Source: ACM Central Europe 1995.

ID for online judge: POJ 1012.

Hint

The problem is also a modified Joseph's problem. In order to avoid time limit exceeded, an offline method is used. First, all solutions to Joseph's problem are calculated. That is, for all possible k, corresponding m is calculated and is stored in an array: $ans[k] = m$. Then, for each test case k, output $ans[k]$ directly.

Because $0 < k < 14$, m can be enumerated.

5.4.4 City Skyline

The best part of the day for Farmer John's cows is when the sun sets. They can see the skyline of the distant city. Bessie wonders how many buildings the city has. Write a program that assists the cows in calculating the minimum number of buildings in the city, given a profile of its skyline.

The city in profile is quite dull architecturally, featuring only box-shaped buildings. The skyline of a city on the horizon is somewhere between 1 and W units wide ($1 \leq W \leq 1,000,000$) and described using N ($1 \leq N \leq 50,000$) successive x and y coordinates ($1 \leq x \leq W$, $0 \leq y \leq 50$), defining at what point the skyline changes to a certain height.

An example skyline could be that shown in Figure 5.4 and would be encoded as (1,1), (2,2), (5,1), (6,3), (8,1), (11,0), (15,2), (17,3), (20,2), (22,1). This skyline requires a minimum of six buildings to form; Figure 5.5 shows a possible set of six buildings that could create the skyline above.

Input

Line 1: Two space-separated integers: N and W

```
............   ....................
.......XX............XXX..........
.XXX. XX..........XXXXXXX.......
XXXXXXXXXX....XXXXXXXXXXXX
```

Figure 5.4 Example skyline.

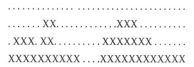

Figure 5.5 One possible set of six buildings.

Lines 2 .. N + 1: Two space-separated integers, the x and y coordinates of a point where the skyline changes. The x coordinates are presented in strictly increasing order, and the first x coordinate will always be 1.

Output

Line 1: The minimum number of buildings to create the described skyline.

Sample Input	Sample Output
10 26	6
1 1	
2 2	
5 1	
6 3	
8 1	
11 0	
15 2	
17 3	
20 2	
22 1	

Source: USACO, November 2005, Silver.

ID for online judge: POJ 3044.

Hint
Because successive x and y coordinates for the skyline are listed from left to right, if the current point's y coordinate is higher than its left adjacent point's y coordinate, the point is a starting position for a building (Figure 5.6a); if the current point's y coordinate is lower than its left adjacent point's y coordinate, the point is an ending position for a building (Figure 5.6b); and if the current point's y coordinate is equal to its left adjacent point's y coordinate, it shows that two points belong to the same building (Figure 5.6c).

A stack is used to store coordinates. Coordinates are used to represent buildings. When a point's coordinate (x, y) is input, it is compared with the coordinate at the top of the stack. The element at the top of the stack is its left adjacent point's coordinate.

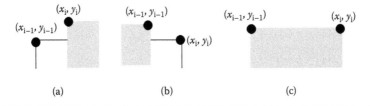

(a) (b) (c)

Figure 5.6 Points and buildings.

If the current point's *y* coordinate is lower than its left adjacent point's *y* coordinate, the point at the top of stack is an ending position for a building and must be popped from the stack, and the number of buildings increases 1; then the point's coordinate (*x*, *y*) is compared with the coordinate at the top of the stack again. The process is repeated until the current point's *y* coordinate isn't lower than its left adjacent point's *y* coordinate.

If the current point's *y* coordinate is higher than its left adjacent point's *y* coordinate, the point is a starting position for a building, and the coordinate (*x*, *y*) is pushed into the stack.

If the current point's *y* coordinate is equal to its left adjacent point's *y* coordinate, it shows that two points belong to the same building, and we needn't do anything.

After *n* coordinates are dealt with, the number of coordinates in the stack is added up to the number of buildings. The result is the solution to the problem.

5.4.5 Anagrams by Stack

How can anagrams result from sequences of stack operations? There are two sequences of stack operators that can convert TROT to TORT:

```
[
i i i i o o o o
i o i i o o i o
]
```

where i stands for push and o stands for pop. Given pairs of words, your program should produce sequences of stack operations that convert the first word to the second.

Input

The input will consist of several lines of input. The first line of each pair of input lines is to be considered a source word (which does not include the end-of-line character). The second line (again, not including the end-of-line character) of each pair is a target word. The end of the input is marked by the end of the file.

Output

For each input pair, your program should produce a sorted list of valid sequences of *i* and *o* that produce the target word from the source word. Each list should be delimited by

```
[
]
```

and the sequences should be printed in dictionary order. Within each sequence, each *i* and *o* is followed by a single space and each sequence is terminated by a new line.

Sample Input	Sample Output
madam	[
adamm	i i i i o o o i o o
bahama	i i i i o o o o i o
bahama	i i o i o i o i o o
long	i i o i o i o o i o

Sample Input	Sample Output
short	]
Eric	[
Rice	i o i i i o o i i o o o
	i o i i i o o o i o i o
	i o i o i o i i i o o o
	i o i o i o i o i o i o
	]
	[
	]
	[
	i i o i o i o o
	]

Source: Zhejiang University Local Contest 2001.

ID for online judge: ZOJ 1004.

Hint
The problem is similar to the rails problem in Section 5.2.1.

Chapter 6

Generalized List Using Indexes

Arrays and generalized lists are indexed data structures. The difference between an array and a generalized list is that the array is directly indexed by an integer index and the generalized list is indexed by a key. In a generalized list, normally one or several items are set as the key to identify records. For example, records for residents of an area use residents' ID numbers as keys to identify residents. Therefore, a generalized list is a set of keyword–data value pairs. There are two kinds of generalized lists:

1. Dictionary
2. Hash table

6.1 Solving Problems Using Dictionaries

Dictionaries are commonly used tools in our lives, such as an English–Chinese dictionary, telephone directories, library catalogs, a computer's file directory, and so on. In computer science, a dictionary can be used as an abstract data type. Such a data type defines a dictionary as a set of ⟨name–attribute⟩ pairs. Based on problems, names can be given different meanings. For example,

Occasion	Name	Attribute
Library catalog	Title	Information for index number, authors, and so on
File list in computer	File name	Information for addresses and sizes of files
Variable list in a program	Variable name	Data type and address for variables

Normally, a file (or a table) contains a set of objects, where a record in the file (or an item in the table) represents an object. In a dictionary, a pair of ⟨name–attribute⟩ will be stored as a record (or an item), and a key (i.e., name in ⟨name–attribute⟩) is used to identify the record (or the item).

Dictionaries can be organized as linear lists or other nonlinear structures. This chapter focuses on dictionaries organized as sequence lists.

6.1.1 References

Editors of an electronic magazine make draft versions of the documents in the form of text files. However, publications should meet some requirements, in particular concerning the rules of reference use. Unfortunately, lots of the draft articles violate some rules. It is desirable to develop a computer program that will make a publication satisfy all the rules from a draft version.

Let's call a *paragraph* a set of lines in the article going one after another, so that paragraphs are separated by at least one empty line (an *empty line* is a line that contains no characters different from spaces). Any paragraph can contain an arbitrary number of references. A reference is a positive integer not greater than 999 enclosed in square brackets (e.g., [23]). There will be no spaces between the brackets and the number. The square brackets are not used in any other context but references.

There can be two types of paragraph: regular and reference description. Reference description differs from the regular paragraph because it begins with the reference it describes, for example:
⟨list⟩[23] It is the description …

The opening square bracket will be at the first position of the first line of the reference description paragraph (i.e., there will be no spaces before it). No reference description paragraph will contain references inside itself.

Each reference will have exactly one corresponding description, and each description will have at least one reference to it.

To convert a draft version to a publication, you have to use the following rules:

■ References should be renumbered by the successive integer numbers, starting from 1 in the order of their first appearance in the regular paragraphs of the source draft version of the document.
■ Reference descriptions should be placed at the end of the article ordered by their number.
■ The order of regular paragraphs in the document should be preserved.
■ Your program should not make any other changes to the paragraphs.

Input

The input file will be a text file containing a draft article your program should process. All lines will be no more than 80 characters long. Any reference description will contain no more than three lines. The input file will contain up to 40,000 lines.

Output

The output file contains the result of processing. All paragraphs should be separated by one true empty line (i.e., a line that contains no characters at all). There should be no empty lines before the first paragraph.

Sample Input	Sample Output
[5] Brownell, D., "Dynamic Reverse Address Resolution Protocol (DRARP)," Work in Progress.	The Reverse Address Resolution Protocol (RARP) [1] (through the extensions defined in the Dynamic RARP (DRARP) [2]) explicitly addresses the problem of network address discovery, and includes an automatic IP address assignment mechanism.

Sample Input	Sample Output
The Reverse Address Resolution Protocol (RARP) [10] (through the extensions defined in the Dynamic RARP (DRARP) [5]) explicitly addresses the problem of network address discovery, and includes an automatic IP address assignment mechanism.	The Trivial File Transfer Protocol (TFTP) [3] provides for transport of a boot image from a boot server. The Internet Control Message Protocol (ICMP) [4] provides for informing hosts of additional routers via "ICMP redirect" messages.
[10] Finlayson, R., Mann, T., Mogul, J., and Theimer, M., "A Reverse Address Resolution Protocol," RFC 903, Stanford, June 1984.	Works [1], [4], and [3] can be obtained via Internet.
[16] Postel, J., "Internet Control Message Protocol," STD 5, RFC 792, USC/Information Sciences Institute, September 1981.	[1] Finlayson, R., Mann, T., Mogul, J., and Theimer, M., "A Reverse Address Resolution Protocol," RFC 903, Stanford, June 1984.
The Trivial File Transfer Protocol (TFTP) [20] provides for transport of a boot image from a boot server. The Internet Control Message Protocol (ICMP) [16] provides for informing hosts of additional routers via "ICMP redirect" messages.	[2] Brownell, D., "Dynamic Reverse Address Resolution Protocol (DRARP)," Work in Progress.
[20] Sollins, K., "The TFTP Protocol (Revision 2)," RFC 783, NIC, June 1981.	[3] Sollins, K., "The TFTP Protocol (Revision 2)," RFC 783, NIC, June 1981.
Works [10], [16], and [20] can be obtained via Internet.	[4] Postel, J., "Internet Control Message Protocol," STD 5, RFC 792, USC/Information Sciences Institute, September 1981.

Source: ACM Northeastern Europe 1997.

IDs for online judges: POJ 1706, UVA 765.

Analysis

In this problem, references are as dictionaries, where names are as reference numbers and attributes are as reference descriptions.

In regular paragraphs, indexes are created by reference numbers. The program should show regular paragraphs and reference descriptions based on rules in the problem description.

Suppose $p[\ ]$ is the sequence of reference descriptions, its length is *refCnt*, its elements' type is *struct*, where $p[i].desc$ is a reference description, $p[i].oldno$ is the old number of the reference in the input, $p[i].newno$ is the sequence number in regular paragraphs ($0 \le i \le refCnt - 1$), and *refSort* is the new current number.

For array p, there are two calculations.

1. Insert the reference whose old number is *oldno* and reference description is *desc* into array p.
2. For the reference whose old number is *oldno*, the new number *newno* is calculated.

Based on it, the algorithm is as follows:

Initially *p* is empty, there is no new number (*refCnt*=0, *refSort*=0). Repeat the following steps until input ends.

1. Ignore empty lines;
2. If the current line s[ ] is a reference description (s[0] == '['), then get *oldno* and *desc* for the reference, and insert it into sequence *p*; else if s[ ] is a regular paragraph, each character is analyzed as follows: if the current character is '[', get *oldno* and compute *newno*; else, output the current character;
3. For each reference in array *p*, the new number *newno* is the sequence number for the reference. Array *p* is sorted based on *newno* and is output.

Program

```
#include <cstdio>
#include <string>
#include <cstring>
#include <cctype>
using namespace std;
const int maxRef = 1000;   //the upper limit of the number of references
const int maxCol = 80 + 5;
struct reference {         // Structure type for reference
    string desc;           // reference description
    int oldno, newno;      // the old and new numbers of the reference
} p[maxRef];               // the sequence of reference descriptions
int refCnt, refSort;
void qsort(int st, int ed)              // Sort references p based on newno
in ascending order
{
    if (st >= ed)
        return;
    int i, j, k = p[(st + ed) / 2].newno;   // the middle newno
    i = st - 1, j = ed + 1;                 // Initialize the left and right
pointers
    do {
        do
            ++i;
        while (p[i].newno < k);    // from left to right, find the first
reference whose newno >=k
        do
            --j;
        while (p[j].newno > k);    // from right to left, find the first
reference whose newno <=k
        if (i < j) {                        // exchange two references
            reference tmp = p[i];
            p[i] = p[j];
            p[j] = tmp;
        }
    } while (i < j);
    qsort(st, j);                           // left and right subintervals
    qsort(j + 1, ed);
}
```

```
bool isBlank(char s[ ])                   // s[ ] is an empty line, return
true, else return false
{
    int k = 0;
    while (s[k] != '\0')
        if (!isspace(s[k++]))
            return false;
    return true;
}
inline bool isReference(char s[ ])     // s[ ] is a reference or not
{
    return s[0] == '[';
}
int searchRef(int oldno)
{ // find the index in p[ ] for oldno, if it doesn't exist, return -1
    for (int i = 0; i < refCnt; i++)
        if (p[i].oldno == oldno)
            return i;
    return -1;
}
int insertRef(string desc, int oldno) // Insert the new reference into p.
   If the reference description hasn't appeared, desc is empty
{ int cur = searchRef(oldno);
    if (cur < 0) {
        cur = refCnt++;
        p[cur].newno = 0;
    }
    p[cur].desc = desc;
    p[cur].oldno = oldno;
    return cur;
}
inline int getRefNo(const char s[ ], int st = 1)
{ // the original number for reference
    int refno;
    sscanf(s + st, "%d", &refno);
    return refno;
}
int getRefNewNo(int oldno)     //find the newno for oldno
{
    int k = searchRef(oldno);
    if (k < 0)
        k = insertRef("", oldno);
    if (!p[k].newno)
        p[k].newno = ++refSort;
    return p[k].newno;
}
void proc(char s[ ])                      // regular paragraph
{
    int len = strlen(s);
    for (int i = 0; i < len; i++) {
        if (s[i] == '[') {
            int oldno = getRefNo(s, i + 1);
            int newno = getRefNewNo(oldno);
            printf("[%d]", newno);
```

```
            while (s[++i] != ']');
        } else
            putchar(s[i]);
    }
    putchar('\n');
}
int main(void)
{
    refCnt = 0;
    refSort = 0;
    char s[maxCol];
    while (gets(s) != NULL) {
        while (isBlank(s)) //
            if (gets(s) == NULL)
                break;
        if (isBlank(s))
            break;
        if (isReference(s)) {         //s[ ] is reference description
            int oldno = getRefNo(s);    // oldno
            string desc;
            do {                      // desc
                desc += s;
                desc += '\n';
            } while (gets(s) != NULL && !isBlank(s));
            insertRef(desc, oldno); // new reference record
        } else { // regular paragraph
            do {
                proc(s);
            } while (gets(s) != NULL && !isBlank(s));
            putchar('\n');
        }
    }
    qsort(0, refCnt - 1);              // Sort references
    for (int i = 0; i < refCnt; i++) {
        printf("[%d]", p[i].newno);
        int k = 0;
        while (p[i].desc[k++] != ']');
        printf("%sn", p[i].desc.c_str( ) + k);
    }
    return 0;
}
```

If sequential search is used in dictionaries, its time complexity for searching an element is $O(n)$. In order to improve the query efficiency, we can sort dictionaries, and then dichotomy is used in the query and its time complexity is $O(\log_2(n))$.

6.1.2 Babelfish

You have just moved from Waterloo to a big city. The people here speak an incomprehensible dialect of a foreign language. Fortunately, you have a dictionary to help you understand them.

Input

The input consists of up to 100,000 dictionary entries, followed by a blank line, followed by a message of up to 100,000 words. Each dictionary entry is a line containing an English word, followed by a space and a foreign language word. No foreign word appears more than once in the dictionary. The message is a sequence of words in the foreign language, one word on each line. Each word in the input is a sequence of at most 10 lowercase letters.

Output

The output is the message translated to English, one word per line. Foreign words not in the dictionary should be translated as "eh".

Sample Input	Sample Output
dog ogday	cat
cat atcay	eh
pig igpay	loops
froot ootfray	
loops oopslay	
atcay	
ittenkay	
oopslay	

Source: Waterloo Local Contest, September 22, 2001.

ID for online judge: POJ 2503.

Analysis
The dictionary is represented as a linear list *dict*, where the length is n, and for the ith dictionary entry, $dict[i][0]$ is used to store the English word and $dict[i][1]$ is used to store the foreign language word, $0 \le i \le n - 1$.

Binary search is used to solve the problem. First, the dictionary is sorted in the foreign language words' alphabet order. Then, after a foreign language word is input, the corresponding English word is searched through binary search. If there is no such foreign language word, "eh" is output; else, the corresponding English word is output.

Because input consists of up to 100,000 dictionary entries, *scanf* and *printf* are used.

Program

```
#include <cstdio>
#include <cstring>
const int maxn = 100000 + 10;   // the upper limit of the number of words
const int maxs = 10 + 5;        // the upper limit of the length of a word
```

```
char dict[maxn][2][maxs]; // dictionary, for the i-th word, English word
is dict[i][0], and foreign language word is dict[i][1]
int n;                    // the number of words
bool isblank(char s[ ])      // the current line is empty line or not
{
    int k = strlen(s);
    while (--k >= 0)
        if (s[k] >= 'a' && s[k] <= 'z')
            return false;
    return true;
}
void swap(char a[ ], char b[ ]) //exchange strings a and b
{
    char t[maxs];
    strcpy(t, a);
    strcpy(a, b);
    strcpy(b, t);
}
void sort(int a, int b, char s[ ][2][maxs]) // sort dictionary in
alphabet order for foreign words
{
    if (a >= b)
        return;
    char t[maxs];
    strcpy(t, s[(a + b) / 2][1]);
    int i, j;
    i = a - 1, j = b + 1;
    do {
        do
            ++i;
        while (strcmp(t, s[i][1]) > 0);
        do
            --j;
        while (strcmp(t, s[j][1]) < 0);
        if (i < j) {
            swap(s[i][0], s[j][0]);
            swap(s[i][1], s[j][1]);
        }
    } while (i < j);
    sort(a, j, s);
    sort(j + 1, b, s);
}
int find(char s[ ])                      // dichotomy is used
{
    int l, r;
    l = 0;
    r = n;
    while (l + 1 < r) {
        int mid = (l + r) / 2;
        if (strcmp(dict[mid][1], s) <= 0)
            l = mid;
        else
            r = mid;
    }
```

```
        if (strcmp(dict[l][1], s))
            return -1;
        return 1;
}
    int main(void)
{
    char s[maxs + maxs];
    n = 0;
    gets(s);                                        //Input the first word
    while (!isblank(s)) {                          // Input all words in the
dictionary
        sscanf(s, "%s%s", dict[n][0], dict[n][1]);  //English word and
foreign language word
        ++n;
        gets(s);                                    //next word
    }
    sort(0, n - 1, dict);
    while (scanf("%s", s) != EOF) {
        int k = find(s);
        if (k < 0)
            printf("%s\n", "eh");
        else
            printf("%s\0n", dict[k][0]);
    }
    return 0;
}
```

6.2 Solving Problems Using a Hash Table and the Hash Method

Like a dictionary, a hash table is also used to search records in an indexed linear list with keys. A hash table uses a hash function that maps keys into addresses in a table; that is, there is a hash function *address* = *hash(key)*. When we need to search element *e* with key *k*, the first step is to calculate hash function *address* = *hash(k)* and get *addresses (hash(k))*, which is the position of the element in the table. Obviously, the ideal situation is that the hash function is an injection function.

Unfortunately, sometimes a hash function maps different keys into a same hash value. Such a situation is called a conflict. There are two methods to eliminate the conflict:

1. Hash with linear open addressing: The data structure of hash table *T* is a one-dimensional array. Hashing with linear open addressing can be used directly. If there exists a conflict, we need to test other addresses until an address can store the element. The hash function is designed based on linear probing, quadratic probing, or double hashing, and so on.
2. Hashing with chains: Hash table *T* uses chains. That is, elements with the same hash value are linked in a chain.

6.2.1 10-20-30

A simple solitaire card game called 10-20-30 uses a standard deck of 52 playing cards in which suit is irrelevant. The value of a face card (king, queen, jack) is 10. The value of an ace is 1. The value of each of the other cards is the face value of the card (2, 3, 4, etc.). Cards are dealt from the top of the deck. You begin by dealing out seven cards, left to right, forming seven piles. After playing a card on the rightmost pile, the next pile upon which you play a card is the leftmost pile.

For each card placed on a pile, check that pile to see if one of the following three card combinations totals 10, 20, or 30:

1. The first two and last one
2. The first one and the last two
3. The last three cards

If so, pick up the three cards and place them on the bottom of the deck. For this problem, always check the pile in the order just described. Collect the cards in the order they appear on the pile and put them at the bottom of the deck. Picking up three cards may expose three more cards that can be picked up. If so, pick them up. Continue until no more sets of three can be picked up from the pile.

For example, suppose a pile contains 5 9 7 3, where the 5 is the first card of the pile, and then a 6 is played. The first two cards plus the last card (5 + 9 + 6) sum to 20. The new contents of the pile after picking up those three cards becomes 7 3. Also, the bottommost card in the deck is now the 6, the card above it is the 9, and the one above the 9 is the 5 (Figure 6.1).

If a queen were played instead of the 6, 5 + 9 + 10 = 24, and 5 + 3 + 10 = 18, but 7 + 3 + 10 = 20, so the last three cards would be picked up, leaving the pile as 5 9 (Figure 6.2).

If a pile contains only three cards when the three sum to 10, 20, or 30, then the pile "disappears" when the cards are picked up. That is, subsequent play skips over the position that the now-empty pile occupied. You win if all the piles disappear. You lose if you are unable to deal a card. It is also possible to have a draw if neither of the previous two conditions ever occurs.

Write a program that will play games of 10-20-30 given initial card decks as input.

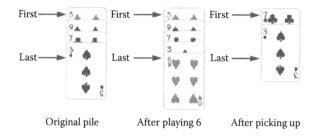

Original pile After playing 6 After picking up

Figure 6.1 An example.

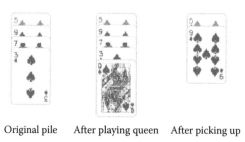

Original pile After playing queen After picking up

Figure 6.2 An example.

Input

Each input set consists of a sequence of 52 integers separated by spaces or ends of line. The integers represent card values of the initial deck for that game. The first integer is the top card of the deck. Input is terminated by a single zero following the last deck.

Output

For each input set, print whether the result of the game is a win, loss, or a draw, and print the number of times a card is dealt before the game results can be determined. (A draw occurs as soon as the state of the game is repeated.) Use the format shown in the "Sample Output" section.

Sample Input	Sample Output
2 6 5 10 10 4 10 10 10 4 5 10 4 5 10 9 7 6 1 7 6 9 5 3 10 10 4 10 9 2 1	Win: 66
10 1 10 10 10 3 10 9 8 10 8 7 1 2 8 6 7 3 3 8 2	Loss: 82
4 3 2 10 8 10 6 8 9 5 8 10 5 3 5 4 6 9 9 1 7 6 3 5 10 10 8 10 9 10 10 7	Draw: 73
2 6 10 10 4 10 1 3 10 1 1 10 2 2 10 4 10 7 7 10	
10 5 4 3 5 7 10 8 2 3 9 10 8 4 5 1 7 6 7 2 6 9 10 2 3 10 3 4 4 9 10 1 1	
10 5 10 10 1 8 10 7 8 10 6 10 10 10 9 6 2 10 10	
0	

Source: ACM-ICPC World Finals 1996.

ID for online judge: UVA 246.

Analysis

A string s is used to represent cards in piles and in the hand.

1. **Intervals are used to represent cards in piles and in the hand.** Cards in piles and in the hand are represented as a string s. Capital letters "ABCDEFGH" are used to separate intervals. Initially values of 52 playing cards are stored in an array $a[1 .. 52]$. That is, $s = $ "A$a[1]$ B$a[2]$C$a[3]$D$a[4]$E$a[5]$F$a[6]$G$a[7]$H$a[8 .. 52]$," where $a[1 .. 7]$ is the first seven cards put into pile 1 ... pile 7, respectively, and $a[8 .. 52]$ are cards in the hand and can be regarded in pile 8. Suppose $sign[i]$ is the flag for pile i ($sign[1] = $ "A," ..., $sign[7] = $ "G," $sign[8] = $ "H"); the front pointer and the rear pointer for pile i are $l[i]$ and $r[i]$, respectively.

2. **Hash technology is used to determine repetition state.** String s is regarded as a state. Obviously, if the current state is the same as a previous state, it is a draw. The hash function

for string s is defined as follows:

$$hash(s)\left(\sum_{i=0}^{s.size-1}(s[i]-\text{'}0\text{'})*13^{s.size-1-i}\right)\%1999997$$

Hashing with open addressing is used to eliminate conflicts. Rejudging the hash function, $hash2(s)$ is defined as follows:

$$hash2(s)=\left(\sum_{i=0}^{s.size-1}(s[i]-\text{'}0\text{'})*13^{s.size-1-i}\right)\%10000009$$

Only when $hash(s_1) == hash(s_2)$ and $hash2(s_1) == hash2(s_2)$, state s_1 and s_2 are same.

3. **Simulate the process putting a card into pile i.** First, the first card in the hand is taken out (the character in $l[8]$) and is inserted into the rear of pile i. Second, the 10-20-30 rule is implemented. If there is no card in pile i, the pile "disappears".

4. **The game process is simulated.**

```
Initially 52 cards are input, and initial state s is gotten;
  The number of pile i is initialized 1;
  Repeat the loop until the result is gotten:
  {      Accumulate the number of cards put into piles (step++);
         Put the first card into pile i, and 10-20-30 rule is implemented;
         If state s exists, output "Draw" and step; and exit;
         If there is no card in the hand, output "Loss" and step; and
exit;
         If all the piles disappear, output "Win" and step; and exit;
         Go to the next pile, and continue the loop;
  }
```

Program

```
#include<iostream>
#include<cstdio>
#include<fstream>
#include<algorithm>
#include<cmath>
#include<vector>
#include<map>
#include<cstring>
#include<math.h>
#include<string>
#include<set>
using namespace std;
const int p=1999997;        // modular for hash function 1
const int p2=10000009;       // modular for hash function 2
```

```
bool h[2100000];      //h[f]: state that hash function 1 is f has appeared
int key[2100000];              // the value for hash function 1 is f, its
corresponding value for hash function 2 is key[f]
string s;                      // the current state
int a[60],n;                   // a: stores face values of cards
char sign[10];                 // sign[i] is the flag for pile i
(sign[1]='A', ..., sign[8]='H')
bool v[10];      // mark whether the pile has been taken. v[i] == true
means pile i has disappeared
int num(char c){   // face values of cards
      if (c=='0') return 10; else return c-'0';
}
int ch(int x){   // face values of cards
      if (x==10) return '0'; else return char('0'+x);
}
int l(int i){                  // pointer for the front
     return s.find(sign[i])+1;
}
int r(int i){                  // pointer for the rear
     return s.find(sign[i+1])-1;
}
int hash(string s)
```
$$\{ \quad // \text{ Hash function 1: } hash\,(s) = \left(\sum_{i=0}^{s.size-1} (s[i] - '0') * 13^{s.size-1-i} \right) \%1999997$$
```
      int f=0;
      for (int i=0;i<=s.size( )-1;i++)  f=(f*13+s[i]-'0')%p;
      return f;
}
int hash2(string s)
```
$$\{ \quad // \text{ Hash function 2: } hash2\,(s) = \left(\sum_{i=0}^{s.size-1} (s[i] - '0') * 13^{s.size-1-i} \right) \%10000009$$
```
      int f=0;
      for (int i=0;i<=s.size( )-1;i++)  f=(f*31+s[i]-'0')%p2;
      return f;
}
bool have(string s)
{              // if state s hasn't appeared, it is added into hash table
   and return 0; else return 1
int f=hash(s), f2=hash2(s);  // calculate hash functions for state s
   while (h[f]&&(key[f]!=f2))  f++;
   if (!h[f]){ // state s hasn't appeard
            h[f]=1;
            key[f]=f2;
            return 0;
      } else return 1;    // state s exists
}
void initialize( ){                    //Initialization
      s.clear( );
      char c='A';
      for (int i=1;i<=n;i++){        //sign[1..8]='ABCDEFGH', state
s='Aa[1]Ba[2]Ca[3]Da[4]Ea[5]Fa[6]Ga[7]Ha[8..52]'
            if (i<=8){
                s+=c;
                sign[i]=c;
```

```
                    c++;
            }
            s+=ch(a[i]);
        }
    memset(v,0,sizeof(v));
    memset(h,0,sizeof(h));            //hash table is initialized empty
    memset(key,0,sizeof(key));
    have(s);
}
void remove1(int i){             // the first combination for pile i
    string t=s.substr(l(i),2);
    t+=s.substr(r(i),1);
    s+=t;
    s.erase(l(i),2);
    s.erase(r(i),1);
}
bool case1(int i){               //determine the first combination in pile
i
    int sum=num(s[l(i)])+num(s[l(i)+1])+num(s[r(i)]);
    if (sum==10||sum==20||sum==30) {remove1(i); return 1;}
    return 0;
}
void remove2(int i){             // the second combination for pile i
    string t=s.substr(l(i),1);
    t+=s.substr(r(i)-1,2);
    s+=t;
    s.erase(l(i),1);
    s.erase(r(i)-1,2);
}
bool case2(int i){               // determine the second combination in
pile i
    int sum=num(s[l(i)])+num(s[r(i)-1])+num(s[r(i)]);
    if (sum==10||sum==20||sum==30) {remove2(i); return 1;}
    return 0;
}
void remove3(int i){             // the third combination for pile i
    string t=s.substr(r(i)-2,3);
    s+=t;
    s.erase(r(i)-2,3);
}
bool case3(int i){               // determine the third combination in
pile i
    int sum=num(s[r(i)-2])+num(s[r(i)-1])+num(s[r(i)]);
    if (sum==10||sum==20||sum==30) {remove3(i); return 1;}
    return 0;
}
void deliver(int i){             //put the card in the hand into pile i, and
then combinations
    string t=s.substr(l(8),1);
    s.insert(r(i)+1,t);
    s.erase(l(8),1);
    if (r(i)-l(i)>=2)
     while (r(i)-l(i)>=2){
        bool flag=0;
```

```
        if (case1(i)) continue; else
        if (case2(i)) continue; else
        if (case3(i)) continue; else flag=1;
        if (flag) break;
      }
      if (r(i)<l(i)) v[i]=1;        // pile i disappears
}
int over( ){                              // determine whether game is
over or not
      if (have(s)) return 3; else        // Draw
      if (l(8)>=s.size( )) return 2; else // Loss
      for (int i=1;i<=7;i++)              // Continue
        if (!v[i]) return 0;
      return 1;                          // Win
}
void game( ){                   // Simulate the game
      int i=1;                  // From pile 1
      int step=7;
      int res;                  // Flag for end
      while (1){
          deliver(i);           // put a card into pile i, and combination
          step++;
          res=over( );
          if (res) break;  // if the result is produced, break the loop
          i=(i%7)+1;           // the next pile
          while (v[i]) i=(i%7)+1;
      }
      // Output the result
      if (res==1) cout<<"Win : "; else
      if (res==2) cout<<"Loss: "; else
      if (res==3) cout<<"Draw: ";
      cout<<step<<endl;         // the number of cards put into piles
}
int main( ){
      n=52;                     // number of cards
      cin>>a[1];                // the first card
      while (a[1]){             // if the first card isn't 0, then the next
51cards
          for (int i=2;i<=n;i++) cin>>a[i];
          initialize( );
          game( );              // simulate the game
          cin>>a[1];            // the first card for the next test case
      }
}
```

6.3 Problems

6.3.1 Spell Checker

You, as a member of a development team for a new spell checking program, are to write a module that will check the correctness of given words using a known dictionary of all correct words in all their forms.

If the word is absent from the dictionary, then it can be replaced by correct words (from the dictionary) that can be obtained by one of the following operations:

- Deleting one letter from the word
- Replacing one letter in the word with an arbitrary letter
- Inserting one arbitrary letter into the word

Your task is to write the program that will find all possible replacements from the dictionary for every given word.

Input

The first part of the input file contains all words from the dictionary. Each word occupies its own line. This part is finished by the single character '#' on a separate line. All words are different. There will be at most 10,000 words in the dictionary.

The next part of the file contains all words that are to be checked. Each word occupies its own line. This part is also finished by the single character '#' on a separate line. There will be at most 50 words that are to be checked.

All words in the input file (words from the dictionary and words to be checked) consist only of small alphabetic characters, and each one contains 15 characters at most.

Output

Write to the output file exactly one line for every checked word in the order of their appearance in the second part of the input file. If the word is correct (i.e., it exists in the dictionary) write the message "is correct." If the word is not correct, then write this word first, and then write the character ':' (colon), and after a single space write all its possible replacements, separated by spaces. The replacements should be written in the order of their appearance in the dictionary (in the first part of the input file). If there are no replacements for this word, then the line feed should immediately follow the colon.

Sample Input	Sample Output
i	me is correct
is	aware: award
has	m: i my me
have	contest is correct
be	hav: has have
my	oo: too
more	or:
contest	i is correct
me	fi: i
too	mre: more me

Sample Input	Sample Output
if	
award	
#	
me	
aware	
m	
contest	
hav	
oo	
or	
i	
fi	
mre	
#	

Source: ACM Northeastern Europe 1998.

IDs for online judges: POJ 1035, ZOJ 2040, UVA 671.

Hint

Because there are at most 10,000 words in the dictionary, a linear list is used to store the dictionary. Suppose *dict*[ ] is used to store the dictionary, where the *i*th word in the dictionary is *dict*[*i*] and the length of the dictionary is *dictSize*.

When *dictSize* words are input, a dictionary is constructed. Then for each checked word *s*, if *s* is in the dictionary, output the correct message (printf("%s is correct\n", *s*)); else, every word in the dictionary is analyzed as follows:

1. If the length of *dict*[*i*] is the same as the length of *s*, and *dict*[*i*] and *s* can be the same by replacing one letter in the word with an arbitrary letter, *s* and *dict*[*i*] are output.
2. If *dict*[*i*] and *s* can be the same by inserting one arbitrary letter into *s*, *s* and *dict*[*i*] are output.
3. If *dict*[*i*] and *s* can be the same by deleting one letter in *s*, *s* and *dict*[*i*] are output.

For steps 2 and 3, function *match*(*s1*[ ], *slen1*, *s2*[ ]) is implemented. The length of *s1*[ ] is *slen1*, and the length of *s2*[ ] is *slen1* + 1. The function *match*(*s1*[ ], *slen1*, *s2*[ ]) is used to determine whether *s1*[ ] and *s2*[ ] can be the same by inserting one letter into *s1*[ ]. The algorithm is as follows:

Search the first position *k* that *s1*[*k*] and *s2*[*k*] are different from left to right. Then we need to determine whether *s1*[*k*]...*s1*[*slen1*] is the same as *s2*[*k*+1]...*s2*[*slen1*+1].

6.3.2 Snowflake Snow Snowflakes

You may have heard that no two snowflakes are alike. Your task is to write a program to determine whether this is really true. Your program will read information about a collection of snowflakes and search for a pair that may be identical. Each snowflake has six arms. For each snowflake, your program will be provided with a measurement of the length of each of the six arms. Any pair of snowflakes that have the same lengths of corresponding arms should be flagged by your program as possibly identical.

Input

The first line of input will contain a single integer n, $0 < n \leq 100{,}000$, the number of snowflakes to follow. This will be followed by n lines, each describing a snowflake. Each snowflake will be described by a line containing six integers (each integer is at least 0 and less than 10 million), the lengths of the arms of the snowflake. The lengths of the arms will be given in order around the snowflake (either clockwise or counterclockwise), but they may begin with any of the six arms. For example, the same snowflake could be described as 1 2 3 4 5 6 or 4 3 2 1 6 5.

Output

If all of the snowflakes are distinct, your program should print the message
    No two snowflakes are alike.
    If there is a pair of possibly identical snowflakes, your program should print the message
    Twin snowflakes found.

Sample Input	Sample Output
2	Twin snowflakes found.
1 2 3 4 5 6	
4 3 2 1 6 5	

Source: Canadian Computing Competition 2007.

ID for online judge: POJ 3349.

Hint
The hash method is used to solve the problem. When a snowflake is input, the value of a hash function is obtained. If there is a same hash value in the hash table, then output "Twin snowflakes found."

6.3.3 Equations

Consider equations having the following form:
    $a_1 x_1^3 + a_2 x_2^3 + a_3 x_3^3 + a_4 x_4^3 + a_5 x_5^3 = 0$
The coefficients are given integers from the interval [−50, 50].
    A solution is considered a system $(x_1, x_2, x_3, x_4, x_5)$ that verifies the equation, $x_i \in [-50, 50]$, $x_i \neq 0$, any $i \in \{1, 2, 3, 4, 5\}$.
    Determine how many solutions satisfy the given equation.

Input

The only line of input contains the five coefficients a_1, a_2, a_3, a_4, a_5, separated by blanks.

Output

The output will contain on the first line the number of solutions for the given equation.

Sample Input	Sample Output
37 29 41 43 47	654

Source: Romania OI 2002.

ID for online judge: POJ 1840.

Hint

The hash method is used to solve the problem. The equation $a_1x_1^3 + a_2x_2^3 + a_3x_3^3 + a_4x_4^3 + a_5x_5^3 = 0$ is transferred into $a_1x_1^3 + a_2x_2^3 + a_3x_3^3 = -(a_4x_4^3 + a_5x_5^3)$. First, solutions to expression $a_1x_1^3 + a2x23 + a_3x_3^3$ are calculated and stored in the hash table. Then solutions to expression $-(a_4x_4^3 + a_5x_5^3)$ are calculated and searched in the hash table.

Chapter 7

Sort of Linear Lists

A sorting algorithm puts elements of a list in a certain order. Sorting algorithms are important for managing data. There are a large number of sorting algorithms, such as bubble sort, insertion sort, selection sort, merge sort, heapsort, and quicksort. Such algorithms have been discussed in many classical books. This chapter focuses on using the sort function in the Standard Template Library (STL).

7.1 Using Sort Function in STL

The STL is a C++ library of container classes, algorithms, and iterators. It provides many algorithms and data structures of computer science, including sorting algorithms. Like many class libraries, the STL includes container classes containing objects. The STL includes the class vector, list, deque, set, multiset, map, multimap, hash_set, hash_multiset, hash_map, and hash_multimap.

For example, map is a sorted associative container that associates objects of type *Key* with objects of type *Data*. Map is a pair associative container, meaning that its value type is pair ⟨const *Key*, *Data*⟩. It is also a unique associative container, meaning that no two elements have the same key.

If there is a mapping relationship between students' names and their scores, map can describe the relationship easily. At the head of the program, there is a preprocessor directive #include ⟨map⟩, and container *mapStudent* is specified as follows:

$$\text{map⟨string, int⟩mapStudent}$$

Based on the above statements, students' names and scores are specified with string and int, respectively. All students' information, *mapStudent[name]=score*, is stored in the container *mapStudent*. The compiling system sorts students' names in alphabetical order.

7.1.1 Hardwood Species

Hardwoods are the botanical group of trees that have broad leaves, produce a fruit or nut, and generally go dormant in the winter. America's temperate climates produce forests with hundreds of hardwood species—trees that share certain biological characteristics. Although oak, maple, and cherry all are types of hardwood trees, for example, they are different species. Together, all the hardwood species represent 40% of the trees in the United States.

On the other hand, softwoods, or conifers, from the Latin word meaning "cone bearing," have needles. Widely available U.S. softwoods include cedar, fir, hemlock, pine, redwood, spruce, and cypress. In a home, the softwoods are used primarily as structural lumber, such as 2 × 4s and 2 × 6s, with some limited decorative applications.

Using satellite imaging technology, the Department of Natural Resources has compiled an inventory of every tree standing on a particular day. You are to compute the total fraction of the tree population represented by each species.

Input

The input to your program consists of a list of the species of every tree observed by the satellite, one tree per line. No species name exceeds 30 characters. There are no more than 10,000 species and no more than 1 million trees.

Output

Print the name of each species represented in the population, in alphabetical order, followed by the percentage of the population it represents, to four decimal places.

Sample Input	Sample Output
Red Alder	Ash 13.7931
Ash	Aspen 3.4483
Aspen	Basswood 3.4483
Basswood	Beech 3.4483
Ash	Black Walnut 3.4483
Beech	Cherry 3.4483
Yellow Birch	Cottonwood 3.4483
Ash	Cypress 3.4483
Cherry	Gum 3.4483
Cottonwood	Hackberry 3.4483
Ash	Hard Maple 3.4483
Cypress	Hickory 3.4483
Red Elm	Pecan 3.4483
Gum	Poplan 3.4483
Hackberry	Red Alder 3.4483
White Oak	Red Elm 3.4483

Sample Input	Sample Output
Hickory	Red Oak 6.8966
Pecan	Sassafras 3.4483
Hard Maple	Soft Maple 3.4483
White Oak	Sycamore 3.4483
Soft Maple	White Oak 10.3448
Red Oak	Willow 3.4483
Red Oak	Yellow Birch 3.4483
White Oak	
Poplar	
Sassafras	
Sycamore	
Black Walnut	
Willow	

Source: Waterloo Local Contest, January 26, 2002.

Hint: This problem has huge input; use *scanf* instead of *cin* to avoid time limit exceeded.

ID for online judge: POJ 2418.

Analysis

Suppose the number of the tree whose species name is *x* is $h[x]$, and the total number of trees is *n*. First, the number of each species and the total number of tree *n* are calculated based on the input, and then in alphabetical order print the name of each species and its percentage of the population, $h[x]/n$.

For this problem, the STL map can be used to store *h* and sort keys.

Program

```
#include<iostream>
#include<string>
#include<map>
using namespace std;
typedef map<string,int> record;
record h;                  // h[x]: the number of the tree whose name is
x
string s;                       // tree name
int n;                          // the number of trees
int main( ){
     n=0;
     while (getline(cin,s)){  // a list of the species of every tree, and
accumulate
```

```
            n++;
            h[s]++;
    }
    for (record::iterator it=h.begin( );it!=h.end( );it++){
        //sequential search (h is sorted in alphabet order)
        string name=(*it).first;
        int k=(*it).second;
        printf("%s %.4lfn", name.c_str( ), double(k)*100/double(n));
        // output name and percentage
    }
}
```

In STL, there is a sort function *sort* to sort elements in an interval. At the head of the program, there must be a preprocessor directive #include ⟨algorithm⟩. There are two kinds of usages: *sort(l, r)* and *sort(l, r, compare)*. For the next two examples, two kinds of usages are used to solve problems.

7.1.2 *Who's in the Middle?*

F.J. is surveying his herd to find the most average cow. He wants to know how much milk this median cow gives: half of the cows give as much or more than the median; half give as much or less.

Given an odd number of cows N ($1 \leq N < 10{,}000$) and their milk output (1 .. 1,000,000), find the median amount of milk given such that at least half the cows give the same amount of milk or more and at least half give the same or less.

Input

Line 1: A single integer N.
Lines 2 .. N + 1: Each line contains a single integer that is the milk output of one cow.

Output

Line 1: A single integer that is the median milk output

Sample Input	Sample Output
5	3
2	
4	
1	
3	
5	

Source: USACO, November 2004.

ID for online judge: POJ 2388.

Analysis

The problem is simple. We only need to sort N cows' milk output. The element in the middle is the median milk output.

The function sort( ) in algorithm.h is used to sort.

Program

```
#include<iostream>
#include<algorithm>
using namespace std;
const int maxn=11000;              //the upper limit of the number of cows
int a[maxn],n;                     // milk outputs and the number of cows
int main( ){
    cin>>n;                             // the number of cows
    for (int i=1; i<=n; i++) cin>>a[i];// milk outputs
    // Sort milk outputs
    sort(a+1, a+n+1);
    cout<<a[(n+1)/2]<<endl;             // the median milk output
}
```

7.1.3 ACM Rank Table

Association for Computing Machinery (ACM) contests, like the one you are participating in, are hosted by the special software. That software, among other functions, performs a job of accepting and evaluating teams' solutions (runs) and displaying results in a rank table. The scoring rules are as follows:

1. Each run is either accepted or rejected.
2. The problem is considered solved by the team if one of the runs submitted for it is accepted.
3. The time consumed for a solved problem is the time elapsed from the beginning of the contest to the submission of the first accepted run for this problem (in minutes) plus 20 minutes for every other run for this problem before the accepted one. For an unsolved problem, consumed time is not computed.
4. The total time is the sum of the time consumed for each problem solved.
5. Teams are ranked according to the number of solved problems. Teams that solve the same number of problems are ranked by the least total time.
6. While the time shown is in minutes, the actual time is measured to the precision of 1 second, and the seconds are taken into account when ranking teams.
7. Teams with equal rank according to the above rules must be sorted by increasing team number.

Given the list of N runs with submission time and the result of each run, your task is to compute the rank table for C teams.

Input

The input contains integer numbers C N, followed by N quartets of integers c_i p_i t_i r_i, where c_i is team number, p_i is problem number, t_i is submission time in seconds, and r_i is 1 if the run was accepted and 0 otherwise.

$$1 \leq C, N \leq 1000, 1 \leq c_i \leq C, 1 \leq p_i \leq 20, 1 \leq t_i \leq 36,000$$

Output

Output file must contain *C* integers—team numbers sorted by rank.

Sample Input	Sample Output
3 3	2 1 3
1 2 3000 0	
1 2 3100 1	
2 1 4200 1	

Source: ACM Northeastern Europe 2004,
Far-Eastern Subregion.

ID for online judge: POJ 2379.

Analysis

The submissions' sequence is represented as an array *a*. For the *i*th submission, $a[i].c$ is the team number, $a[i].t$ is the submission time, and $a[i].p$ is the problem number, $1 \leq i \leq n$.

The teams' sequence is represented as an array *t*. For the *i*th team, $t[i].id$ is the team number, $t[i].ac$ is the number of solved problems, and $t[i].p[j]$ is the number of wrong runs for problem *j*; the flag that problem *k* is accepted is $t[i].sol[k]$, $1 \leq i \leq C$, $1 \leq k, j \leq n$.

The problem is a simulation problem. In the problem description, rules for ACM rank are shown. The program should implement these rules.

First, *n* submissions are sorted in the submission time's ascending order.

Then every submission is processed based on rules.

Finally, arrays *t* are sorted based on rules.

Function sort in STL can be used to simplify the program.

Program

```
#include<iostream>
#include<algorithm>
using namespace std;
const int maxn=1100;
struct judgement{              // submission structure
        int c, t, p, r;         // team number, submission time, problem
number, and run result
};
struct team{                   // team structure
        int id, ac, t;          // team number, the number of solved
problems, time
        int p[25];             //p[i] is the number of wrong runs for problem
i
        bool sol[25];          //sol[i] is he flag that problem i is accepted
};
bool cmp_t(const judgement &a, const judgement &b){   //compare a and b
with time
        return a.t<b.t;
};
```

```
bool cmp_ac(const team &a, const team &b){        // array t are sorted
based on rules
     if (a.ac!=b.ac) return a.ac>b.ac;
     if (a.t!=b.t) return a.t<b.t;
     return a.id<b.id;
};
judgement a[maxn];         // submissions
team t[maxn];              //teams
int n, m;                   //numbers of teams and submissions
     int main( ){
     memset(a, 0, sizeof(a));
     memset(t, 0, sizeof(t));
     cin>>n>>m;            // numbers of teams and submissions
     for (int i=1; i<=m; i++) cin>>a[i].c>>a[i].p>>a[i].t>>a[i].r; //
Each submission
     for (int i=1; i<=n; i++) t[i].id=i;
     sort(a+1, a+m+1, cmp_t);      // Sort array a based on submission time
     for (int i=1; i<=m; i++){          // For each submission
          int x=a[i].c, y=a[i].p;
          if (t[x].sol[y]) continue;
          if (a[i].r){                     // if accepted
               t[x].t+=1200*t[x].p[y]+a[i].t;
               t[x].sol[y]=1;
               t[x].ac++;
          } else t[x].p[y]++;
     }
     sort(t+1, t+n+1, cmp_ac);                    // sort
     for (int i=1; i<n; i++) cout<<t[i].id<<' ';         // rank
     cout<<t[n].id<<endl;
}
```

7.2 Using Sort Algorithms

STL is encapsulated for programmers. But some problems require you to implement sort algorithms. For example, a problem requires you to calculate the number of exchanges in sorting.

7.2.1 Flip Sort

Sorting in computer science is an important part. Almost every problem can be solved efficiently if sorted data are found. There are some excellent sorting algorithms that have already achieved the lower bound nlgn. In this problem, we will also discuss a new sorting approach. In this approach only one operation (flip) is available, and that is you can exchange two adjacent terms. If you think a while, you will see that it is always possible to sort a set of numbers in this way.

A set of integers will be given. Now using the above approach, we want to sort the numbers in ascending order. You have to find the minimum number of flips required. For example, to sort 1 2 3, we need no flip operation, whereas to sort 2 3 1, we need at least two flip operations.

Input

The input will start with a positive integer N ($N \leq 1000$). In the next few lines there will be N integers. Input will be terminated by the end of the file (EOF).

Output

For each data set print "Minimum exchange operations: *M*," where *M* is the minimum flip operations required to perform sorting. Use a separate line for each case.

Sample Input	Sample Output
3	Minimum exchange operations : 0
1 2 3	Minimum exchange operations : 2
3	
2 3 1	

ID for online judge: UVA 10327.

Analysis

Suppose the given set of integers is stored in an array *a*. Initially, the minimum flip operations *ans* is 0. Minimum exchange operations are simulated by using bubble sort. For each exchange, *ans*++.

When there is no exchange operation, *ans* is the solution to the problem.

Program

```
#include<iostream>
using namespace std;
const int maxn=1100;              // the upper limit of the number of
integers
int n,a[maxn];        // the number of integers n, the given set of
integers is stored in an array a
int main( ){
    cin>>n;
    while (!cin.eof( )){
        for (int i=1;i<=n;i++) cin>>a[i];
        bool flag=1;
        int ans=0;                          // initialize the number of
exchange operations
        while (flag){                          //Simulate bubble sort
            flag=0;
            for (int i=1;i<n;i++) if (a[i]>a[i+1]) {   //adjacent
elements
                swap(a[i],a[i+1]);                 //exchange
                flag=1;
                ans++;            //accumulate the number of exchange
operations
            }
        }
        cout<<"Minimum exchange operations : "<<ans<<endl; //output
result
        cin>>n;
    }
    system("pause");
}
```

7.2.2 Ultra-Quicksort

In this problem, you have to analyze a particular sorting algorithm. The algorithm processes a sequence of n distinct integers by swapping two adjacent sequence elements until the sequence is sorted in ascending order. For the input sequence,

$$9\ 1\ 0\ 5\ 4$$

Ultra-quicksort produces the output

$$0\ 1\ 4\ 5\ 9$$

Your task is to determine how many swap operations Ultra-quicksort needs to perform in order to sort a given input sequence.

Input

The input contains several test cases. Every test case begins with a line that contains a single integer $n < 500,000$—the length of the input sequence. Each of following n lines contains a single integer $0 \leq a[i] \leq 999,999,999$, the ith input sequence element. Input is terminated by a sequence of length $n = 0$. This sequence must not be processed.

Output

For every input sequence, your program prints a single line containing an integer number op, the minimum number of swap operations necessary to sort the given input sequence.

Sample Input	Sample Output
5	6
9	0
1	
0	
5	
4	
3	
1	
2	
3	
0	

Source: Waterloo Local Contest, February 5, 2005.

IDs for online judges: POJ 2299, ZOJ 2386, UVA 10810.

Analysis

For n integers, if each pair (A_i, A_j) is enumerated to determine whether A_i is larger than A_j, $i < j$, and to calculate the number of inversions. The time complexity is $O(n^2)$.

Divide and conquer can be used to improve the time complexity. The new time complexity is $O(n\log n)$.

Suppose the number of inversions for a series of numbers $A[l .. r]$ is $d(l, r)$. $A[l .. r]$ is divided into two parts, $A[l .. mid]$ and $A[mid + 1 .. r]$, where

$$mid = \left\lfloor \frac{l+r}{2} \right\rfloor$$

If two numbers in an inversion are from $A[l .. mid]$ and $A[mid + 1 .. r]$, respectively, the number of such inversions is $f(l, mid, r)$. Obviously, $d(l, r) = d(l, mid) + d(mid + 1, r) + f(l, mid, r)$.

After $d(l, r)$ is calculated, $A[l .. r]$ is also sorted. That is, after $d(l, mid)$ and $d(mid + 1, r)$ are calculated, $A[l, mid]$ and $A[mid + 1, r]$ are sorted. Suppose i and j are pointers for $A[l, mid]$ and $A[mid + 1, r]$ respectively, where $A[j - 1] < A[i]$ and $A[j] \geq A[i]$. Therefore, in $A[mid + 1 .. r]$, there are $j - mid - 1$ numbers that are less than $A[i]$, as shown in Figure 7.1. Therefore, $j - mid - 1$ will be accumulated into $f(l, mid, r)$.

Because $A[l, mid]$ and $A[mid + 1, r]$ are all sorted, $f(l, mid, r)$ is calculated through moving pointer i and pointer j. The process is like merge sort.

Program

```
#include<iostream>
#define lolo long long
using namespace std;
const lolo maxn=510000;          // the upper limit of the length of the
input sequence
lolo n, a[maxn], ans, t[maxn];    // the length of the input sequence
n, the input sequence a, Merge sequence t, the number of inversions ans
void Sort(lolo l, lolo r){        // Merge Sort
        if (l==r) return;         // Sort finish
        lolo mid=(l+r)/2;
        Sort(l, mid);             // Sort the left sub interval
        Sort(mid+1, r);           // Sort the right sub interval
        lolo i=l, j=mid+1, now=0;
        while (i<=mid&&j<=r){
            if (a[i]>a[j]){
                ans+=mid-i+1; //if a[i]>a[j], a[i] and numbers after a[i]
and a[j] are inversions
```

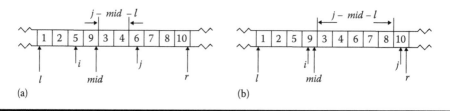

(a)   (b)

Figure 7.1 Dichotomy.

```
                t[++now]=a[j++];
        } else {
                t[++now]=a[i++];
        }
    }
    while (i<=mid) t[++now]=a[i++];
    while (j<=r)   t[++now]=a[j++];
    now=0;
    for (lolo k=l; k<=r; k++) a[k]=t[++now];
}
int main( ){
    cin>>n;
    while (n){
        for (lolo i=1; i<=n; i++) cin>>a[i]; // the input sequence
        ans=0;
        Sort(1, n);
        cout<<ans<<endl;      // the minimum number of swap operations
        cin>>n;
    }
}
```

7.3 Problems

7.3.1 Ananagrams

Most crossword puzzle fans are used to *anagrams*—groups of words with the same letters in different orders—for example, OPTS, SPOT, STOP, POTS, and POST. Some words, however, do not have this attribute; no matter how you rearrange their letters, you cannot form another word. Such words are called *ananagrams*; an example is QUIZ.

Obviously, such definitions depend on the domain within which we are working; you might think that ATHENE is an ananagram, whereas any chemist would quickly produce ETHANE. One possible domain would be the entire English language, but this could lead to some problems. One could restrict the domain to, say, music, in which case SCALE becomes a *relative ananagram* (LACES is not in the same domain), but NOTE is not since it can produce TONE.

Write a program that will read in the dictionary of a restricted domain and determine the relative ananagrams. Note that single-letter words are, ipso facto, relative ananagrams since they cannot be rearranged' at all. The dictionary will contain no more than 1000 words.

Input

The input will consist of a series of lines. No line will be more than 80 characters long, but may contain any number of words. Words consist of up to 20 upper- and lowercase letters and will not be broken across lines. Spaces may appear freely around words, and at least one space separates multiple words on the same line. Note that words that contain the same letters but of differing case are considered to be anagrams of each other; thus, tIeD and EdiT are anagrams. The file will be terminated by a line consisting of a single #.

Output

The output will consist of a series of lines. Each line will consist of a single word that is a relative ananagram in the input dictionary. Words must be output in lexicographic (case-sensitive) order. There will always be at least one relative ananagram.

Sample Input	Sample Output
ladder came tape soon leader acme RIDE lone Dreis peat	Disk
ScAIE orb eye Rides dealer NotE derail LaCeS drled	NotE
noel dire Disk mace Rob dries	derail
#	drled
	eye
	ladder
	soon

Source: New Zealand Contest 1993.

ID for online judge: UVA 156.

Hint

If the current word's ascending order is the same as another word's ascending order, the word is a *relative anagram*, and if a word's ascending order isn't the same as any other word's ascending order, the word is a *relative ananagram*. The algorithm is as follows.

```
Suppose the words' list is a, where the ith word is a[i].s and its
ascending order is a[i].t, 1<=i<=n. First, a is sorted in alphabet order
for a[i].s, and a[i].t is calculated. Then for each word, determine
whether the word is a relative ananagram or not.
```

7.3.2 Grandpa Is Famous

The whole family was excited by the news. Everyone knew grandpa had been an extremely good bridge player for decades, but when it was announced he would be in the *Guinness Book of World Records* as the most successful bridge player ever, that was astonishing.

The International Bridge Association (IBA) has maintained, for several years, a weekly ranking of the best players in the world. Considering that each appearance in a weekly ranking constitutes a point for the player, grandpa was nominated the best player ever because he got the highest number of points.

Having many friends who were also competing against him, grandpa is extremely curious to know which player(s) took second place. Since the IBA rankings are now available on the Internet, he turned to you for help. He needs a program that, when given a list of weekly rankings, finds out which player(s) got second place according to the number of points.

Input

The input contains several test cases. Players are identified by integers from 1 to 10,000. The first line of a test case contains two integers N and M, indicating, respectively, the number of rankings available ($2 \leq N \leq 500$) and the number of players in each ranking ($2 \leq M \leq 500$). Each of the next N lines contains the description of one weekly ranking. Each description is composed of a sequence of M integers, separated by a blank space, identifying the players who figured in that weekly ranking. You can assume that

■ In each test case there is exactly one best player and at least one second-best player.

■ Each weekly ranking consists of M distinct player identifiers.

The end of input is indicated by $N = M = 0$.

Output

For each test case in the input, your program must produce one line of output, containing the identification number of the player who is second best in number of appearances in the rankings. If there is a tie for second best, print the identification numbers of all second-best players in increasing order. Each identification number produced must be followed by a blank space.

Sample Input	Sample Output
4 5	32 33
20 33 25 32 99	1 2 21 23 31 32 34 36 38 67 76 79 88 91 93 100
32 86 99 25 10	
20 99 10 33 86	
19 33 74 99 32	
3 6	
2 34 67 36 79 93	
100 38 21 76 91 85	
32 23 85 31 88 1	
0 0	

Source: ACM South America 2004.

ID for online judge: POJ 2092.

Hint

Linear list a is used to represent bridge players, where $a[i].id$ is the ith bridge player's identification number, and $a[i].p$ is the number of appearances in weekly rankings for the bridge player.

Obviously, $a[i].id == i$. And $a[i].p$ is calculated when a test case is input. Then linear list a is sorted in $a[i].p$'s descending order (as the first key) and in $a[i].id$'s ascending order (as the second key). Finally, the identification numbers of all second-best players are output in increasing order.

7.3.3 *Word Amalgamation*

In millions of newspapers across the United States there is a word game called Jumble. The object of this game is to solve a riddle, but in order to find the letters that appear in the answer, it is necessary to unscramble four words. Your task is to write a program that can unscramble words.

Input

The input contains four parts: (1) a dictionary, which consists of at least 1 and at most 100 words, one per line; (2) a line containing XXXXXX, which signals the end of the dictionary; (3) one or more scrambled words that you must unscramble, each on a line by itself; and (4) another line containing XXXXXX, which signals the end of the file. All words, including both dictionary words and scrambled words, consist only of lowercase English letters and will be at least one and at most six characters long. (Note that the sentinel XXXXXX contains uppercase X's.) The dictionary is not necessarily in sorted order, but each word in the dictionary is unique.

Output

For each scrambled word in the input, output an alphabetical list of all dictionary words that can be formed by rearranging the letters in the scrambled word. Each word in this list must appear on a line by itself. If the list is empty (because no dictionary words can be formed), output the line "NOT A VALID WORD" instead. In either case, output a line containing six asterisks to signal the end of the list.

Sample Input	Sample Output
tarp	score
given	******
score	refund
refund	******
only	part
trap	tarp
work	trap
earn	******
course	NOT A VALID WORD
pepper	******
part	course
XXXXXX	******
resco	
nfudre	
aptr	
sett	
oresuc	
XXXXXX	

Source: ACM Mid-Central United States 1998.

ID for online judge: POJ 1318.

Hint

Suppose the dictionary is represented by a linear list a, and the number of words is n. After a test case is input, a dictionary is set up.

For each word $a[i]$ in a, letters in $a[i]$ are sorted in alphabetical order and stored in $b[i]$, $1 \le i \le n$.

After a scrambled word is input, its letters are also sorted in alphabetical order and stored in a string t.

If there exists $b[i]$ such that $b[i] == t$, output $a[i]$; else, output the line "NOT A VALID WORD."

7.3.4 Questions and Answers

The database of the Pentagon contains top-secret information. We don't know what the information is—you know, it's top secret—but we know the format of its representation. It is extremely simple. We don't know why, but all the data is coded by the natural numbers from 1 up to 5000. The size of the main base (we'll denote it be N) is rather big—it may contain up to 100,000 of those numbers. The database is to quickly process every query. The most often query is "Which element is ith by its value?"—with i being a natural number in a range from 1 to N.

Your program is to play a role of a controller of the database. In other words, it should be able to quickly process queries like this.

Input

The standard input of the problem consists of two parts. At first, a database is written, and then there's a sequence of queries. The format of the database is very simple: in the first line there's a number N, and in the next N lines there are numbers of the database, one in each line, in an arbitrary order. A sequence of queries is written simply as well: in the first line of the sequence a number of queries K ($1 \le K \le 100$) are written, and in the next K lines there are queries, one in each line. The query "Which element is ith by its value?" is coded by the number i. A database is separated from a sequence of queries by the string of three '#' symbols.

Output

The output should consist of K lines. In each line there should be an answer to the corresponding query. The answer to the query i is an element from the database, which is ith by its value (in the order from the least to the greatest element).

Sample Input	Sample Output
5	121
7	121
121	7
123	123
7	
121	
###	
4	

(Continued)

Sample Input	Sample Output
3	
3	
2	
5	

Source: Ural State University Internal Contest, October 2000, Junior Session.

ID for online judge: POJ 2371.

Hint

For n integers $a[1] \ldots a[n]$ in the database, array a is sorted in ascending order. Then for query i, $a[i]$ is the answer.

7.3.5 Find the Clones

Doubleville, a small town in Texas, was attacked by aliens. They have abducted some of the residents and taken them to a spaceship orbiting around earth. After some (quite unpleasant) human experiments, the aliens cloned the victims and released multiple copies of them back in Doubleville. So now it might happen that there are six identical persons named Hugh F. Bumblebee: the original person and his five copies. The Federal Bureau of Unauthorized Cloning (FBUC) charged you with the task of determining how many copies were made from each person. To help you in your task, FBUC has collected a DNA sample from each person. All copies of the same person have the same DNA sequence, and different people have different sequences (we know that there are no identical twins in the town, so this is not an issue).

Input

The input contains several blocks of test cases. Each case begins with a line containing two integers: the number $1 \le n \le 20{,}000$ people and the length $1 \le m \le 20$ of the DNA sequences. The next n lines contain the DNA sequences: each line contains a sequence of m characters, where each character is either 'A', 'C', 'G', or 'T'.

The input is terminated by a block with $n = m = 0$.

Output

For each test case, you have to output n lines, each line containing a single integer. The first line contains the number of different people that were not copied. The second line contains the number of people that were copied only once (i.e., there are two identical copies for each such person.) The third line contains the number of people that are present in three identical copies, and so on: the ith line contains the number of persons that are present in i identical copies. For example, if there are 11 samples and one of them is from John Smith and all the others are from copies of Joe Foobar, then you have to print '1' in the 1st and the 10th lines and '0' in all the other lines.

Sample Input	Sample Output
9 6	1
AAAAAA	2
ACACAC	0
GTTTTG	1
ACACAC	0
GTTTTG	0
ACACAC	0
ACACAC	0
TCCCCC	0
TCCCCC	
0 0	

Source: ACM Central Europe 2005.

Hint: The problem has a huge input file; scanf is recommended to avoid time limit exceeded.

ID for online judge: POJ 2945.

Hint

Suppose DNA sequences are stored in s, where the ith person's DNA sequence is stored in $s[i]$, $1 \leq i \leq n$, and the number of persons who share the kth same DNA sequence is stored in $ans[k]$.

First, s is sorted in alphabetical order. Then s is searched in order; accumulate the number of persons who share the same DNA sequence. Finally, output the result.

7.3.6 487-3279

Businesses like to have memorable telephone numbers. One way to make a telephone number memorable is to have it spell a memorable word or phrase. For example, you can call the University of Waterloo by dialing the memorable TUT-GLOP. Sometimes only part of the number is used to spell a word. When you get back to your hotel tonight, you can order a pizza from Gino's by dialing 310-GINO. Another way to make a telephone number memorable is to group the digits in a memorable way. You could order your pizza from Pizza Hut by calling their "three tens" number, 3-10-10-10.

The standard form of a telephone number is seven decimal digits with a hyphen between the third and fourth digits (e.g., 888-1200). The keypad of a phone supplies the mapping of letters to numbers, as follows:

A, B, and C map to 2
D, E, and F map to 3
G, H, and I map to 4
J, K, and L map to 5
M, N, and O map to 6

P, R, and S map to 7
T, U, and V map to 8
W, X, and Y map to 9

There is no mapping for Q or Z. Hyphens are not dialed and can be added and removed as necessary. The standard form of TUT-GLOP is 888-4567, the standard form of 310-GINO is 310-4466, and the standard form of 3-10-10-10 is 310-1010.

Two telephone numbers are equivalent if they have the same standard form (they dial the same number).

Your company is compiling a directory of telephone numbers from local businesses. As part of the quality control process, you want to check that no two (or more) businesses in the directory have the same telephone number.

Input

The input will consist of one case. The first line of the input specifies the number of telephone numbers in the directory (up to 100,000) as a positive integer alone on the line. The remaining lines list the telephone numbers in the directory, with each number alone on a line. Each telephone number consists of a string composed of decimal digits, uppercase letters (excluding Q and Z), and hyphens. Exactly seven of the characters in the string will be digits or letters.

Output

Generate a line of output for each telephone number that appears more than once in any form. The line should give the telephone number in standard form, followed by a space, followed by the number of times the telephone number appears in the directory. Arrange the output lines by telephone number in ascending lexicographical order. If there are no duplicates in the input, print the line "No duplicates."

Sample Input	Sample Output
12	310-1010 2
4873279	487-3279 4
ITS-EASY	888-4567 3
888-4567	
3-10-10-10	
888-GLOP	
967-11-11	
967-11-11	
310-GINO	
F101010	
888-1200	

Sample Input	Sample Output
-4-8-7-3-2-7-9- 487-3279	

Source: ACM East Central North America 1999.

ID for online judge: POJ 1002.

Hint

First, n telephone numbers in the directory are transferred into their standard forms. Then the n telephone numbers are sorted in ascending order; calculate numbers for each telephone number that appears. Finally, output the result.

7.3.7 Holiday Hotel

Mr. and Mrs. Smith are going to the seaside for their holiday. Before they start off, they need to choose a hotel. They got a list of hotels from the Internet and want to choose some candidate hotels that are cheap and close to the seashore. A candidate hotel M meets two requirements:

1. Any hotel that is closer to the seashore than M will be more expensive than M.
2. Any hotel that is cheaper than M will be farther away from the seashore than M.

Input

There are several test cases. The first line of each test case is an integer N ($1 \leq N \leq 10{,}000$), which is the number of hotels. Each of the following N lines describes a hotel, containing two integers D and C ($1 \leq D$, $C \leq 10000$). D means the distance from the hotel to the seashore, and C means the cost of staying in the hotel. You can assume that there are no two hotels with the same D and C. A test case with $N = 0$ ends the input and should not be processed.

Output

For each test case, you should output one line containing an integer, which is the number of all the candidate hotels.

Sample Input	Sample Output
5	2
100 300	
100 300	
400 200	
200 400	

(Continued)

Sample Input	Sample Output
100 500	
0	

Source: ACM Beijing 2005.

ID for online judge: POJ 2726.

Hint

The list of hotels is represented as linear list a; for the ith hotel, $a[i].d$ is the distance from the hotel to the seashore, and $a[i].c$ is the cost of staying in the hotel, $1 \le i \le n$. Based on two requirements for candidate hotels, a is sorted so that the distance from the hotel to the seashore is the first key, and the cost of staying in the hotel is the second key. Suppose the number of candidate hotels is ans, and the last candidate hotel is pre.

Initially, the first hotel is a candidate hotel ($ans = pre = 1$).

From hotel 2 to hotel n, if $a[i].c < a[pre].c$, then hotel i is also a candidate hotel (ans++; $pre = i$).

Finally, output the number of candidate hotels ans.

7.3.8 Train Swapping

At an old railway station, you may still encounter one of the last remaining "train swappers." A train swapper is an employee of the railroad whose sole job it is to rearrange the carriages of trains.

Once the carriages are arranged in the optimal order, all the train driver has to do is drop the carriages off, one by one, at the stations for which the load is meant.

The title *train swapper* stems from the first person who performed this task, at a station close to a railway bridge. Instead of opening up vertically, the bridge rotated around a pillar in the center of the river. After rotating the bridge 90°, boats could pass left or right.

The first train swapper had discovered that the bridge could be operated with at most two carriages on it. By rotating the bridge 180°, the carriages switched place, allowing him to rearrange the carriages (as a side effect, the carriages then faced the opposite direction, but train carriages can move either way, so who cares).

Now that almost all train swappers have died out, the railway company would like to automate its operation. Part of the program to be developed is a routine that decides for a given train the least number of swaps of two adjacent carriages necessary to order the train. Your assignment is to create that routine.

Input

The input contains on the first line the number of test cases (N). Each test case consists of two input lines. The first line of a test case contains an integer L, determining the length of the train ($0 \le L \le 50$). The second line of a test case contains a permutation of the numbers 1 through L, indicating the current order of the carriages. The carriages should be ordered such that carriage 1 comes first, then 2, and so forth, with carriage L coming last.

Output

For each test case, output the sentence "Optimal train swapping takes S swaps," where S is an integer.

Sample Input	Sample Output
3	Optimal train swapping takes 1 swap.
3	Optimal train swapping takes 6 swaps.
1 3 2	Optimal train swapping takes 1 swap.
4	
4 3 2 1	
2	
2 1	

Source: ACM North Western European Regional Contest 1994.

ID for online judge: UVA 299.

Hint

Bubble sort is used to solve the problem. The number of exchanges is the solution to the problem.

7.3.9 Unix ls

The computer company you work for is introducing a brand new computer line and is developing a new Unix-like operating system to be introduced along with the new computer. Your assignment is to write the formatter for the ls function.

Your program will eventually read input from a pipe (although for now your program will read from the input file). Input to your program will consist of a list of (F) filenames that you will sort (ascending based on the ASCII character values) and format into (C) columns based on the length (L) of the longest filename. Filenames will be between 1 and 60 (inclusive) characters in length and will be formatted into left-justified columns. The rightmost column will be the width of the longest filename, and all other columns will be the width of the longest filename plus 2. There will be as many columns as will fit in 60 characters. Your program should use as few rows (R) as possible, with rows being filled to capacity from left to right.

Input

The input file will contain an indefinite number of lists of filenames. Each list will begin with a line containing a single integer ($1 \leq N \leq 100$). There will then be N lines, each containing one left-justified filename, and the entire line's contents (between 1 and 60 characters) are considered to be part of the filename. Allowable characters are alphanumeric (a to z, A to Z, and 0 to 9) and from the following set {._-} (not including the curly braces). There will be no illegal characters in any of the filenames and no line will be completely empty.

Immediately following the last filename will be the N for the next set or the end of the file. You should read and format all sets in the input file.

Output

For each set of filenames, you should print a line of exactly 60 dashes (-), followed by the formatted columns of filenames. The sorted filenames 1 to R will be listed down column 1, filenames $R + 1$ to $2R$ listed down column 2, and so forth.

Sample Input	Sample Output
10	---
tiny	12345678.123 size-1
2short4me	2short4me size2
very_long_file_name	mid_size_name size3
shorter	much_longer_name tiny
size-1	shorter very_long_file_name
size2	---
size3	Alfalfa Cotton Joe Porky
much_longer_name	Buckwheat Darla Mrs._Crabapple Stimey
12345678.123	Butch Froggy P.D. Weaser
mid_size_name	---
12	Alice Chris Jan Marsha Ruben
Weaser	Bobby Cindy Jody Mike Shirley
Alfalfa	Buffy Danny Keith Mr._French Sissy
Stimey	Carol Greg Lori Peter
Buckwheat	
Porky	
Joe	
Darla	
Cotton	
Butch	
Froggy	
Mrs._Crabapple	
P.D.	
19	
Mr._French	
Jody	
Buffy	
Sissy	
Keith	

Sample Input	Sample Output
Danny	
Lori	
Chris	
Shirley	
Marsha	
Jan	
Cindy	
Carol	
Mike	
Greg	
Peter	
Bobby	
Alice	
Ruben	

Source: ACM South Central Regional 1995.

ID for online judge: UVA 400.

Hint

A string array s is used to store N filenames. First, s is sorted in alphabetical order, and

$$\max l = \max_{1 \le i \le N} \{s[i].\text{size}\}$$

is calculated. Obviously, the number of columns

$$c = \left\lfloor \frac{62}{\max l + 2} \right\rfloor$$

and the number of rows

$$r = \left\lfloor \frac{n-1}{c} + 1 \right\rfloor.$$

Suppose $ans[j]$ is to store formatted output in the jth row, $1 \le j \le r$. Finally, output $ans[1] \dots ans[r]$.

7.3.10 Children's Game

There are lots of number games for children. These games are pretty easy to play but not so easy to make. We will discuss an interesting game here. Each player will be given N positive integer. He or she can make a big integer by appending those integers after one another. For example, if there are four integers 123, 124, 56, and 90, then the following integers can be made: 1231245690, 1241235690, 5612312490, 9012312456, 9056124123, and so forth. In fact, 24 such integers can be made. But one thing is sure: 9056124123 is the largest possible integer that can be made.

You may think that it's very easy to find the answer, but will it be easy for a child who has just gotten the idea of number?

Input

Each input starts with a positive integer N (≤ 50). In the next lines there are N positive integers. Input is terminated by $N = 0$, which should not be processed.

Output

For each input set, you have to print the largest possible integer that can be made by appending all the N integers.

Sample Input	Sample Output
4	9056124123
123 124 56 90	99056124123
5	99999
123 124 56 90 9	
5	
9 9 9 9 9	
0	

Source: 4th IIUC Interuniversity Programming Contest 2005, Problemsetter: Md. Kamruzzaman.

ID for online judge: UVA 10905.

Hint

Note that whether we should swap two subsequent strings depends only on these two strings. If the strings are x and y, their possible concatenations are xy and yx. If and only if $yx > xy$ should we swap them.

Given any starting sequence, make such swaps until no more swaps are possible. It can be proved that the result is optimal.

7.3.11 DNA Sorting

One measure of unsortedness in a sequence is the number of pairs of entries that are out of order with respect to each other. For instance, in the letter sequence "DAABEC", this measure is 5, since D is greater than four letters to its right and E is greater than one letter to its right. This measure is called the number of inversions in the sequence. The sequence "AACEDGG" has only one inversion (E and D)—it is nearly sorted—while the sequence "ZWQM" has six inversions (it is as unsorted as can be—exactly the reverse of sorted).

You are responsible for cataloging a sequence of DNA strings (sequences containing only the four letters A, C, G, and T). However, you want to catalog them, not in alphabetical order, but rather in order of "sortedness," from "most sorted" to "east sorted." All the strings are of the same length.

Input

The first line contains two integers: a positive integer n ($0 < n \leq 50$), giving the length of the strings, and a positive integer m ($0 < m \leq 100$), giving the number of strings. These are followed by m lines, each containing a string of length n.

Output

Output the list of input strings, arranged from "most sorted" to "least sorted". Since two strings can be equally sorted, output them according to the original order.

Sample Input	Sample Output
10 6	CCCGGGGGGA
AACATGAAGG	AACATGAAGG
TTTTGGCCAA	GATCAGATTT
TTTGGCCAAA	ATCGATGCAT
GATCAGATTT	TTTTGGCCAA
CCCGGGGGGA	TTTGGCCAAA
ATCGATGCAT	

Source: ACM East Central North America 1998.

ID for online judge: POJ 1007.

Hint

A linear list a is used to represent the sequence of DNA strings. For the ith DNA string, $a[i].s$ is the string and $a[i].x$ is the number of inversions.

Bubble sort is used to calculate the numbers of inversions. Then a is sorted in ascending order for numbers of inversions. Finally, output a.

7.3.12 Exact Sum

Peter received money from his parents this week and wants to spend it all buying books. But he does not read a book fast because he likes to enjoy every single word while he is reading. In this way, it takes him a week to finish a book.

As Peter receives money every 2 weeks, he decided to buy two books; then he can read them until he receives more money. As he wishes to spend all the money, he should choose two books whose prices summed up are equal to the money that he has. It is a little bit difficult to find these books, so Peter asks for your help to find them.

Input

Each test case starts with $2 \leq N \leq 10,000$, the number of available books. The next line will have N integers, representing the price of each book; a book costs less than 1,000,001. Then there is another line with an integer M, representing how much money Peter has. There is a blank line after each test case. The input is terminated by the end of the file (EOF).

Output

For each test case you must print the message "Peter should buy books whose prices are i and j," where i and j are the prices of the books whose sum is equal to M and $i \leq j$. You can consider that it is always possible to find a solution; if there are multiple solutions, print the solution that minimizes the difference between the prices i and j. After each test case, you must print a blank line.

Sample Input	Sample Output
2	Peter should buy books whose prices are 40 and 40.
40 40	
80	Peter should buy books whose prices are 4 and 6.
5	
10 2 6 8 4	
10	

Source: ACM-ICPC: UFRN Qualification Contest 2006.

ID for online judge: UVA 11057.

Hint by the Problemsetter (http://www.algorithmist.com/index.php/Main_Page)

Peter wants to spend his allowance on books. It takes him a week to read a book because he likes to savor every word. Peter receives his allowance every two weeks, so he'd like to buy two books that he can read until he gets his allowance again.

Given the value of his allowance and the prices of a list of books that he wants, find two books whose prices summed up are equal to his allowance exactly. Where there are multiple answers, output the pair that minimizes the difference in price between the two books.

There can be up to 10,000 books in the worst case, so we can't afford to check every pair of books. So, we need to be a bit more clever. If our target is T and you choose to buy a book of price P, the price of the other book must be $T - P$. If we sort the books, then we can find the other book in logarithmic time using binary search. Keep track of the books that add up to the target with the least difference.

Note: The problem statement specifically states that there will always be a solution. In other words, there will always be at least two books that add up to the target value.

7.3.13 Shellsort

He made each turtle stand on another one's back
And he piled them all up in a nine-turtle stack.

And then Yertle climbed up. He sat down on the pile.
What a wonderful view! He could see 'most a mile!*

King Yertle wishes to rearrange his turtle throne to place his highest-ranking nobles and clos-est advisors nearer to the top. A single operation is available to change the order of the turtles in the stack: a turtle can crawl out of its position in the stack and climb up over the other turtles to sit on the top.

Given an original ordering of a turtle stack and a required ordering for the same turtle stack, your job is to determine a minimal sequence of operations that rearranges the original stack into the required stack.

Input

The first line of the input consists of a single integer K giving the number of test cases. Each test case consists of an integer n giving the number of turtles in the stack. The next n lines specify the original ordering of the turtle stack. Each of the lines contains the name of a turtle, starting with the turtle on the top of the stack and working down to the turtle at the bottom of the stack. Turtles have unique names, each of which is a string of no more than 80 characters drawn from a character set consisting of the alphanumeric characters, the space character, and the period. The next n lines in the input give the desired ordering of the stack, once again by naming turtles from top to bottom. Each test case consists of exactly $2n + 1$ lines in total. The number of turtles (n) will be less than or equal to 200.

Output

For each test case, the output consists of a sequence of turtle names, one per line, indicating the order in which turtles are to leave their positions in the stack and crawl to the top. This sequence of operations should transform the original stack into the required stack and should be as short as possible. If more than one solution of shortest length is possible, any of the solutions may be reported. Print a blank line after each test case.

Sample Input	Sample Output
2	Duke of Earl
3	
Yertle	Sir Lancelot
Duke of Earl	Richard M. Nixon
Sir Lancelot	Yertle
Duke of Earl	
Yertle	
Sir Lancelot	
9	

(Continued)

---

* Dr. Seuss. *Yertle the Turtle and Other Stories*. Random House, 1958.

Sample Input	Sample Output
Yertle	
Duke of Earl	
Sir Lancelot	
Elizabeth Windsor	
Michael Eisner	
Richard M. Nixon	
Mr. Rogers	
Ford Perfect	
Mack	
Yertle	
Richard M. Nixon	
Sir Lancelot	
Duke of Earl	
Elizabeth Windsor	
Michael Eisner	
Mr. Rogers	
Ford Perfect	
Mack	

ID for online judge: UVA 10152.

Hint

Suppose the original ordering of the turtle stack is b, and the desired ordering of the stack is a, where $b[i]$ and $a[i]$ are the ith turtle top down in their stacks.

Suppose i is the first position that $b[i] \neq a[i]$ bottom up. It can be obtained through the following statements: ($i = n$, while ($b[i] == a[i]$ && $i >= 1$) i––;).

Then turtle $b[k]$ that crawls out of its position in the stack and climbs up over the other turtles to sit on the top is found.

```
for (j=i; j>=1; j--)
for (k=i; k>=j; k--)
if (a[j]==b[k]){
Output b[k];
temp= b[k];
for (int t=k-1; t>=1; t--) b[t+1]=b[t];
b[1]=temp;
break;
}
```

7.3.14 Tell Me the Frequencies!

Given a line of text, you will have to find the frequencies of the ASCII characters present in it. The given lines will contain none of the first 32 or last 128 ASCII characters. Of course, lines may end with/n and/r, but always keep those out of consideration.

Input

Several lines of text are given as input. Each line of text is considered a single input. The maximum length of each line is 1000.

Output

Print the ASCII value of the ASCII characters that are present and their frequency according to the given format below. A blank line should separate each set of output. Print the ASCII characters in the ascending order of their frequencies. If two characters are present at the same time, print the information of the ASCII character with the higher ASCII value first. Input is terminated by the end of the file.

Sample Input	Sample Output
AAABBC	67 1
122333	66 2
	65 3
	49 1
	50 2
	51 3

Source: Bangladesh Programming Contest 2001.

ID for online judge: UVA 10062.

Hint
First, the frequencies of all ASCII characters are calculated. Then results are obtained through a loop statement.

7.3.15 Anagrams (II)

One of the preferred kinds of entertainment of people living in the final stages of the twentieth century was filling in crosswords. Almost every newspaper and magazine has a column dedicated to entertainment, but only amateurs have enough after solving one crossword. Real professionals require more than one crossword a week. And it is so dull—just crosswords and crosswords—while so many other riddles are waiting out there. For those people there are special, dedicated magazines. There are also quite a few competitions to take part in, even reaching the level of world championships.

You were taken on by such a professional for whom competing to solve riddles is just a job. He had a brilliant idea to use a computer at work not just to play games. Somehow, anagrams found themselves first in line. You are to write a program that searches for anagrams of given words, using a given vocabulary, tediously filled with new words by your employer.

Input

The structure of input data is given below:

⟨number of words in vocabulary⟩
⟨word 1⟩
....
⟨word N⟩
⟨test word 1⟩
....
⟨test word k⟩
END

⟨number of words in vocabulary⟩ is an integer number $N < 1000$. ⟨word 1⟩ up to ⟨word N⟩ are words from the vocabulary. ⟨test word 1⟩ up to ⟨test word k⟩ are the words to find anagrams for. All words are lowercase (word END means end of data—it is NOT a test word). You can assume all words are not longer than 20 characters.

Output

For each ⟨test word⟩ list the found anagrams in the following way:

Anagrams for: ⟨test word⟩
⟨No)) ⟨anagram⟩
....
⟨No⟩ should be printed on 3 chars.
In case of failing to find any anagrams, your output should look like this:
Anagrams for ⟨test word⟩
No anagrams for ⟨test word⟩

Sample Input	Sample Output
1	Anagrams for: tola
	1) atol
8	2) lato
atol	3) tola
lato	Anagrams for: kola
microphotographics	No anagrams for: kola
rata	Anagrams for: aatr
rola	1) rata
tara	2) tara

Sample Input	Sample Output
tola	Anagrams for: photomicrographics
pies	1) microphotographics
tola	
kola	
aatr	
photomicrographics	
END	

ID for online judge: UVA 630.

Hint

First, a dictionary of words is given. Then a set of words is given, and for each word, you are required to find all its permutations in the dictionary.

The solution is like that of the find the clones problem in Section 7.3.5. That is, for each word in the directory, its letters are sorted in alphabetical order and stored in another list t. Then, for each word in the set of words, its letters are also sorted in alphabetical order; search whether it exists in t.

7.3.16 *Flooded!*

To enable home buyers to estimate the cost of flood insurance, a real-estate firm provides clients with the elevation of each 10×10–meter square of land in regions where homes may be purchased. Water from rain, melting snow, and burst water mains will collect first in those squares with the lowest elevations, since water from squares of higher elevation will run downhill. For simplicity, we also assume that storm sewers enable water from high-elevation squares in valleys (completely enclosed by still higher-elevation squares) to drain to lower-elevation squares, and that water will not be absorbed by the land.

From weather data archives, we know the typical volume of water that collects in a region. As prospective home buyers, we wish to know the elevation of the water after it has collected in low-lying squares and also the percentage of the region's area that is completely submerged (i.e., the percentage of 10-meter squares whose elevation is strictly less than the water level). You are to write the program that provides these results.

Input

The input consists of a sequence of region descriptions. Each begins with a pair of integers, m and n, each less than 30, giving the dimensions of the rectangular region in 10-meter units. Immediately following are m lines of n integers giving the elevations of the squares in row-major order. Elevations are given in meters, with positive and negative numbers representing elevations above and below sea level, respectively. The final value in each region description is an integer that indicates the number of cubic meters of water that will collect in the region. A pair of zeros follows the description of the last region.

Output

For each region, display the region number (1, 2, ...), the water level (in meters above or below sea level), and the percentage of the region's area under water, each on a separate line. The water level and percentage of the region's area under water are to be displayed accurately to two fractional digits. Follow the output for each region with a blank line.

Sample Input	Sample Output
3 3	Region 1
25 37 45	Water level is 46.67 meters.
51 12 34	66.67 percent of the region is under water.
94 83 27	
10000	
0 0	

Source: ACM World Finals 1999.

ID for online judge: POJ 1877.

Hint

Based on the problem description, the area of each square of land is 10*10 = 100 square meters. $n*m$ elevations of all squares are stored in an array $a[\ ]$ and sorted in ascending order.

The difference of elevations of $a[i + 1]$ and $a[i]$ is $a[i + 1] - a[i]$. The area for the first i squares is $i*100$; that is, water is increased $100*(a[i + 1] - a[i])*i$ from square i to square $i + 1$. Suppose the elevation of water is between $a[k]$ and $a[k + 1]$, that is,

$$\sum_{i=1}^{k} 100 * (a[i+1] - a[i]) * i \le w < \sum_{i=1}^{k+1} 100 * (a[i+1] - a[i]) * i.$$

The number of cubic meters of water over $a[k]$ is

$$w_k = w - \sum_{i=1}^{k} 100 * (a[i+1] - a[i]) * i.$$

Therefore, the water level is $a[k] + (w_k/100*k)$, and the percentage of the region's area under water is $100*(k/n*m)\%$ ($1 \le k < n*m$).

7.3.17 Football Sort

Write a program that, given the fixtures of a football championship, outputs the corresponding classification following the format specified below. Win, draw, and loss earn three, one, and zero points, respectively.

The criteria of classification are the number of points scored, followed by goal average and then scored goals. When more than one team have exactly the same number of points, goal average, and scored goals, these are considered as having the same position in the classification.

Input

The input will consist of a series of tests. Each test starts with a line containing two positive integers, $28 \geq T \geq 1$ and $G \geq 0$. T is the number of teams, and G is the number of games played. The following T lines each contain the name of a squad. Squad names have up to 15 characters and may only contain letters and dash characters (-). Finally, the following G lines contain the score of each game. The scores are output with the following format: name of the home team, number of goals scored by the home team, a dash, number of goals scored by the away team, and name of the away team.

The input ends with a test case where $T = G = 0$ and should not be processed.

Output

The program shall output the classification tables corresponding to each input test separated by blank lines. In each table, the teams appear in order of classification, or alphabetically when they have the same position. The statistics of each team are displayed on a single line: team position, team name, number of points, number of games played, number of scored goals, number of suffered goals, goal average, and percentage of earned points, when available. Note that if several teams are in a draw, only the position of the first is printed. Fields shall be formatted and aligned as shown in the sample output.

Sample Input	Sample Output							
6 10	1.	tA	4	4	1	1	0	33.33
tA		tB	4	4	0	0	0	33.33
tB	3.	tC	4	4	0	0	0	33.33
tC		td	4	4	0	0	0	33.33
td		tE	4	4	0	0	0	33.33
tE	6.	tF	0	0	0	0	0	N/A
tF								
tA 1 - 1 tB	1.	Botafogo	6	2	6	4	2	100.00
tC 0 - 0 td	2.	Flamengo	0	2	4	6	-2	0.00
tE 0 - 0 tA								
tC 0 - 0 tB	1.	tA	4	4	0	0	0	33.33
td 0 - 0 tE		tB	4	4	0	0	0	33.33
tA 0 - 0 tC		tC	4	4	0	0	0	33.33
tB 0 - 0 tE		tD	4	4	0	0	0	33.33

(Continued)

Sample Input	Sample Output						
td 0 - 0 tA	tE	4	4	0	0	0	33.33
tE 0 - 0 tC							
tB 0 - 0 td	1. Quinze-Novembro	3	1	6	0	6	100.00
2 2	2. Santo-Andre	3	1	2	0	2	100.00
Botafogo	3. Flamengo	0	2	0	8	–8	0.00
Flamengo							
Botafogo 3 - 2 Flamengo							
Flamengo 2 - 3 Botafogo							
5 10							
tA							
tB							
tC							
tD							
tE							
tA 0 - 0 tB							
tC 0 - 0 tD							
tE 0 - 0 tA							
tC 0 - 0 tB							
tD 0 - 0 tE							
tA 0 - 0 tC							
tB 0 - 0 tE							
tD 0 - 0 tA							
tE 0 - 0 tC							
tB 0 - 0 tD							
3 2							
Quinze-Novembro							
Flamengo							

Sample Input	Sample Output
Santo-Andre	
Quinze-Novembro 6 - 0 Flamengo	
Flamengo 0 - 2 Santo-Andre	
0 0	

Source: Federal University of Rio Grande do Norte Classifying Contest 2004, Round 2.

ID for online judge: UVA 10698.

Hint

The problem is a simulation problem. Following the instruction, the points for each team are calculated, and then teams are sorted according to the criteria of classification.

A linear list p is used to represent teams; for the ith team, the team name is $p[i].name$, the number of points is $p[i].pts$, the number of games played is $p[i].gms$, the number of scored goals is $p[i].goal$, and the number of suffered goals is $p[i].suffer$, $(0 \leq i \leq n - 1)$.

1. **Initialization.** First, n teams' information is input. For team i, $p[i].pts$, $p[i].gms$, $p[i].goal$, and $p[i].suffer$ are initialized to 0. Sort p in team names by alphabetical order.

   Second, m games' information is input. For the current game, suppose the sequence number of the home team is x, the number of goals scored by the home team is u; the sequence number of the away team is y, and the number of goals scored by the away team is v. Then $++p[x].gms$; $++p[y].gms$; $p[x].goal += u$; $p[x].suffer += v$; $p[y].goal += v$; $p[y].suffer += u$; if $u > v$, $p[x].pts += 3$; else, if $u < v$, $p[y].pts += 3$; else, if $u == v$, $++ p[x].pts$, $++ p[y].pts$.

2. **Teams are sorted.** The number of points is the first key, the goal average is the second key, and the number of scored goals is the third key. Teams are sorted based on these keys.

3. **Output the classification table.**

7.3.18 Trees

The road off the east gate of Peking University used to be decorated with a lot of trees. However, because of the construction of a subway, many of them have been cut down or moved away. Now please help to count how many trees are left.

Let's only consider one side of the road. Assume that trees were planted every 1 meter from the beginning of the road. Now some sections of the road are assigned for the subway station, crossover, or other buildings, so trees in those sections will be moved away or cut down. Your job is to give the number of trees left.

For example, the road is 300 meter long and trees are planted every 1 meter from the beginning of the road (0 meter). That is, there used to be 301 trees on the road. Now the section from 100 to 200 meters is assigned for the subway station, so 101 trees need to be moved away and only 200 trees are left.

Input

There are several test cases in the input. Each case starts with an integer L ($1 \le L < 200{,}000{,}000$), representing the length of the road, and M ($1 \le M \le 5000$), representing the number of sections that are assigned for other use.

The following M lines each describe a section. A line is in the following format:
Start End
Here Start and End ($0 \le$ Start $\le$ End $\le L$) are both nonnegative integers representing the start point and the end point of the section. It is confirmed that these sections do not overlap with each other.

A case with $L = 0$ and $M = 0$ ends the input.

Output

Output the number of trees left in one line for each test case.

Sample Input	Sample Output
300 1	200
100 200	300
500 2	
100 200	
201 300	
0 0	

Source: ACM Beijing 2005, Preliminary.

ID for online judge: POJ 2665.

Hint

The problem is a simulation problem. Based on the problem description, for each test case, the number of moved trees is calculated.

SUMMARY OF SECTION II

A linear list is a data structure containing a finite ordered set of data elements. First, experiments for applications of linear lists were shown: applications of arrays, strings, stacks, queues, dictionaries, hash tables, and so on. Then practices for sort algorithms of linear lists were given.

In Section II, experiments for applications of STL were also introduced. Through experiments, students can become familiar with STL.

Also in this section, object-oriented programming was introduced.

EXPERIMENTS FOR TREES

Section III focuses on trees. A tree is a data structure representing hierarchical data. In this section, experiments are organized in two fields: trees and binary trees. In Chapter 8, experiments for hierarchical problems, tree storage structures, and union–find sets are shown. In Chapters 9 and 10, experiments for binary trees, such as traversal of binary trees, binary search trees, Huffman trees, and heaps, are shown.

Chapter 8

Programming by Tree Structure

A tree can be defined recursively. A tree is a collection of n vertices. The collection can be empty ($n == 0$); otherwise, a tree constitutes a distinguished vertex r, called the root, and zero or more nonempty subtrees that the root of each subtree is a child of r, and r is the parent of each subtree root.

In Chapter 8, there are two parts of the experiments:

1. Solving hierarchical problems by tree traversal
2. Union–find sets supported by tree structure

8.1 Solving Hierarchical Problems by Tree Traversal

A hierarchical structure can be modeled mathematically as a rooted tree: the root of the tree forms the top level, and the children of the root are at the same level, under their common parent. Vertices in a rooted tree constitute a partially ordered set, and the relations between vertices constitute relations of partial orders.

Hierarchical problems can be represented as tree structures and can be solved by tree traversal.

Tree traversal (also known as tree search) refers to the process of visiting (examining or updating) each vertex in a tree exactly once, in a systematic way.

There are two ways to traverse a tree:

Preorder traversal
    Visit the root
    Preorder traversal for subtrees from left to right
Postorder traversal
    Postorder traversal for subtrees from left to right
    Visit the root

The algorithm for preorder traversal is as follows:

```
void preorder(int v);
{  visit vertex v;
   for (i ∈ the set of adjacent vertices for v)      // Pre-order traverse
all adjacent unvisited vertices for v
   if  (vertex i isn't visited)
     preorder(i);
};
```

The algorithm for postorder traversal is as follows:

```
void postorder(int v);
{  for (i ∈ the set of adjacent vertices for v)      // Post-order traverse
all adjacent unvisited vertices for v
   if (vertex i isn't visited)
       postorder (i);
  visit vertex v;
};
```

Different storage structures for trees can affect tree traversal's efficiency. There are many storage representations for trees. The most common representation methods are as follows:

1. **Representation of generalized list.** A generalized list can be used to represent a tree. In a tree there are three kinds of vertices: root, leaf, and inner node. In a generalized list, there are also three kinds of corresponding vertices: ATOM, HEAD, and LST. There are two kinds of representations for a tree: bracket and linked list.
2. **Representation of parents.** Representation of parents is suitable for postorder traversal for a tree. Representation of parents for a tree uses an array to store vertices in the tree. There are two fields for a vertex: data field, to store data, and pointer field, to store the pointer pointing to its parent.
3. **Representation of multiple linked list.** Representation of multiple linked list is suitable for preorder traversal for a tree. An element in a multiple linked list is to store its data and its pointers pointing to its children. It is shown as

data	(child₁)	(child₂)	...	(child_d)

The shortage for representation of multiple linked lists is to waste storage space.

A Standard Template Library (STL) container can be used to define a multiple linked list for a tree. For example, a multiple linked list *adj[n]* can be defined as a vector, where *adj[x].push_back(y)* is to push *y* into the list for the children of *x*, and *y = adj[x].pop_back( )* is to get a vertex from the list for the children of *x*.

8.1.1 Nearest Common Ancestor

A rooted tree is a well-known data structure in computer science and engineering. An example is shown in Figure 8.1.

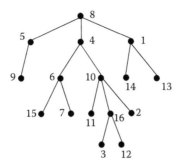

Figure 8.1　A rooted tree.

In the figure, each node is labeled with an integer from {1, 2, ..., 16}. Node 8 is the root of the tree. Node x is an ancestor of node y if node x is in the path between the root and node y. For example, node 4 is an ancestor of node 16. Node 10 is also an ancestor of node 16. In fact, nodes 8, 4, 10, and 16 are the ancestors of node 16. Remember that a node is an ancestor of itself. Nodes 8, 4, 6, and 7 are the ancestors of node 7. A node x is called a common ancestor of two different nodes y and z if node x is an ancestor of node y and an ancestor of node z. Thus, nodes 8 and 4 are the common ancestors of nodes 16 and 7. A node x is called the nearest common ancestor of nodes y and z if x is a common ancestor of y and z and nearest to y and z among their common ancestors. Hence, the nearest common ancestor of nodes 16 and 7 is node 4. Node 4 is nearer to nodes 16 and 7 than node 8 is.

For other examples, the nearest common ancestor of nodes 2 and 3 is node 10, the nearest common ancestor of nodes 6 and 13 is node 8, and the nearest common ancestor of nodes 4 and 12 is node 4. In the last example, if y is an ancestor of z, then the nearest common ancestor of y and z is y.

Write a program that finds the nearest common ancestor of two distinct nodes in a tree.

Input

The input consists of T test cases. The number of test cases (T) is given in the first line of the input file. Each test case starts with a line containing an integer N, the number of nodes in a tree, $2 \le N \le 10{,}000$. The nodes are labeled with integers 1, 2, ..., N. Each of the next $N - 1$ lines contains a pair of integers that represent an edge—the first integer is the parent node of the second integer. Note that a tree with N nodes has exactly $N - 1$ edges. The last line of each test case contains two distinct integers whose nearest common ancestor is to be computed.

Output

Print exactly one line for each test case. The line should contain the integer that is the nearest common ancestor.

Sample Input	Sample Output
2	4
16	3

(Continued)

Sample Input	Sample Output
1 14	
8 5	
10 16	
5 9	
4 6	
8 4	
4 10	
1 13	
6 15	
10 11	
6 7	
10 2	
16 3	
8 1	
16 12	
16 7	
5	
2 3	
3 4	
3 1	
1 5	
3 5	

Source: ACM Taejon 2002.

ID for online judge: POJ 1330.

Analysis
From each node in a tree, there is only one path to the root. Therefore, any pair of nodes has common ancestors in a tree. A tree is represented with representation of parents and representation of multiple linked lists. Each node's level number can be obtained by preorder traversal (the root's level number is 0, its children's level number is 1, and so on). The multiple linked list is represented by class *vector*. And an integer array is used to represent parents and hierarchical data.

The algorithm finding the nearest common ancestor for nodes x and y is as follows:

```
while (x≠y)
{ if (the level number of x is great than the level number of y)
```

```
x=the parent of x;
else
y= the parent of y;
}
```

When the loop ends, *x* is the nearest common ancestor.

Program

```
#include <iostream>
#include <vector>
using namespace std;
const int N = 10000;
vector<int> a[N];          // multiple linked list, the list of children
    for node i is a vector
int f[N], r[N];            // representations of parents and hierarchy,
    the parent and hierarchy for node i is f[i] and r[i]
void DFS (int u, int dep) // Pre-order Traversal from node u
{
    r[u]=dep;              // node u is at hierarchy dep
    for (vector<int>::iterator it = a[u].begin(); it != a[u].end(); ++it)
        DFS(*it, dep + 1);// Recursion for every child of u
}
int main()
{
    int casenum, num, n, i, x, y;
    scanf("%d", &casenum); //      number of test cases
    for (num = 0; num < casenum; num++)
    {
        scanf("%d", &n);      // number of nodes
        for (i = 0; i < n; i++) a[i].clear(); //initialization
        memset(f, 255, sizeof(f));
        for (i = 0; i < n - 1; i++)
        {
            scanf("%d %d", &x, &y);     // edge (x, y)
            a[x - 1].push_back(y - 1);  // push node (y-1) into the list of
(x-1)'s children
            f[y - 1] = x - 1;            //node(y-1)'s parent is (x-1)
        }
        for (i = 0; f[i] >= 0; i++);   // search the root i
        DFS(i, 0);                     // calculate every nodes' hierarchy
from the root
        scanf("%d %d", &x, &y);        // a pair of nodes
        x--; y--;
        while (x != y)                 // to find the Nearest Common Ancestors
        {
            if (r[x]>r[y]) x = f[x];
            else y = f[y];
        }
        printf("%dn", x + 1);// Output
    }
    return 0;
}
```

8.1.2 Hire and Fire

In this problem, you are asked to keep track of the hierarchical structure of an organization's changing staff. As the first event in the life of an organization, the chief executive officer (CEO) is named. Subsequently, any number of hires and fires can occur. Any member of the organization (including the CEO) can hire any number of direct subordinates, and any member of the organization (including the CEO) can be fired. The organization's hierarchical structure can be represented by a tree. Consider the example shown by Figure 8.2.

VonNeumann is the CEO of this organization. VonNeumann has two direct subordinates: Tanenbaum and Dijkstra. Members of the organization who are direct subordinates of the same member are ranked by their respective seniority. In the diagram, the seniority of such members decreases from left to right. For example, Tanenbaum has higher seniority than Dijkstra.

When a member hires a new direct subordinate, the newly hired subordinate has lower seniority than any other direct subordinates of the same member. For example, if VonNeumann (in Figure 8.2) hires Shannon, then VonNeumann's direct subordinates are Tanenbaum, Dijkstra, and Shannon in order of decreasing seniority.

When a member of the organization gets fired, there are two possible scenarios. If the victim (the person who gets fired) had no subordinates, then he or she will simply be dropped from the organization's hierarchy. If the victim had any subordinates, then his or her highest-ranking (by seniority) direct subordinate will be promoted to fill the resulting vacancy. The promoted person will also inherit the victim's seniority. Now, if the promoted person also had some subordinates, then his or her highest-ranking direct subordinate will similarly be promoted, and the promotions will cascade down the hierarchy until a person having no subordinates has been promoted. In Figure 8.2, if Tanenbaum gets fired, then Stallings will be promoted to Tanenbaum's position and seniority, and Knuth will be promoted to Stallings's previous position and seniority.

Figure 8.3 shows the hierarchy resulting from Figure 8.2 after (1) VonNeumann hires Shannon and (2) Tanenbaum gets fired.

Input

The first line of the input contains only the name of the person who is initially the CEO. All names in the input file consist of 2–20 characters, which may be upper- or lowercase letters, apostrophes, and hyphens (but no blank spaces). Each name contains at least one uppercase and at least one lowercase letter.

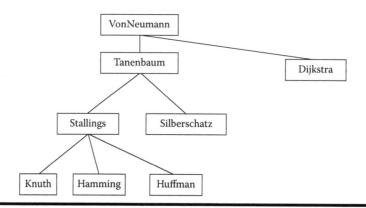

Figure 8.2   The hierarchical structure of an organization.

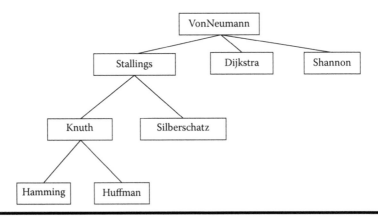

Figure 8.3   The hierarchy resulting from Figure 8.2.

The first line will be followed by one or more additional lines. The format of each of these lines will be determined by one of the following three rules of syntax:

- [Existing member] hires [new member]
- Fire [existing member]
- Print

Here [existing member] is the name of any individual who is already a member of the organization, and [new member] is the name of an individual who is not a member of the organization yet. The three types of lines (hires, fire, and print) can appear in any order, any number of times.

You may assume that at any time there is at least one member (who is the CEO) and no more than 1000 members in the organization.

Output

For each print command, print the current hierarchy of the organization, assuming all hires and fires since the beginning of the input have been processed as explained above. Tree diagrams (such as those in Figures 8.2 and 8.3) are translated into textual format according to the following rules:

- Each line in the textual representation of the tree will contain exactly one name.
- The first line will contain the CEO's name, starting in column 1.
- The entire tree, or any subtree, having the form shown in Figure 8.4 will be represented in textual form as shown in Figure 8.5.

The output resulting from each print command in the input will be terminated by one line consisting of exactly 60 hyphens. There will not be any blank lines in the output.

Sample Input	Sample Output
VonNeumann	VonNeumann
VonNeumann hires Tanenbaum	+Tanenbaum

(Continued)

Sample Input	Sample Output
VonNeumann hires Dijkstra	++Stallings
Tanenbaum hires Stallings	+++Knuth
Tanenbaum hires Silberschatz	+++Hamming
Stallings hires Knuth	+++Huffman
Stallings hires Hamming	++Silberschatz
Stallings hires Huffman	+Dijkstra
print	---
VonNeumann hires Shannon	VonNeumann
fire Tanenbaum	+Stallings
print	++Knuth
fire Silberschatz	+++Hamming
fire VonNeumann	+++Huffman
print	++Silberschatz
	+Dijkstra
	+Shannon

	Stallings
	+Knuth
	++Hamming
	+++Huffman
	+Dijkstra
	+Shannon

Source: ACM Rocky Mountain 2004.

IDs for online judges: POJ 2003, ZOJ 2348, UVA 3048.

Analysis

The hierarchical structure of an organization is a rooted tree, where CEO is the root; and can be represented as a multiple linked list. When members are hired and fired, the hierarchical structure of an organization is changed. Each node's parents and its level number should be recorded. Each subtree is preceded by one more '+' than its root.

The hierarchical structure of an organization and commands are analyzed as follows:

1. Because a member's seniority should be considered, the tree is an ordered tree. The seniority of children decreases from left to right. A multiple linked list is used as the storage mode. All children for a node are stored in a queue. And the queue is defined as class *list* in STL.

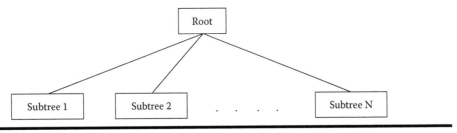

Figure 8.4 Tree diagram.

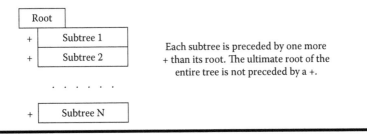

Each subtree is preceded by one more + than its root. The ultimate root of the entire tree is not preceded by a +.

Figure 8.5 Textual form.

2. x hires y: y is added into the queue for x's children, and y's parent pointer points to x.
3. fire y: y's highest-ranking (by seniority) direct subordinate will be promoted to fill the resulting vacancy. The promoted person will also inherit y's seniority. And if the promoted person also had some subordinates, then his or her highest-ranking direct subordinate will similarly be promoted, and the promotions will cascade down the hierarchy until a person having no subordinates has been promoted.
4. print: A multiple linked list is set up to represent the tree. CEO is the root, and it is at level 0. Preorder traversal is used for the command. If the current node i is at level p, print p '+' and the member name, and then children of node i are visited recursively.

STL containers, such as string, map, and list, are used to set up relationships between members' names and nodes and to implement operations.

Program

```
#include <string>
#include <iostream>
#include <list>
#include <map>
using std::list;
using std::string;
using std::cin;
using std::cout;
using std::endl;
using std::map;
struct Tman                        // Struct for multiple linked list
Tman
{
  string name;                     // Member name name
  Tman *f;                         // Pointer for parent *f
```

```
  list<Tman *> s;                    // Pointers list for children s
  Tman( ) {f = NULL;}
};
map<string, Tman*> hash;            //hash[x] stores pointers for subtree
whose root is x
Tman *root;                          // pointer for root
void print(long dep, Tman *now)     //print a tree from node pointer now
(level dep)
{
  if (now == NULL) return;          // if pointer is NULL, backtracking
  for (long i=1; i<=dep; ++i)       // output dep '+'
    cout<<'+';
    cout<<now->name<<endl;           // output name, and then output children
at dep+1 level
  for(list<Tman *>::iterator j=now->s.begin(); j!=now->s.end(); ++j)
    print(dep+1, *j);
}
void hires(string n1, string n2)  // n1 hires new member n2
{
  Tman *f = hash[n1];                // pointer for the subtree whose root
is n1
  Tman *s = new Tman( );             // New node s: name is n2 and parent
is f
  s->name = n2;
  s->f = f;
  f->s.push_back(s);                 //Add new node s into a queue for n1's
children
  hash[n2] = s;
}
void fire(string n1)                 //fire a member whose name is n1
{
  Tman *s = hash[n1]; // pointer for the subtree whose root is n1 is *s,
and pointer for parent of s is *f
  Tman *f = s->f;
  hash.erase(n1);
  while (s->s.size( ) != 0)          // Promotion from s
{
    s->name = s->s.front( )->name; // the name for the first child of s
is adjusted as the name for s
    hash[s->name] = s;               // s is adjusted as a subtree whose
root is s->name
    s = s->s.front( );
  }
  s->f->s.remove(s);
  delete s;                          //delete s
}
void solve( )                        //Calculate the change of a tree's struct
{
  string s1, s2;
  long i;
  cin>>s1;                           //Input name of CEO
  root = new Tman( );                //Initialize a subtree whose root
is CEO
  hash[s1] = root;
```

```
      root->name = s1;
      while (cin>>s1)                  // Input the first string for the command
      {
        if (s1 == "print")             // Print command
        {
            print(0, root);            // Output a tree's textual format
            for (i=1; i<=60; ++i) cout<<'-';
            cout<<endl;
        }
        else if (s1 == "fire")         // Fire command
        {
            cin>>s2;                    // Fired person's name
            fire(s2);
        }
        else                            // Hire command
        {
            cin>>s2;                    // input "hires"
            cin>>s2;                    // new member's name s2 as the new
child of s1
            hires(s1, s2);
        }
      }
}
int main( )
{
  solve( );
  return 0;
}
```

8.2 Union–Find Sets Supported by Tree Structure

In some applications, n elements are divided into several groups. Each group is a set. Because such problems are mainly related to union and search for sets, they are called union–find sets.

Union–find sets are disjoint sets $S = \{S_1, S_2, ..., S_r\}$, where set S_i has an element $rep[S_i]$, called a representative. Any two different sets in S are disjoint. There are three operations for union–find sets:

1. *Make_Set(x)*: For union–find sets $S = \{S_1, S_2, ..., S_r\}$, a set containing only one element $\{x\}$ is added into union–find sets S, and $rep[\{x\}] = x$, where x is not in any S_i, $1 \le i \le r$. Initially, for each element x, *Make_Set(x)* is called.
2. *join(x, y)*: Merge two different sets containing x and y, respectively. That is, S_x and S_y are deleted from S, and $S_x \cup S_y$ is added into S.
3. *set_find(x)*: Return representative $rep[S_x]$ for set S_x containing x.

There are two storage structures for union–find sets:

1. Linear list: A set is represented as a doubly linked list, where $rep[S_i]$ is the front of the list. Each element has a pointer pointing to $rep[S_i]$ (Figure 8.6).

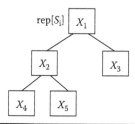

Figure 8.6   A set is represented as a doubly linked list $S_i = \{X_i, X_2, ..., X_k\}$.

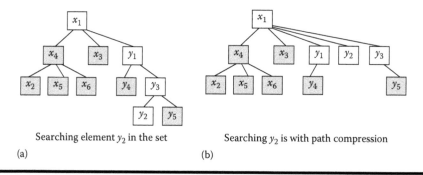

Figure 8.7   A set is represented as a tree $S_i = \{X_i, X_2, ..., X_5\}$.

Searching element y_2 in the set   Searching y_2 is with path compression

(a)   (b)

Figure 8.8   Search with path compression.

If union–find sets are represented as linear lists, the program becomes complex and its time efficiency of merging two sets is low. If two different lists containing x and y, respectively, are merged into a new linear list, we need to let *rep* pointers for all elements in S_y point to $rep[S_x]$, and the time complexity is $O(n)$.

2. **Tree structure:** A set is represented as a tree, where the root is the representative for the set (Figure 8.7).

Each node p has a pointer $set[p]$ pointing to its parent. If $set[p] < 0$, p is the root node. Initially, a set is constructed for each element, that is, $set[x] = -1$ ($1 \le x \le n$).

In search operation, we make use of the method that the search is with "path compression," to reduce the depth of the tree in the search process. For example, in Figure 8.8a, we need to search element y_2 in the set. The path is y_2-y_3-y_1-x_1 from y_2. So set pointers for y_2, y_3, and y_1 point to x_1 (Figure 8.8b).

The algorithm that the search is with "path compression" is as follows:

First, from node x, through set pointers the root of the tree f ($set[f] < 0$) is found. Then set pointers for all nodes on the path from x to f point to f to compress the path. The search process is as follows:

```
int set_find(int p) // Search the representative of the set
containing p, and compress the path
{
    if (set[p]<0)
        return p;
    return set[p]=set_find(set[p]);
}
```

Merging two sets is to connect roots for the two corresponding trees. That is, merging the set containing *x* (the tree root is *fx*) and the set containing *y* (the tree root is *fy*) is to let the set pointer for *fx* point to *fy* (Figure 8.9).

The merging algorithm is as follows:

Calculate root *fx* in the tree for the union–find set containing *x*, and calculate root *fy* in the tree for the union–find set containing *y*. If *fx*==*fy*, then *x* and *y* are in the same union–find set; else, the set containing *x* is merged into the set containing *y*, that is, the set pointer for *fx* points to *fy*:

```
void join(int p, int q)     // Merging the set containing p into the set
containing q
{
    p=set_find(p);
    q=set_find(q);
    if (p!=q)
        set[p]=q;
}
```

Search with "path compression" can reduce the length of a tree and improve the time complexity. In algorithm complexity, a union–find set represented as a tree is better than one represented as a linear list.

8.2.1 Find Them, Catch Them

The police department in Tadu City decides to end the chaos and launch actions to root up the two gangs in the city, Gang Dragon and Gang Snake. However, the police first need to identify which gang a criminal belongs to. The present question is, given two criminals, do they belong to the same clan? You must give your judgment based on incomplete information (since the gangsters are always acting secretly).

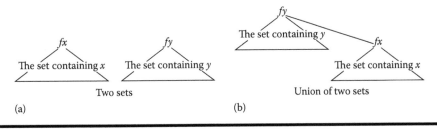

(a)   (b)

Figure 8.9 Merging two sets.

Assume N ($N \leq 10^5$) criminals are currently in Tadu City, numbered from 1 to N. And of course, at least one of them belongs to Gang Dragon, and the same for Gang Snake. You will be given M ($M \leq 10^5$) messages in sequence, which are of the following two kinds:

1. D [a] [b], where [a] and [b] are the numbers of two criminals, and they belong to different gangs.
2. A [a] [b], where [a] and [b] are the numbers of two criminals. This requires you to decide whether a and b belong to the same gang.

Input

The first line of the input contains a single integer T ($1 \leq T \leq 20$), the number of test cases. Then T cases follow. Each test case begins with a line with two integers, N and M, followed by M lines, each containing one message as described above.

Output

For each message "A [a] [b]" in each case, your program should give the judgment based on the information obtained before. The answers might be one of "In the same gang," "In different gangs," and "Not sure yet."

Sample Input	Sample Output
1	Not sure yet.
5 5	In different gangs.
A 1 2	In the same gang.
D 1 2	
A 1 2	
D 2 4	
A 1 4	

Source: POJ Monthly, July 18, 2004.

ID for online judge: POJ 1703.

Analysis

Criminals in Gang Dragon and criminals in Gang Snake are two different sets, respectively. Suppose $set[d]$ is the representative of the set containing d, and $set[d + n]$ is the representative of the "opposite" set, for the set containing d, $1 \leq d \leq n$. Function $set\_find(i)$ is used to find the representative of the set containing i, and compress the path, $1 < i < 2n$.

Initially, $set[i] = -1$, $1 < i < 2n$. That is, each criminal constitutes a gang. Then messages are processed as follows:

1. Decide whether a and b belong to the same gang ($s[0] == $ 'A'). If a and b don't belong to the same gang ($set\_find(a)! = set\_find(b)$), and the gang that a belongs to and the "opposite" gang for b are different ($set\_find(a)! = set\_find(b + n)$), then we can't decide whether a and b belong to the same gang; else if the representative of the set containing a is the same as the

representative of the set containing *b* (*set_find(a)* == *set_find(b)*), *a* and *b* belong to the same gang; else, *a* and *b* belong to different gangs.

2. Set up *a* and *b* as belonging to two different gangs (*s*[0] == 'D'). If the gang that *a* belongs to isn't the "opposite" gang for *b* (*set_find(a)*! = *set_find(b + n)*), then set up the gang that *a* belongs to as the "opposite" gang for *b*, and the gang that *b* belongs to is also the "opposite" gang for *a* (*set[set_find(a)]* = *set_find(b + n)*; *set[set_find(b)]* = *set_find(a + n)*).

Program

```cpp
#include <cstdio>
#include <cstring>
const int maxn = 100000 + 5;     // upper limit for the number of criminals
int n, m;
int set[maxn + maxn];    // the gang that k belongs to is set[k], and the
opposite gang is set[k+n]
int set_find(int d)                          // Search with path compress
{
  if (set[d] < 0)                            // if d is the representative
     return d;
  return set[d] = set_find(set[d]);                    //calculate the
representative of the set that d belongs to
}
int main(void)
{
    int loop;
    scanf("%d", &loop);               // number of test cases
    while (loop--) {
    scanf("%d%d", &n, &m);            //numbers of criminals and messages
    memset(set, -1, sizeof(set));     // each criminal constitutes a gang
    for (int i = 0; i < m; i++) {      // each time one message
        int a, b;
        char s[5];
        scanf("%s%d%d", s, &a, &b);    // input the i-th message
        if (s[0] == 'A') {             // Decide whether a and b belong
to a same gang
            if (set_find(a) != set_find(b)
                && set_find(a) != set_find(b + n))       // the case
not sure
                printf("%s/n", "Not sure yet.");
            else if (set_find(a) == set_find(b))     // a and b belong
to a same gang
                printf("%s/n", "In the same gang.");
            else                        // a and b are in two different gangs
                printf("%s/n", "In different gangs.");
        } else {                        // a and b are in two different gangs
            if (set_find(a) != set_find(b + n)) {
                set[set_find(a)] = set_find(b + n);
                set[set_find(b)] = set_find(a + n);
            }
        }
        }
      }
    }
    return 0;
}
```

The above problem only judges whether two elements belong to the same set. Sometimes the number of elements and the permutation of elements in a set are required for calculation.

8.2.2 Cube Stacking

Farmer John and Betsy are playing a game with N ($1 \leq N \leq 30,000$) identical cubes labeled 1 through N. They start with N stacks, each containing a single cube. Farmer John asks Betsy to perform P ($1 \leq P \leq 100,000$) operations. There are two types of operations: moves and counts.

- In a move operation, Farmer John asks Bessie to move the stack containing cube X on top of the stack containing cube Y.
- In a count operation, Farmer John asks Bessie to count the number of cubes on the stack with cube X that are under cube X and report that value.

Write a program that can verify the results of the game.

Input

- Line 1: A single integer, P.
- Lines 2 .. P + 1: Each of these lines describes a legal operation. Line 2 describes the first operation, and so on. Each line begins with an 'M' for a move operation or a 'C' for a count operation. For move operations, the line also contains two integers: X and Y. For count operations, the line also contains a single integer: X.

Note that the value for N does not appear in the input file. No move operation will request to move a stack onto itself.

Output

Print the output from each of the count operations in the same order as the input file.

Sample Input	Sample Output
6	1
M 1 6	0
C 1	2
M 2 4	
M 2 6	
C 3	
C 4	

Source: USACO 2004, U.S. Open.

ID for online judge: POJ 1988.

Analysis

Each stack is a set. Elements in the set are cubes in the stack. Initially, there are N stacks, each containing a single cube. Suppose $set[k]$ is the number of the cube at the bottom of the stack containing cube k and is also the representative of the set; $cnt[k]$ is the number of elements in interval $[k \ldots set[k]]$; and $top[k]$ is the number of the cube at the top of the stack containing cube k.

1. Count operation: Function $set\_find(p)$ is used to calculate the number of cubes in the stack and under the cube p, and calculate the element at the bottom of the stack. Compressing the path is used.

    If there are elements under $set[p]$ ($set[set[p]] \geq 0$), there are elements in the stack before moving elements in interval $[p \ldots set[p]]$ (Figure 8.10).

    The number of elements under p is modified $cnt[p] \mathrel{+}= cnt[set[p]]$, and the element at the bottom of the stack is modified $set[p] = set\_find(set[p])$.

2. Move operation: Function $set\_join(x, y)$ is used to move the stack containing cube x onto the top of the stack containing cube y.

    First, elements at the bottom of stacks containing x and y are found ($x = set\_find(x)$; $y = set\_find(y)$). Then the element at the bottom of the stack containing x is modified ($set[x] = y$). The number of elements of the stack containing y is calculated ($set\_find(top[y])$). Finally, the element at the top of the stack containing y is renewed ($top[y] = top[x]$), and the number of elements in the stack under x is adjusted ($cnt[x] = cnt[top[y]]$).

Program

```
#include <cstdio>
#include <cstring>
const int maxn = 100000 + 5;      //upper limit of elements
int set[maxn], cnt[maxn], top[maxn]; //set[k] is the serial number of the
element at the bottom of the stack containing k; cnt[k] is the number of
elements in [k…set[k]]; top[k] is the serial number of the element at the
top of the stack containing k
int set_find(int p)
{
    if (set[p] < 0) // element p constitutes a set
        return p;
    if (set[set[p]] >= 0) { // If there are elements under set[p]
```

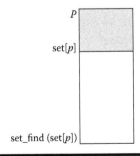

Figure 8.10   Stack.

```
        int fa = set[p];
        set[p] = set_find(fa);
        cnt[p] += cnt[fa];
    }
    return set[p];
}
void set_join(int x, int y)
{   x = set_find(x);
    y = set_find(y);
    set[x] = y;
    set_find(top[y]);
    cnt[x] = cnt[top[y]]+1;
    top[y] = top[x];
}
int main(void)
{
    int p;
    scanf("%d", &p);                  //number of operations
    memset(set, -1, sizeof(set));
    memset(cnt, 0, sizeof(cnt));
    for (int i = 0; i < maxn; i++)
        top[i] = i;                   //Initialization
    while (p--) {
        char s[5];
        scanf("%s", s);               //current operation
        if (s[0] == 'M') {            //move operation
            int x, y;
            scanf("%d%d", &x, &y);    // Input elements x and y
            set_join(x, y);
        } else {
            int x;
            scanf("%d", &x);          // Input element x
            set_find(x);
            printf("%dn", cnt[x]);
        }
    }
    return 0;
}
```

8.3 Calculation of Sum of Weights of Subtrees by Binary Indexed Trees

Sometimes a node in a tree is assigned with a weight, and the problem requires you to dynamically calculate sums of weights of subtrees. The time complexity of postorder traversal is $O(n)$. When weights of nodes are changed, sums of weights of dynamic subtrees are also changed. Obviously, it is not efficient that postorder traversal is used again and again to calculate the sum of weights of dynamic subtrees.

A binary indexed tree (BIT) is used to solve the problem. Suppose a is an array and its length is n; *sum* is the sum of weights of subintervals,

$$sum[i, j] = \sum_{k=i}^{j} a[k];$$

and s is a prefix sum for array a, that is,

$$s[i] = \sum_{k=1}^{i} a[k].$$

Obviously, $sum[i, j] = s[j] - s[i - 1]$.

Function $lowbit(k)$ returns the number of the last 1 in binary representation that k corresponds to. For example, $lowbit(34)$ returns 2. The binary representation of 34 is 00100010. The last 1 in the binary representation of 34 is the 1 at digit 2^1, that is, 2. $lowbit(k) = k \ \& \ (-k)$, or $lowbit(k) = k \ \& \ (k \wedge (k - 1))$.

Array c is used to represent a binary indexed tree, where $c[k]$ is used to store the sum of elements from $a[k - lowbit(k) + 1]$ to $a[k]$, that is,

$$c[k] = \sum_{i=k-lowbit(k)+1}^{k} a[i]$$

It is shown in Figure 8.11.

Modifying and searching elements in array a is implemented in array c. For example, if $a[2]$ is modified, $c[2]$, $c[4]$, and $c[8]$ will also be modified, and the time complexity is $O(logN)$.

For example, $s[7]$ is required for calculation. The binary representation of 7 is 0111. The last 1 in binary representation of 7 corresponds to 1. Therefore, from $a[7]$, only one element ($a[7]$) is obtained, that is, $c[7]$. Then this 1 is deleted and 6 is obtained. The binary representation of 6 is 0110, and the last 1 in the binary representation of 6 corresponds to 2. Therefore, from $a[6]$ two

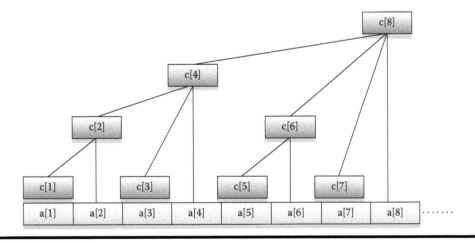

Figure 8.11   A BIT.

elements ($a[6]$ and $a[5]$) are obtained, that is, $c[6]$. Finally, this 1 is deleted and 4 is obtained. The binary representation of 4 is 0100, and the last 1 in the binary representation of 4 corresponds to 4. Therefore, from $a[4]$ four elements ($a[4]$, $a[3]$, $a[2]$, and $a[1]$) are obtained, that is, $c[4]$. Obviously, $s[7] = c[7] + c[6] + c[4]$.

When k is added to $a[x]$, a binary indexed tree c is adjusted as follows:

$$\text{for}(i = x;\ i < cnt;\ i+ = lowbit(i))\ c[i] + = k;$$

The process calculating $s[x] = \Sigma_{k=1}^{x} a[k]$ is as follows:

$$s[x] = 0;$$

$$\text{for}(i = x;\ i > 0;\ i- = lowbit(i))\ s[x] + = c[i];$$

The time complexity for calculating $s[x]$ is $\log_2 n$. It takes $2\log_2 n$ to compute $sum[x, y] = \Sigma_{i=x}^{y} a[i] = s[y] - s[x-1]$.

8.3.1 Apple Tree

There is an apple tree outside of Kaka's house. Every autumn, a lot of apples will grow on the tree. Kaka likes apples very much, so he has been carefully nurturing the big apple tree.

The tree has N forks that are connected by branches. Kaka numbers the forks from 1 to N and the root is always numbered by 1. Apples will grow on the forks, and two apples won't grow on the same fork. Kaka wants to know how many apples are in a subtree for his study of the produce ability of the apple tree (Figure 8.12).

The trouble is that a new apple may grow on an empty fork some time and Kaka may pick an apple from the tree for his dessert. Can you help Kaka?

Input

The first line contains an integer N ($N \le 100,000$), which is the number of the forks in the tree.

The following $N-1$ lines each contain two integers, u and v, which means fork u and fork v are connected by a branch.

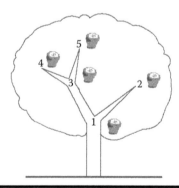

Figure 8.12   Apple tree.

The next line contains an integer M ($M \leq 100{,}000$).

The following M lines each contain a message that is one of the following:

C *x*, which means the existence of the apple on fork x has been changed, that is, if there is an apple on the fork, then Kaka picks it; otherwise, a new apple has grown on the empty fork.

Q *x*, which means an inquiry for the number of apples in the subtree above the fork x, including the apple (if it exists) on the fork x.

Note the tree is full of apples at the beginning.

Output

For every inquiry, output the corresponding answer per line.

Sample Input	Sample Output
3	3
1 2	2
1 3	
3	
Q 1	
C 2	
Q 1	

Source: POJ Monthly.

ID for online judge: POJ 3321.

Hint

The apple tree can be represented as a tree. Forks can be regarded as tree nodes. And the number of apples at a fork can be regarded as the weight of the node. Message **Q** *x* returns the weight of the subtree whose root is x. Message **C** *x* means the weight of node *x* has been changed from 1 to 0 (Kaka picked an apple on fork *x*), or from 0 to 1 (a new apple has grown on empty fork *x*). In order to calculate the weight of a subtree, a binary indexed tree is introduced.

1. Calculate subintervals produced by post-order traversals of all subtrees. The result of post-order traversal for any subtree in the tree is a continuous subinterval. The root of the subtree is the element at the right end of the subinterval. Suppose array *apple* is used to represent the apple tree, and for node *i*, the result of postorder traversal for the subtree whose root is node *i* is a subinterval [*apple[i].l*, *apple[i].r*]. Such a subinterval can be obtained through postorder traversal. The time that the postorder traversal takes is *cnt*.

```
void DFS(int u);
{
  apple[u].l=cnt;    // The left pointer for the subinterval
  for ( each node v connecting with node u)
  DFS(v);
  apple[u].r=cnt++;  // The right pointer for the subinterval
}
Initially cnt=1, and DFS( root) is called.
```

2. Adjust the BIT from $a[x]$. Suppose array a is the result of postorder traversal for a tree. If $a[x]$ is 1, in the path from $a[x]$ to the root, all weights of subtrees increase 1; otherwise, all weights of subtrees decrease 1.

```
void change(int x)
{
    int i;
    if (a[x])
        for (i=x; i<cnt; i+=lowbit(i)) c[i]++;
    else
        for (i=x; i<cnt; i+=lowbit(i)) c[i]--;
}
```

3. Construct the BIT c. Initially, the tree is full of apples. Based on it, BIT c is constructed as follows:

```
for (i=1; i<=n; i++){
    a[i]=1;
    change(i);
}
```

4. Every message is implemented as follows:
   Q x: Output the number of apples in the subtree above the fork x, including the apple (if it exists) on the fork x:

$$sum\big(apple[x].r\big) - sum\big(apple[x].l - 1\big) \ (sum(i) = \sum_{k=1}^{i} a[k]);$$

C x: The existence of the apple on fork x has been changed $(a[apple[x].r] = (a[apple[x].r] + 1)$ %2); from $a[apple[x].r]$ the BIT c will be adjusted $(change(apple[t].r))$.

Program

```
#include<iostream>
#include<cstring>
#define max 100002        // the upper limit of the number of nodes
using namespace std;
struct node1            //a list of edges is edge, where the node
connected by the i-th edge is edge[i]. tail, and the serial number of the
next connected edge is edge[i]. next
{
    int next, tail;
} edge[max];
struct node2 //In apple tree apple, the interval of post-order traversal
for the subtree whose root is node i is [apple[i].l, apple[i].r]
{
    int r, l;
} apple[max];
int s[max], cnt, c[max], a[max]; //The weight of the i-th node in the
post-order traversal is a[i]; the serial number for the post-order
traversal is cnt; BIT is c;
```

```
void DFS (int u)              //Calculate the interval of subtree [apple[ ].l,
apple[ ].r]
{
    int i;
    apple[u].l=cnt;
    for(i=s[u]; i!=-1; i=edge[i].next)
        DFS(edge[i].tail);
    apple[u].r=cnt++;
}
    inline int lowbit (int x)      //Function lowbit(k)
{
    return x&(-x);
}
void change (int x)           // Adjust BIT from a[x]
{
    int i;
    if(a[x])
        for(i=x; i<cnt; i+=lowbit(i))
            c[i]++;
    else
        for(i=x; i<cnt; i+=lowbit(i))
            c[i]--;
}
int sum (int x)               //Calculate ∑_{k=1}^{x} a[k]
{
    int i,res=0;
    for(i=x; i>0; i-=lowbit(i))
        res+=c[i];
    return res;
}
int main( )
{
    int i, n, m, t1, t2, t;
    char str[3];
    scanf("%d", &n);                    // The number of nodes in the tree
    memset(s, -1, sizeof(s[0])*(n+1));            //Initialization
    memset(c, 0, sizeof(c[0])*(n+1));
    memset(apple, 0, sizeof(apple[0])*(n+1));
    for(i=0; i<n-1; i++){
        scanf("%d%d", &t1, &t2);        // The i-th edge (t1, t2)
        edge[i].tail=t2;
        edge[i].next=s[t1];
        s[t1]=i;
    }
    cnt=1;
    DFS(1);              // DFS starts from node 1
    scanf("%d",&m);     //number of messages
    for(i=1;i<=n;i++){   // Construct BIT c full of apples
        a[i]=1;
        change(i);
    }
    while(m--){
        scanf("%s%d", &str, &t);    // Input a message
```

```
        if(str[0]=='Q')                  //Output the number of apples in the
sub-tree above the fork t
            printf("%dn",sum(apple[t].r)-sum(apple[t].l-1));
        else{        // the existence of the apple on fork t has been changed
            a[apple[t].r]=(a[apple[t].r]+1)%2;        // the number of apple
on fork t
            change(apple[t].r);
        }
    }
    return 0;
}
```

8.4 Problems

8.4.1 Friends

There is a town with N citizens. It is known that some pairs of people are friends. According to the famous saying that "the friends of my friends are my friends, too," it follows that if A and B are friends and B and C are friends, then A and C are friends, too.

Your task is to count how many people there are in the largest group of friends.

Input

The first line of the input consists of N and M, where N is the number of the town's citizens ($1 \leq N \leq 30{,}000$) and M is the number of pairs of people ($0 \leq M \leq 500{,}000$) that are known to be friends. Each of the following M lines consists of two integers, A and B ($1 \leq A \leq N$, $1 \leq B \leq N$, $A \neq B$), that describe that A and B are friends. There could be repetitions among the given pairs.

Output

The output should contain one number denoting how many people there are in the largest group of friends.

Sample Input	Sample Output
2	3
3 2	6
1 2	
2 3	
10 12	
1 2	
3 1	
3 4	
5 4	
3 5	

Sample Input	Sample Output
4 6	
5 2	
2 1	
7 10	
1 2	
9 10	
8 9	

Source: Bulgarian National Olympiad in Informatics 2003.

ID for online judge: UVA 10608.

Hint

A group of friends is a set. Calculating friends' relationships is the union operation of sets. Suppose $set[k]$ represents the representative in the group of friends containing k, $s[k]$ is the number of friends in the group of friends containing k, and *max* is the number of friends in the largest group of friends.

Function $set\_find(p)$ is used to find the representative in the group of friends containing p. That is, $set[p]$ is to be calculated.

Function $join(p, q)$ is used to merge two groups of friends containing p and q, respectively. First, representatives in the group of friends containing p and q are found ($p = set\_find(p)$; $q = set\_find(q)$). If the two groups are different ($p \mathrel{!=} q$), the group of friends containing p is merged into the group of friends containing q ($set[p] = q$); accumulate the number of friends in the group containing q ($s[q] \mathrel{+}= s[p]$), and *max* is adjusted ($max = s[q] > max? s[q]: max$).

Next, each citizen is a group of friends; that is, $set[i] = -1$, $s[i] = 1$, $max = 1$. Then, while a pair of friends A and B is input, function $join(A, B)$ is called. Finally, *max* is the solution to the problem.

8.4.2 Wireless Network

An earthquake takes place in Southeast Asia. The Asia Cooperated Medical team have set up a wireless network with laptop computers, but an unexpected aftershock occurred, and all computers in the network were broken. The computers were repaired one by one, and the network gradually began to work again. Because of the hardware restrictions, each computer can only directly communicate with the computers that are not farther than d meters from it. But every computer can be regarded as the intermediary of the communication between two other computers; that is, computer A and computer B can communicate if computer A and computer B can communicate directly or there is a computer C that can communicate with both A and B.

In the process of repairing the network, workers can take two kinds of operations at every moment, repairing a computer or testing if two computers can communicate. Your job is to answer all the testing operations.

Input

The first line contains two integers N and d ($1 \le N \le 1001$, $0 \le d \le 20{,}000$). Here N is the number of computers, which are numbered from 1 to N, and d is the maximum distance two computers can communicate directly. In the next N lines, each contains two integers, x_i and y_i ($0 \le x_i, y_i \le 10,000$), which are the coordinates of N computers. From the $(N+1)$th line to the end of input, there are operations, which are carried out one by one. Each line contains an operation in one of following two formats:

1. O p $(1 \le p \le N)$, which means repairing computer p
2. S p q $(1 \le p, q \le N)$, which means testing whether computers p and q can communicate

The input will not exceed 300,000 lines.

Output

For each testing operation, print "SUCCESS" if the two computers can communicate or "FAIL" if they cannot.

Sample Input	Sample Output
4 1	FAIL
0 1	SUCCESS
0 2	
0 3	
0 4	
O 1	
O 2	
O 4	
S 1 4	
O 3	
S 1 4	

Source: POJ Monthly, HQM.

ID for online judge: POJ 2236.

Hint

Working computers that can communicate constitute a set. Suppose the representative of the set containing computer p is $set[p]$; the sign whether computer p works is $valid[p]$; function $join(p, q)$ is used to merge the set containing p into the set containing q; and function $set\_find(p)$ is used to find the representative of the set containing p.

1. O p $(1 \le p \le N)$; repairing computer p. Computer p works ($valid[p] = $ true); Find each working computer i that can communicate with computer p ($valid[i]$ && $((x_i - x_p)^2 + (y_i - y_p)^2) \le d^2$, $1 \le i \le n$) and merge the set containing i into the set containing q ($join$ (i, p)).

2. $S\ p\ q$ $(1 \le p, q \le N)$; testing whether computers p and q can communicate. If computer p and computer q belong to the same set ($set\_find(p)$ == $set\_find(q)$), then computer p and computer q can communicate; else, they can't.

8.4.3 War

A war is being fought between two countries, A and B. As a loyal citizen of C, you decide to help your country's espionage by attending the peace talks taking place these days (incognito, of course). There are n people at the talks (not including you), but you do not know which person belongs to which country. You can see people talking to each other, and through observing their behavior during their occasional one-to-one conversations, you can guess if they are friends or enemies. In fact, what your country would need to know is whether certain pairs of people are from the same country or are enemies. You may receive such questions from C's government even during the peace talks, and you have to give replies on the basis of your observations so far. Fortunately, nobody talks to you, as nobody pays attention to your humble appearance.

Now, more formally, consider a black box with the following operations:

setFriends(x, y) shows that x and y are from the same country.
setEnemies(x, y) shows that x and y are from different countries.
areFriends(x, y) returns true if you are sure that x and y are friends.
areEnemies(x, y) returns true if you are sure that x and y are enemies.

The first two operations should signal an error if they contradict your former knowledge. The two relations "friends" (denoted by ~) and "enemies" (denoted by *) have the following properties:

~ is an equivalence relation, that is,
1. If $x \sim y$ and $y \sim z$, then $x \sim z$ (the friends of my friends are my friends as well).
2. If $x \sim y$, then $y \sim x$ (friendship is mutual).
3. $x \sim x$ (everyone is a friend of himself).
* is symmetric and irreflexive,
4. If $x*y$, then $y*x$ (hatred is mutual).
5. Not $x*x$ (nobody is an enemy of himself).
Also,
6. If $x*y$ and $y*z$, then $x \sim z$ (a common enemy makes two people friends).
7. If $x \sim y$ and $y*z$, then $x*z$ (an enemy of a friend is an enemy).
Operations setFriends(x, y) and setEnemies(x, y) must preserve these properties.

Input

The first line contains a single integer, n, the number of people.

Each of the following lines contains a triple of integers, $c\ x\ y$, where c is the code of the operation:

$c = 1$, setFriends
$c = 2$, setEnemies
$c = 3$, areFriends
$c = 4$, areEnemies

x and y are its parameters, which are integers in the range $[0, n)$, identifying two (different) people. The last line contains 0 0 0.

All integers in the input file are separated by at least one space or line break.

For every 'areFriends' and 'areEnemies' operation, write 0 (meaning no) or 1 (meaning yes) to the output. Also, for every 'setFriends' or 'setEnemies' operation that contradicts previous knowledge, output a −1 to the output; note that such an operation should produce no other effect and execution should continue. A successful 'setFriends' or 'setEnemies' gives no output.

All integers in the output file must be separated by at least one space or line break.

Constraints

If $n < 10,000$, the number of operations is unconstrained.

Sample Input	Sample Output
10	1
1 0 1	0
1 1 2	1
2 0 5	0
3 0 2	0
3 8 9	−1
4 1 5	0
4 1 2	
4 8 9	
1 8 9	
1 5 2	
3 5 2	
0 0 0	

Source: Programming Contest for Newbies 2005.

ID for online judge: UVA 10158.

Hint

Suppose *set*[*k*] is the representative of the set containing *k*, and *set*[*k* + *n*] is the representative of the set containing *k*'s enemy. Initially, *set*[*i*] is −1 ($1 \leq i \leq 2n$); that is, initially you can't determine any pair of people that are friends or enemies.

Function *set_find*(*p*) is used to find the representative of the set containing *p*, that is, to calculate *set*[*p*].

Function *areFriends*(*x*, *y*) is used to calculate the relationship between *x* and *y*. If the relationship can't be determined ((*set_find*(*x*) != *set_find*(*y*)) && (*set_find*(*x*) != *set_find*(*y* + *n*))), then return −1; if *x* and *y* are friends (*set_find*(*x*) = *set_find*(*y*)), then return 1; and if *x* and *y* are enemies (*set_find*(*x*) != *set_find*(*y*)), then return 0.

1. Operation 1 x y. If we know x and y are from the same country: if x and y are enemies $(areFriends(x, y) == 0)$, then there exists a conflict; if $areFriends(x, y) == -1$, the set containing x is merged into the set containing y, and the set containing x's enemies is merged into the set containing y's enemies $(set[set\_find(x)] = set\_find(y); set[set\_find(x + n)] = set\_find(y + n))$.

2. Operation 2 x y. If we know x and y are from different countries: if x and y are friends $(areFriends(x, y) == 1)$, then there exists a conflict; if $areFriends(x, y) == -1$, then the set containing x's enemies is merged into the set containing y, and the set containing x is merged into the set containing y's enemies $(set[set\_find(x + n)] = set\_find(y); set[set\_find(x)] = set\_find(y + n))$.

3. Operation 3 x y. Function $areFriends(x, y)$ is used to determine whether x and y are friends.

4. Operation 4 x y. The opposite value for function $areFriends(x, y)$ is used to determine whether x and y are enemies.

8.4.4 Ubiquitous Religions

There are so many different religions in the world today that it is difficult to keep track of them all. You are interested in finding out how many different religions students in your university believe in.

You know that there are n students in your university ($0 < n \leq 50{,}000$). It is infeasible for you to ask every student their religious beliefs. Furthermore, many students are not comfortable expressing their beliefs. One way to avoid these problems is to ask m ($0 \leq m \leq [n(n - 1)/2]$) pairs of students and ask them whether they believe in the same religion (e.g., they may know if they both attend the same church). From this data, you may not know what each person believes in, but you can get an idea of the upper bound of how many different religions can possibly be represented on campus. You may assume that each student subscribes to at most one religion.

Input

The input consists of a number of cases. Each case starts with a line specifying the integers n and m. The next m lines each consist of two integers, i and j, specifying that students i and j believe in the same religion. The students are numbered 1 to n. The end of input is specified by a line in which $n = m = 0$.

Output

For each test case, print on a single line the case number (starting with 1), followed by the maximum number of different religions that the students in the university believe in.

Sample Input	Sample Output
10 9	Case 1: 1
1 2	Case 2: 7
1 3	
1 4	
1 5	
1 6	

(Continued)

Sample Input	Sample Output
1 7	
1 8	
1 9	
1 10	
10 4	
2 3	
4 5	
4 8	
5 8	
0 0	

Source: Alberta Collegiate Programming Contest, October 18, 2003.

IDs for online judges: POJ 2524, UVA 10583.

Hint

Students who believe in the same religion constitute a set. The problem requires calculating the maximum number of different religions that the students in the university believe in. A union–find set is used to solve the problem. Initially, each student constitutes a set. When a pair of students x and y who believe in the same religion is input, the set containing x is merged into the set containing y; that is, the representative of the set containing y is the representative of the set containing x.

Finally, the maximum number of different religions that the students in the university believe in can be obtained.

8.4.5 Network Connections

Bob, who is a network administrator, supervises a network of computers. He is keeping a log of connections between the computers in the network. Each connection is bidirectional. Two computers are interconnected if they are directly connected or if they are interconnected with the same computer. Occasionally, Bob has to decide, quickly, whether two given computers are connected, directly or indirectly, according to the log information.

Write a program that, based on information input from a text file, counts the number of successful and unsuccessful answers to questions such as "Is $computer_i$ interconnected with $computer_j$?"

Input and Output

The program reads data from a text file, as follows:

1. The number of computers in the network (a strictly positive integer)
2. A list of pairs of the following forms
   a. c $computer_i$ $computer_j$, where $computer_i$ and $computer_j$ are integers from 1 to $no\_of\_computers$. A pair of this form shows that $computer_i$ and $computer_j$ get interconnected.

b. q *computer_i* *computer_j*, where *computer_i* and *computer_j* are integers from 1 to *no_of_computers*. A pair of this form stands for the question "Is *computer_i* interconnected with *computer_j*?"

Each pair is on a separate line. Pairs can appear in any order, regardless of their type. The log is updated after each pair of type a, and each pair of type b is processed according to the current network configuration.

For example, the input file illustrated in the sample below corresponds to a network of 10 computers and 7 pairs. There are N_1 successfully answered questions and N_2 unsuccessfully answered questions. The program prints these two numbers to the standard output on the same line, in the order of successful answers and unsuccessful answers, as shown in the sample output.

Sample Input	Sample Output
1	1, 2
10	
c 1 5	
c 2 7	
q 7 1	
c 3 9	
q 9 6	
c 2 5	
q 7 5	

Source: ACM Southeastern European Regionals 1997.

ID for online judge: UVA 793.

Hint

This is a problem for union–find sets. Interconnected computers constitute a set. Initially, each computer constitutes a set.

Command c *computer_i* *computer_j* is to merge the set containing *computer_i* into the set containing *computer_j*; that is, the representative of the set containing *computer_i* is the representative of the set containing *computer_j*, and computers in the two sets are interconnected.

Command q *computer_i* *computer_j* is to ask whether *computer_i* and *computer_j* belong to the same set, that is, whether the representative of the set containing *computer_i* is the representative of the set containing *computer_j*; then the numbers of successfully answered questions or unsuccessfully answered questions are accumulated.

After all commands are finished, the numbers of successfully answered questions and unsuccessfully answered questions are output.

8.4.6 Building Bridges

The City Council of New Altonville plans to build a system of bridges connecting all of its downtown buildings together so people can walk from one building to another without going outside. You must write a program to help determine an optimal bridge configuration.

New Altonville is laid out as a grid of squares. Each building occupies a connected set of one or more squares. Two occupied squares whose corners touch are considered to be a single building and do not need a bridge. Bridges may be built only on the grid lines that form the edges of the squares. Each bridge must be built in a straight line and must connect exactly two buildings.

For a given set of buildings, you must find the minimum number of bridges needed to connect all the buildings. If this is impossible, find a solution that minimizes the number of disconnected groups of buildings. Among possible solutions with the same number of bridges, choose the one that minimizes the sum of the lengths of the bridges, measured in multiples of the grid size. Two bridges may cross, but in this case they are considered to be on separate levels and do not provide a connection from one bridge to the other.

Figure 8.13 illustrates four possible city configurations. City 1 consists of five buildings that can be connected by four bridges with a total length of 4. In City 2, no bridges are possible, since no buildings share a common grid line. In City 3, no bridges are needed because there is only one building. In City 4, the best solution uses a single bridge of length 1 to connect two buildings, leaving two disconnected groups (one containing two buildings and one containing a single building).

Input

The input data set describes several rectangular cities. Each city description begins with a line containing two integers, r and c, representing the size of the city on the north–south and east–west axes measured in grid lengths ($1 \le r \le 50$ and $1 \le c \le 50$). These numbers are followed by exactly r lines, each consisting of c hash (#) and dot (.) characters. Each character corresponds to one square of the grid. A hash character corresponds to a square that is occupied by a building, and a dot character corresponds to a square that is not occupied by a building.

The input data for the last city will be followed by a line containing two zeros.

Output

For each city description, print two or three lines of output as shown below. The first line consists of the city number. If the city has fewer than two buildings, the second line is the sentence "No bridges are needed." If the city has two or more buildings but none of them can be connected by bridges, the second line is the sentence "No bridges are possible." Otherwise, the second line is "*N* bridges of total length *L*," where *N* is the number of bridges and *L* is the sum of the lengths of the bridges of the best solution. (If *N* is 1, use the word *bridge* rather than *bridges*.) If the solution

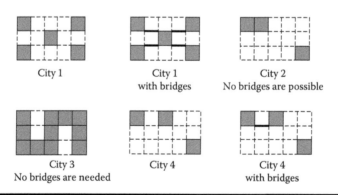

City 1   City 1   City 2
                with bridges   No bridges are possible

City 3   City 4   City 4
No bridges are needed   with bridges

Figure 8.13   City configurations.

leaves two or more disconnected groups of buildings, print a third line containing the number of disconnected groups.

Print a blank line between cases. Use the output format shown in the example.

Sample Input	Sample Output
3 5	City 1
#...#	
..#..	4 bridges of total length 4
#...#	City 2
3 5	No bridges are possible.
##...	2 disconnected groups
.....	
....#	City 3
3 5	No bridges are needed.
#.###	
#.#.#	City 4
###.#	1 bridge of total length 1
3 5	2 disconnected groups
#.#..	
.....	
....#	
0 0	

Source: ACM World Finals 2002–2003, Beverly Hills (United States).

ID for online judge: UVA 2721.

Hint

Union–find sets are used to solve the problem. A building is a set, and a square is the representative of the set.

First, all squares are as nodes and are numbered from left to right and from top to down, as follows:

1	2	...	m
$m + 1$	$m + 2$		$m*2$
	......		
$m*(n - 1) + 1$	$m*(n - 1) + 2$	...	$m*n$

Union–find sets are used to calculate the number of buildings. Squares occupied by a building constitute a set. Initially, $m*n$ squares are as $m*n$ sets.

1. Based on four directions (top, down, left, and right), connected squares covered by buildings are calculated. A square (i, j) occupied by a building is searched from top to bottom and from left to right, and the representative *Tmp* of the set containing the square is calculated. Search its adjacent squares occupied by the building.
2. The number of buildings *part* is calculated, and the list of distances between contracted nodes *Elist* is constructed. Because two occupied squares whose corners touch are considered to be a single building, the process is as follows.

    Rows under row i are analyzed. For row k ($i + 1 \le k \le n$), if square $(k, j + det)$ ($-1 \le det \le 1$) is occupied by a building, and square $(k, j + det)$ and square (i, j) aren't in the same building, then the shortest distance for the two building is $|k - i| - 1$. The two buildings are contracted into two nodes: the representative of the set containing square $(k, j + det)$ and *Tmp* are as contracted nodes, and the shortest distance $|k - i| - 1$ for the two nodes is put into list *Elist*.

    In the same way, columns after column j are also analyzed. For column k ($j + 1 \le k \le m$), if square $(i + det, k)$ ($-1 \le det \le 1$) is occupied by a building, and square $(i + det, k)$ and square (i, j) aren't in the same building, then the shortest distance for the two buildings is $|k - j| - 1$. The two buildings are contracted into two nodes: the representative of the set containing square $(i + det, k)$ and *Tmp* are as contracted nodes, and the shortest distance $|k - i| - 1$ for the two nodes is put into list *Elist*.

    Then the shortest distances for contracted nodes are sorted in ascending order in list *Elist*.
3. The result is analyzed based on *Elist*. Every element in list *Elist* is analyzed. If two contracted nodes don't belong to the same set, then the two sets are combined, the number of bridges *bridge*++, and the shortest distance between is added to the sum of the lengths of the bridges *len*.

    After all elements in list *Elist* are analyzed, if the number of bridges is 0, then output "No bridges are possible"; else, output *bridge* and *len*. If the number of buildings *part* ≠ bridge + 1, then output the number of unconnected buildings *part–bridge*.

8.4.7 Family Tree

A professor of anthropology was interested in people living on isolated islands and their history. He collected their family trees to conduct an anthropological experiment. For the experiment, he needed to process the family trees with a computer. For that purpose, he translated them into text files. The following is an example of a text file representing a family tree.

```
John
  Robert
    Frank
    Andrew
  Nancy
    David
```

Each line contains the given name of a person. The name in the first line is the oldest ancestor in this family tree. The family tree contains only the descendants of the oldest ancestor. Their husbands and wives are not shown in the family tree. The children of a person are indented with one more space than the parent. For example, Robert and Nancy are the children of John, and Frank and Andrew are the children of Robert. David is indented with one more space than Robert, but he is not a child of Robert, but of Nancy. To represent a family tree in this way, the professor excluded some people from the family trees so that no one had both parents in a family tree.

For the experiment, the professor also collected documents of the families and extracted the set of statements about relations of two persons in each family tree. The following are some examples of statements about the family above.

John is the parent of Robert.
Robert is a sibling of Nancy.
David is a descendant of Robert.

For the experiment, he needs to check whether each statement is true or not. For example, the first two statements above are true and the last statement is false. Since this task is tedious, he would like to check it by a computer program.

Input

The input contains several data sets. Each data set consists of a family tree and a set of statements. The first line of each data set contains two integers, n ($0 < n < 1000$) and m ($0 < m < 1000$), which represent the number of names in the family tree and the number of statements, respectively. Each line of the input has less than 70 characters.

As a name, we consider any character string consisting of only alphabetic characters. The names in a family tree have less than 20 characters. The name in the first line of the family tree has no leading spaces. The other names in the family tree are indented with at least one space; that is, they are descendants of the person in the first line. You can assume that if a name in the family tree is indented with k spaces, the name in the next line is indented with at most $k + 1$ spaces.

This guarantees that each person except the oldest ancestor has his or her parent in the family tree. No name appears twice in the same family tree. Each line of the family tree contains no redundant spaces at the end.

Each statement occupies one line and is written in one of the following formats, where X and Y are different names in the family tree.

X is a child of Y.
X is the parent of Y.
X is a sibling of Y.
X is a descendant of Y.
X is an ancestor of Y.

Names not appearing in the family tree are never used in the statements. Consecutive words in a statement are separated by a single space. Each statement contains no redundant spaces at the beginning and at the end of the line.

The end of the input is indicated by two zeros.

Output

For each statement in a data set, your program should output one line containing "True" or "False."

The first letter of "True" or "False" in the output must be a capital. The output for each data set should be followed by an empty line.

Sample Input	Sample Output
6 5	True
John	True
Robert	True
Frank	False
Andrew	False
Nancy	
David	True
Robert is a child of John.	
Robert is an ancestor of Andrew.	
Robert is a sibling of Nancy.	
Nancy is the parent of Frank.	
John is a descendant of Andrew.	
2 1	
abc	
xyz	
xyz is a child of abc.	
0 0	

Source: Asia 2000, Tsukuba (Japan).

IDs for online judges: ZOJ 1674, UVA 2146.

Hint
Obviously, a family tree can be represented as a rooted tree. The oldest ancestor in this family tree is the root. Representation of parents is used to represent the family tree. Suppose the parent of x is $parent[x]$:

- If $parent[s1] == s2$, then $s1$ is a child of $s2$.
- If $parent[s2] == s1$, then $s1$ is the parent of $s2$.
- If $((parent[s1] \neq NIL)$ && $(parent[s2] \neq NIL)$ && $(parent[s1] == parent[s2]))$, then $s1$ is a sibling of $s2$.
- Using parent pointers, $s1$ can be reached from $s2$, and then $s1$ is an ancestor of $s2$ and $s2$ is a descendant of $s1$.

List *parent* is constructed when a family tree is input. Then, based on *parent*, each statement is analyzed. In a statement, there are six substrings: $s1\_t1\_t2\_relation\_t2\_s2'.$, where $s1$, $s2$, and *relation*("child", "parent", "sibling", "descendant", or "ancestor") are useful information. Based on $parent[s1]$, $parent[s2]$, and *relation*, the statement can be determined true or false.

8.4.8 Directory Listing

Given a tree of UNIX directories and file and directory sizes, you are supposed to list them as a tree with proper indention and sizes.

Input

The input consists of several test cases. Each case consists of several lines that represent the levels of the directory tree. The first line contains the root file or directory. If it is a directory, then its children will be listed in the second line, inside a pair of parentheses. Similarly, if any of its children is a directory, then the contents of that directory will be listed in the next line, inside a pair of parentheses. The format of a file or directory is

   name size or ***name size**

where **name**, the name of the file or directory, is a string of no more than 10 characters; **size** > 0 is the integer size of the file or directory; * means the **name** is a directory. It is guaranteed that **name** will not contain characters (', '), [', '], and *. There are no more than 10 levels for each case and no more than 10 files or directories on each level.

Output

For each test case, list the tree in the format shown by the sample. Files and directories that are of depth *d* will have their names indented by $8d$ spaces. Do *not* print tabs to indent the output. The size of a directory *D* is the sum of the sizes of all the files and directories in *D*, plus its own size.

Sample Input	Sample Output
*/usr 1	|_*/usr[24]
(*mark 1 *alex 1)	|_*mark[17]
(hw.c 3 *course 1) (hw.c 5)	| |_hw.c[3]
(aa.txt 12)	| |_*course[13]
*/usr 1	| |_aa.txt[12]
()	|_*alex[6]
	|_hw.c[5]
	|_*/usr[1]

Source: Zhejiang University Local Contest 2003.

ID for online judge: ZOJ 1635.

Hint

In a directory tree, a file or directory is represented as a vertex. A directory is as a root or an internal vertex, and a file is a leaf. If a directory contains files and directories, these files and directories are children for the directory.

There are three main parts for the solution:

1. A directory tree is constructed. An array *a* is used to represent a directory tree, where *tot* is its length, and for vertex *i*, *a*[*i*].*name* is to store the name of the file or directory, *a*[*i*].*size* is to store size of the file or directory, *a*[*i*].*up* is a pointer pointing to its parent, and there are two pointers for its children: *a*[*i*].*first* points to its first child, *a*[*i*].*last* points to its last child; and *a*[*i*].*next* points to its next sibling.
   A directory tree is constructed as a test case is input.
2. Calculate the size for each vertex. The size of a directory *D* is the sum of the sizes of all the files and directories in *D*, plus its own size. That is, the size for each vertex is the sum of the sizes of all children, plus its own size. Postorder traversal can be used to calculate the size of each vertex in a tree.

```
makesize(x) {
for (t=a[x].first; t!=0; t=a[t].next) makesize(t); //x is the root
a[a[x].up].size+= a[x].size;                        //accumulation
}
```

   Obviously in the main program *makesize*(1) calculates sizes of all directories.
3. Output a directory tree. Based on the requirement "Files and directories that are of depth *d* will have their names indented by 8*d* spaces," the program segment is as follows:

```
output( x, pre) {
output pre|_a[x].name[a[x].size];
if (a[x].next==0)                      //child is a file or a directory
     pre+="    "; else pre+="|    ";
   for (int t=a[x].first; t!=0; t=a[t].next) output(t, pre);    // all
children for x
}
```

   In the main program, *output*(1,"") outputs a directory tree.

8.4.9 Closest Common Ancestors

Write a program that takes as input a rooted tree and a list of pairs of vertices. For each pair (*u*, *v*) the program determines the closest common ancestor of *u* and *v* in the tree. The closest common ancestor of two nodes *u* and *v* is the node *w* that is an ancestor of both *u* and *v* and has the greatest depth in the tree. A node can be its own ancestor (e.g., in Figure 8.14 the ancestors of node 2 are 2 and 5).

Input

The data set, which is read from the standard input, starts with the tree description, in the form

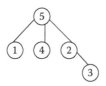

Figure 8.14   A tree.

nr_of_vertices
vertex:(nr_of_successors) successor₁ successor₂ ... successorₙ
...

where *vertices* are represented as integers from 1 to *n* ($n \leq 900$). The tree description is followed by a list of pairs of vertices, in the form

nr_of_pairs
(u v) (x y) ...

The input file contents include several data sets (at least one).
Note that white spaces (tabs, spaces and line breaks) can be used freely in the input.

Output

For each common ancestor the program prints the ancestor and the number of the pair for which it is an ancestor. The results are printed on the standard output on separate lines, into the ascending order of the vertices, in the format

ancestor:times

For example, for the tree in Figure 8.14, see the following table:

Sample Input	Sample Output
5	2:1
5:(3) 1 4 2	5:5
1:(0)	
4:(0)	
2:(1) 3	
3:(0)	
6	
(1 5) (1 4) (4 2) (2 3)(1 3) (4 3)	

Source: ACM Southeastern Europe 2000.

IDs for online judges: POJ 1470, ZOJ 1141, UVA 2045.

Hint
The analysis and solution to the problem refers to the nearest common ancestor problem in Section 8.1.1. The two problems are similar.

8.4.10 Who's the Boss?

Several surveys indicate that the taller you are, the higher you can climb the corporate ladder. At TALL Enterprises, Inc. this de facto standard has been properly formalized: your boss is always at least as tall as you are. Furthermore, you can safely assume that your boss earns a bit more than you do. In fact, you can be absolutely sure that your immediate boss is the person who earns the least among all the employees that earn more than you and are at least as tall as you are. Furthermore, if you are the immediate boss of someone, that person is your subordinate, and all his subordinates are your subordinates as well. If you are nobody's boss, then you have no subordinates. As simple as these rules are, many people working for TALL are unsure of to whom they should be turning in their weekly progress report and how many subordinates they have. Write a program that will help in determining for any employee who the immediate boss of that employee is and how many subordinates they have. Quality assurance at TALL has devised a series of tests to ensure that your program is correct. These tests are described below.

Input

On the first line of the input is a single positive integer n, telling the number of test scenarios to follow. Each scenario begins with a line containing two positive integers, m and q, where m (at most 30,000) is the number of employees and q (at most 200) is the number of queries. The following m lines each list an employee by three integers on the same line: employee ID number (six decimal digits, the first one of which is not zero), yearly salary in euros, and finally height in micrometers (1 micrometer = 10^{-6} meters—accuracy is important at TALL). The chairperson is the employee that earns more than anyone else and is also the tallest person in the company. Then there are q lines listing queries. Each query is a single legal employee ID.

The salary is a positive integer which is at most 10 million. No two employees have the same ID, and no two employees have the same salary. The height of an employee is at least 1 million micrometers and at most 2.5 million micrometers.

Output

For each employee ID x in a query, output a single line with two integers y k, separated by one space character, where y is the ID of x's boss and k is the number of subordinates of x. If the query is the ID of the chairperson, then you should output 0 as the ID of his or her boss (since the chairperson has no immediate boss except, possibly, God).

Sample Input	Sample Output
2	123457 0
3 3	0 2
123456 14323 1700000	123458 1
123458 41412 1900000	200001 2
123457 15221 1800000	200004 0
123456	200004 0
123458	0 3
123457	

Sample Input	Sample Output
4 4	
200002 12234 1832001	
200003 15002 1745201	
200004 18745 1883410	
200001 24834 1921313	
200004	
200002	
200003	
200001	

Source: ACM Northwestern Europe 2003.

IDs for online judges: POJ 1634, ZOJ 1989, UVA 2934.

Hint
The hierarchy for TALL can be represented as a tree. Employees are represented as vertices, and the CEO is the root at level 0. The level number for an employee is based on his yearly salary and height. An employee's immediate boss is his parent, and the number of his subordinates is the number of vertices in the subtree whose root is the employee.

Based on each employee's yearly salary and height, his immediate boss and subordinates can be calculated.

1. Employees are sorted in descending order of their yearly salaries. A linear list *emp* is used to store employees sorted in descending order of their yearly salaries. For the ith employee, $emp[i].ID$ is his ID, $emp[i].sal$ is his salary, $emp[i].hte$ is his height, $emp[i].boss$ is his immediate boss's number, and $emp[i].nsub$ is the number of his subordinates. Initially, $emp[i].boss = -1$, $emp[i].nsub = 0$ ($0 <= i <= m - 1$).

2. Search each employee's immediate boss. Based on the rule "your immediate boss is the person who earns the least among all the employees that earn more than you and are at least as tall as you are," the algorithm searching each employee's immediate boss is as follows:

```
for (int i = m-2; i >= 0; --i) {
int b=i+1;
while(emp[i].hte>emp[b].hte) b=emp[b].boss;
emp[i].boss = b;
}
```

3. Calculate the number of subordinates for each employee:

```
for (int i=0; i<=m-1; ++i) emp[emp[i].boss].nsub+=(1+emp[i].nsub);
```

For each query ID x, find his sequence number ix in *emp* ($emp[ix].ID = x$).
If he has no boss ($emp[ix].boss == -1$), then output '0'; else, output $emp[emp[ix].boss].ID$.
Output the number of his subordinates ($emp[ix].nsub$).

8.4.11 Disk Tree

Hacker Bill has accidentally lost all the information from his workstation's hard drive and he has no backup copies of its contents. He does not regret the loss of the files themselves, but the very nice and convenient directory structure that he had created and cherished during years of work. Fortunately, Bill has several copies of directory listings from his hard drive. Using those listings, he was able to recover full paths (like WINN\TSYSTEM32\CERTSRV\CERTCO~1\X86) for some directories. He put all of them in a file by writing each path he has found on a separate line. Your task is to write a program that will help Bill restore his state-of-the-art directory structure by providing a nicely formatted directory tree.

Input

The first line of the input file contains single integer number N ($1 \leq N \leq 500$) that denotes a total number of distinct directory paths. Then N lines with directory paths follow. Each directory path occupies a single line and does not contain any spaces, including leading or trailing ones. No path exceeds 80 characters. Each path is listed once and consists of a number of directory names separated by a back slash "(\)".

Each directory name consists of one to eight uppercase letters, numbers, or special characters from the following list: exclamation mark, number sign, dollar sign, percent sign, ampersand, apostrophe, opening and closing parentheses, hyphen sign, commercial at, circumflex accent, underscore, grave accent, opening and closing curly brackets, and tilde ("!#$%&'()-@^_`{}~").

Output

Write to the output file the formatted directory tree. Each directory name shall be listed on its own line preceded by a number of spaces that indicate its depth in the directory hierarchy. The subdirectories shall be listed in lexicographic order immediately after their parent directories, preceded by one more space than their parent directory. Top-level directories shall have no spaces printed before their names and shall be listed in lexicographic order. See the sample below for clarification of the output format.

Sample Input	Sample Output
7	GAMES
WINN\TSYSTEM32\CONFIG	DRIVERS
GAMES	HOME
WINNT\DRIVERS	WIN
HOME	SOFT
WIN\SOFT	WINNT
GAMES\DRIVERS	DRIVERS
WINNT\SYSTEM32\CERTSRV\CERTCO~1\X86	SYSTEM32
	CERTSRV
	CERTCO~1

Sample Input	Sample Output
	X86
	CONFIG

Source: ACM Northeastern Europe 2000.

IDs for online judges: POJ 1760, ZOJ 2057, UVA 2223.

Hint

First, a directory tree whose root is " " is constructed. Then N directory paths are input. When a directory path is input, the directory tree is extended. Finally, preorder traversal is used to output the directory tree.

8.4.12 Marbles on a Tree

n boxes are placed on the vertices of a rooted tree, which are numbered from 1 to n, $1 \le n \le 10,000$. Each box is either empty or contains a number of marbles; the total number of marbles is n.

The task is to move the marbles such that each box contains exactly one marble. This is to be accomplished be a sequence of moves; each move consists of moving one marble to a box at an adjacent vertex. What is the minimum number of moves required to achieve the goal?

Input

The input contains a number of cases. Each case starts with the number n, followed by n lines. Each line contains at least three numbers: v, the number of a vertex, followed by the number of marbles originally placed at vertex v, followed by a number d that is the number of children of v, followed by d numbers giving the identities of the children of v.

The input is terminated by a case where $n = 0$, and this case should not be processed.

Output

For each case in the input, output the smallest number of moves of marbles resulting in one marble at each vertex of the tree.

Sample Input	Sample Output
9	7
1 2 3 2 3 4	14
2 1 0	20
3 0 2 5 6	
4 1 3 7 8 9	
5 3 0	
6 0 0	
7 0 0	

(Continued)

Sample Input	Sample Output
8 2 0	
9 0 0	
9	
1 0 3 2 3 4	
2 0 0	
3 0 2 5 6	
4 9 3 7 8 9	
5 0 0	
6 0 0	
7 0 0	
8 0 0	
9 0 0	
9	
1 0 3 2 3 4	
2 9 0	
3 0 2 5 6	
4 0 3 7 8 9	
5 0 0	
6 0 0	
7 0 0	
8 0 0	
9 0 0	
0	

Source: Waterloo Local Contest, June 12, 2004.

IDs for online judges: POJ 1909, ZOJ 2374, UVA 10672.

Hint
Suppose the number of descendants for vertex c is $child[c]$, and the number of marbles for the subtree whose root is c is $tot[c]$. Then the subtree contains $child[c] + 1$ marbles finally, and the smallest number of moving marble is $|tot[c] - (child[c] + 1)|$.

Obviously, through postorder traversal used for a subtree whose root is c, $child[c]$ and $tot[c]$ can be calculated.

1. Rooted trees are constructed when a test case is input. Suppose *ele* is the sequence for vertices' numbers; for vertex v, $start[v]$ is the pointer pointing to its first child in *ele*, $child[v]$ is the number of its children, $flag[v]$ means whether vertex v has parents or not. The reason why

flag[ ] is set is that *n* vertices can constitute a forest. If *flag*[*v*] == false, vertex *v* is the root for a rooted tree. The process is as follows:

```
     cnt=0;
for (int i = 0; i<N; i++) {
   Input the number of the i-th vertex v;
   Input the number of marbles tot[v] originally placed at vertex v;
start[v]= cnt;
Input the number of children of v child[v];
for ( k=child[v]; k>0; k--) {
   Input the identify p for the kth child;
   Put it into list ele (ele[cnt++]=p-1);
   flag[ele[cnt-1]]=true;
}
```

2. Postorder traversal is used for each rooted tree to calculate the smallest number of moves of marbles.

```
dfs(c) { // Post-order traversal is from vertex c and calculate the
smallest number of moves of marbles for each vertex
for (int i=child[c]; i > 0; i--) { //each child for c
   dfs(ele[i + start[c]  - 1]);
  child[c] += child[ele[i + start[c] - 1]]; //accumulate the number
of descendants for c
   tot[c]+= tot[ele[i+ start[c]-1]];          //accumulate the number
of marbles
}
   ans += Math.abs(tot[c]-(child[c]+1));       //accumulate the number
of moves
}
```

Postorder traversal is used for the vertex whose *flag* is false. The smallest number of moves of marbles resulting in one marble at each vertex of the tree can be obtained.

8.4.13 *This Sentence Is False*

The court of King Xeon 2.4 is plagued with intrigue and conspiracy. A document recently discovered by the king's secret service is thought to be part of some mischievous scheme. The document contains simply a set of sentences that state the truth or falsehood of each other. Sentences have the form "Sentence *X* is true/false," where *X* identifies one sentence in the set. The king's secret service suspects the sentences in fact refer to another, yet uncovered, document.

While they try to establish the origin and purpose of the document, the king ordered you to find whether the set of sentences it contains is consistent, that is, if there is a valid truth assignment for the sentences. If the set is consistent, the king wants you to determine the maximum number of sentences that can be made true in a valid truth assignment for the document.

Input

The input contains several instances of documents. Each document starts with a line containing a single integer, *N*, which indicates the number of sentences in the document ($1 \leq N \leq 1000$). The following *N* lines each contain a sentence. Sentences are numbered sequentially, in the order they appear in the input (the first is sentence 1, the second is sentence 2, and so on). Each sentence has the form "Sentence X is true" or "Sentence X is false," where $1 \leq X \leq N$.

Output

For each document in the input, your program should output one line. If the document is consistent, your program should print the maximum number of sentences in a valid truth assignment for the document. Otherwise, your program should print "Inconsistent."

Sample Input	Sample Output
1	Inconsistent
Sentence 1 is false.	1
1	3
Sentence 1 is true.	
5	
Sentence 2 is false.	
Sentence 1 is false.	
Sentence 3 is true.	
Sentence 3 is true.	
Sentence 4 is false.	
0	

Source: ACM South America 2002.

IDs for online judges: POJ 1291, ZOJ 1518, UVA 2612.

Hint

The goal of the problem is to determine whether there exists inconsistency in the document, and if the document is consistent, the program should print the maximum number of sentences in a valid truth assignment for the document.

Suppose $i \rightarrow j$ means sentence i is "Sentence j is true," and $i \,!\!> j$ means sentence i is "Sentence j is false."

If $i \rightarrow j$, the ith sentence and the jth sentence are true or false simultaneously, and if $i \,!\!> j$, the ith sentence and the jth sentence are opposite. Therefore, a union–find set can be used to solve the problem.

For sentence i ($1 \leq i \leq n$), an opposite sentence $n + i$ is set. Then the document is inconsistent if and only if sentence j is not only in the set containing sentence $i + n$, but also in the set containing sentence i. Therefore,

If $i \rightarrow j$: If sentence i and sentence $j + n$ are in the same set, or sentence $i + n$ and sentence j are in the same set, then the document is inconsistent; else, the set containing sentence i and the set containing sentence j are combined, and the set containing sentence $i + n$ and the set containing sentence $j + n$ are combined.

If $i \mathrel{!>} j$: If sentence i and sentence j are in the same set, or sentence $i + n$ and sentence $j + n$ are in the same set, then the document is inconsistent; else, the set containing sentence i and the set containing sentence $j + n$ are combined, and the set containing sentence $i + n$ and the set containing sentence j are combined.

After union–find sets are set up, for each set and its opposite set, get its maximal cardinal number and accumulate it, that is, *sum* += max{$v[i]$, $v[opt[i]]$}, where $v[i]$ is the number of vertices of the tree (the cardinal number of the set) containing i, $v[opt[i]]$ is the number of vertices of the opposite tree for i, and the two trees aren't processed before. Finally, *sum* is the maximum number of sentences in a valid truth assignment for the document.

Program

```cpp
#include<iostream>
#include<cstdio>
#include<cstring>
#include<string>
#include<algorithm>
#include<cmath>
using namespace std;
#define MAXN 10000          // The upper limit of the number of vertices
int set[MAXN];              // the root of the subtree containing vertex
i is set[i]
int n, ans;                 //n: the number of sentences, ans: the
maximum number of sentences which can be made true in a valid truth
assignment for the document
void init(int n)            // set up 2*n sets with one vertex
{
    for(int i=1;i<=n*2;i++) set[i]=i;
}
int find(int a)         // find the representative of the set containing a
{
    int root=a;         // find the root root from a
    int temp;
    while(set[root]!=root) root=set[root];
    while(set[a]!=root)     // representative for all vertices on the path
from a to root is root
    {
        temp=a; a=set[a]; set[temp]=root;
    }
    return root;
}
bool unions(int a,int b)            // if a and b belong to a same set,
then return false; else the two sets containing a and b are combined and
return true
{
    int x=find(a);int y=find(b);  // find representatives for sets
containing a and b
    if(x==y) return false;      // if a and b belong to a same set, then
return false
    set[x]=y;                   // the two sets containing a and b are
combined and return true
    return true;
}
```

```
char str[50];
int d;
bool jud(int i, int j)          // Determine whether i and j belong to a
same union-find set or not
{
    return find(i)==find(j);
}
int opt[MAXN], cnt[MAXN];          // cnt[i]: the size of the subtree
  with root i, opt[i]: the root of the subtree for the opposite set,
  vis[i]: visited mark
bool vis[MAXN];
void solve( )                      //Calculate and output the maximum
number of sentences which can be made true in a valid truth assignment
for the document
{
  memset(cnt, 0, sizeof(cnt));
  for(int i=1; i<=n; i++)
  {
    cnt[find(i)]++;
    opt[find(i)]=find(i+n);
    opt[find(i+n)]=find(i);
  }
  int ans=0;
  memset(vis, 0, sizeof(vis));
  for(int i=1; i<=n; i++)
  {
    if(vis[find(i)]||vis[find(i+n)]) continue;
    ans+=max(cnt[find(i)],cnt[find(i+n)]);
    vis[find(i)]=vis[find(i+n)]=1;
  }
  printf("%\dn", ans);             // Output
}
int main( )
{
  while(~scanf("%d", &n)&& n)      // the number of sentences
  {
    bool flag=0;
    init(n);                       // set up 2*n sets with one vertex
    for(int i=1; i<=n; i++)        //Input
    {
      scanf("%s", str);            //Input 'Sentence'
      scanf("%d", &d);
      scanf("%s", str);            //Input 'is'
      scanf("%s", str);
      if (flag) continue;
      if (strcmp(str,"false.")==0) //false
      {
// if i and d belong to a same union-find set or i+n and d+n belong to a
same union-find set,
// else combine union-find sets containing i and d+n, union-find sets
containing i+n and d
              if (jud(i, d) || jud(i+n, d+n)) flag=1;
                else unions(i, d+n),unions(i+n, d);
          }
```

```
            else                    //true
            {
                if(jud(i,d+n) || jud(i+n,d)) flag=1;
                    else unions(i,d),unions(i+n,d+n);
            }
        }
        if (flag) printf("Inconsistent\n");
        else solve( );
    }
    return 0;
}
```

Chapter 9

Applications of Binary Trees

A binary tree is a tree in which no node can have more than two children (referred to as the left child and the right child). Based on the definition of binary tree, there are a lot of important data structures, such as heap, Huffman tree, and binary search tree.

Any ordered tree can be transferred into a binary tree. A binary tree can be transferred into a linear sequence by traversing it. There are four ways to traverse a binary tree: preorder, inorder, postorder, and level order.

This chapter focuses on the following:

1. Converting ordered trees to binary trees
2. Paths of binary trees
3. Traversal of binary trees

9.1 Converting Ordered Trees to Binary Trees

In the real world, some problems can be modeled as ordered trees. In order to save memory and make programming convenient, ordered trees should be converted to corresponding binary trees. *A corresponding binary tree* refers to the preorder traversal and postorder traversal for an ordered tree being the same as the preorder traversal and postorder traversal for a converted binary tree.

The process of converting an ordered tree T to a binary tree T' is as follows:

1. The root of T is as the root of T'.
2. The first child of T' is as the root of the left subtree of T' x.
3. If x has children, its first child is as the root of its left subtree, and if x has siblings, its first sibling is as the root of its right subtree.

An example is shown in Figure 9.1.

A forest can also be transferred into a binary tree. First, all trees are transferred into binary trees. Then all binary trees are transferred into a corresponding binary tree: from the first binary tree, the next tree's root is the right child of the previous tree's root. The process is shown in Figure 9.2.

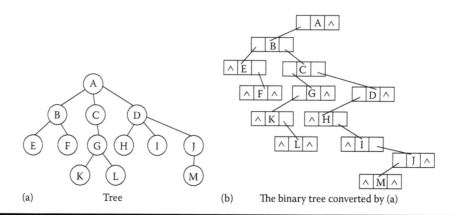

    (a)   Tree   (b)   The binary tree converted by (a)

Figure 9.1   Converting an ordered tree to a binary tree.

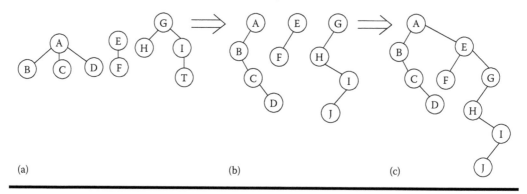

    (a)   (b)   (c)

Figure 9.2   A forest is transferred into a binary tree.

9.1.1 Tree Grafting

Trees have many applications in computer science. Perhaps the most commonly used trees are rooted binary trees, but there are other types of rooted trees that may be useful as well. One example is ordered trees, in which the subtrees for any given node are ordered. The number of children of each node is variable, and there is no limit on the number. Formally, an ordered tree consists of a finite set of nodes T such that

- There is one node designated as the root, denoted root(T).
- The remaining nodes are partitioned into subsets T_1, T_2, ..., T_m, each of which is also a tree (subtrees).

Also, define root(T_1), ..., root(T_m) to be the children of root(T), with root(T_i) being the ith child. The nodes root(T_1), ..., root(T_m) are siblings.

It is often more convenient to represent an ordered tree as a rooted binary tree, so that each node can be stored in the same amount of memory. The conversion is performed by the following steps:

1. Remove all edges from each node to its children.
2. For each node, add an edge to its first child in T (if any) as the left child.
3. For each node, add an edge to its next sibling in T (if any) as the right child.

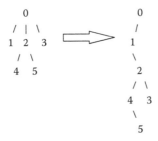

Figure 9.3   Conversion.

This is illustrated by Figure 9.3.

In most cases, the height of the tree (the number of edges in the longest root-to-leaf path) increases after the conversion. This is undesirable because the complexity of many algorithms on trees depends on their height.

You are asked to write a program that computes the height of the tree before and after the conversion.

Input

The input is given by a number of lines giving the directions taken in a depth-first traversal of the trees. There is one line for each tree. For example, the tree above would give dudduduudu, meaning 0 down to 1, 1 up to 0, 0 down to 2, and so forth. The input is terminated by a line whose first character is #. You may assume that each tree has at least 2 and no more than 10,000 nodes.

Output

For each tree, print the heights of the tree before and after the conversion specified above. Use the format

    Tree *t*: *h*1 => *h*2

where *t* is the case number (starting from 1), *h*1 is the height of the tree before the conversion, and *h*2 is the height of the tree after the conversion.

Sample Input	Sample Output
dudduduudu	Tree 1: 2 => 4
ddddduuuuu	Tree 2: 5 => 5
ddddduduuu	Tree 3: 4 => 5
ddddduuduu	Tree 4: 4 => 4
#	

Source: ACM Rocky Mountain 2007.

IDs for online judges: POJ 3437, UVA 3821.

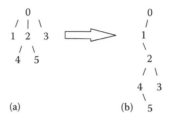

(a)   (b)

Figure 9.4   Transformation.

Analysis

The problem is about the transformation from an ordered tree into its corresponding binary tree. The transformation is illustrated as follows. For the figure shown in the problem, the depth-first traversal for the tree (Figure 9.4a) is "dudduduudu," and the transferred binary tree is shown in Figure 9.4b.

Suppose the root is at level 0, its children are at level 1, and so on; the heights of the tree before and after the conversion are *height*1 and *height*2, respectively, and the current level numbers for the tree before and after the conversion are *level*1 and *level*2, respectively.

In order to calculate the heights of the tree before and after the conversion, the key to the problem is to calculate level numbers for all nodes in the tree before and after the conversion.

The height of the tree before the conversion can be calculated based on the direction taken in a depth-first traversal of the tree. Each 'd' in a direction taken in a depth-first traversal of a tree means the current level number for the tree increases 1, that is, *level*1++. In Figure 9.4a, the first 'd' visits vertex 1 (at level 1) from vertex 0 (at level 0), and the current level number for the tree is 1. The second 'd' visits vertex 2 (at level 1) from vertex 0, and the current level number for the tree is 1. The third 'd' visits vertex 4 (at level 2) from vertex 2, and the current level number for the tree is 2. And so on.

The structure of the tree before the conversion can also be obtained based on the direction taken in a depth-first traversal of the tree. Each 'd' means the number of children increases 1 for its parent. In Figure 9.4a, the first 'd' visits the first child for vertex 0, and the second 'd' visits the second child for vertex 0.

The height of the tree after the conversion can be calculated based on the following formula: for node x and its parent in the tree before the conversion y, $level2(x) = level2(y) +$ the sequence number for x as a child in the tree before the conversion.

For example, in Figure 9.4a, for the tree before the conversion, vertex 0 is the parent for vertex 3, and vertex 3 is the third child for vertex 0. $level2(\text{vertex } 3) = level2(\text{vertex } 0) +$ the sequence number for vertex 3 as a child in the tree before the conversion $= 0 + 3 = 3$. Vertex 2 is the parent for vertex 5, and vertex 5 is the second child for vertex 2. $level2(\text{vertex } 5) = level2(\text{vertex } 2) +$ the sequence number for vertex 5 as a child in the tree before the conversion $= 2 + 2 = 4$. It is shown in Figure 9.4b.

Program

```
#include <iostream>
#include<string>
using namespace std;
string s;                   // direction taken in a depth-first traversal of
a tree
```

```
int i,n=0, height1, height2;   //character pointer i, the number of test
case n, the heights of the tree before and after the conversion are
height1 and height2 respectively
void work(int level1, int level2){
// recursively calculate heights of the tree before and after the
conversion, where the current level numbers for the tree before and after
the conversion are level1 and level2 respectively
      int tempson=0;              // Number of children
   while (s[i]=='d'){             // while the current character is 'd',
character pointer i is moved right, and the number of children+1
         i++;  tempson++;
         work(level1+1, level2+tempson);       //recursion
   }
   // adjust the heights of the tree before and after the conversion
   height1=level1>height1?level1:height1;
   height2=level2>height2?level2:height2;
   i++;                          //character pointer i is moved
right
}
int main ( )
{
   while (cin>>s && s!="#"){      //Input direction taken in a depth-
first traversal of a tree
      i=height1=height2=0;                // Initialization
      work(0, 0);            // calculate heights of the tree before and
after the conversion
      cout<<"Tree "<<++n<<": "<<height1<<" => "<<height2<<endl;
   }
   return 0;
}
```

9.2 Paths of Binary Trees

Paths in a tree are paths from the root to other nodes. In a tree there is no cycle; therefore, there is unique path from the root to any node.

Normally there should be pointers between parents and children. In a complete binary tree with n nodes, node numbers are from 0 to $n-1$, and nodes are numbered top down, and from left to right. For a node i, if it has a parent, then its parent is node $\lfloor (i-1)/2 \rfloor$ ($1 \leq i \leq n-1$); if it has a left child ($2*i+1 \leq n-1$), then its left child is node $2*i+1$; if it has a right child ($2*i+2 \leq n-1$), then its right child is node $2*i+2$. If i is even and $i \neq 0$, the left sibling for node i is node $i-1$, and if i is odd and $i \neq n-1$, the right sibling for node i is node $i+1$. The level where node i is at $\lfloor \log_2(i+1) \rfloor$, that is, the length of the path from node i to the root, is $\lfloor \log_2(i+1) \rfloor$. Based on it, a complete binary tree can be stored in an array.

9.2.1 Binary Tree

Binary trees are a common data structure in computer science. In this problem, we will look at an infinite binary tree where the nodes contain a pair of integers. The tree is constructed like this:

■ The root contains the pair (1, 1).
■ If a node contains (a, b), then its left child contains ($a + b$, b) and its right child (a, $a + b$).

Given the contents (*a*, *b*) of some node of the binary tree described above, suppose you are walking from the root of the tree to the given node along the shortest possible path. Can you find out how often you have to go to a left child and how often to a right child?

Input

The first line contains the number of scenarios. Every scenario consists of a single line containing two integers *i* and *j* ($1 \leq i, j \leq 2*10^9$) that represent a node (*i*, *j*). You can assume that this is a valid node in the binary tree described above.

Output

The output for every scenario begins with a line containing "Scenario #*i*:," where *i* is the number of the scenario starting at 1. Then print a single line containing two numbers *l* and *r* separated by a single space, where *l* is how often you have to go left and *r* is how often you have to go right when traversing the tree from the root to the node given in the input. Print an empty line after every scenario.

Sample Input	Sample Output
3	Scenario #1:
42 1	41 0
3 4	
17 73	Scenario #2:
	2 1
	Scenario #3:
	4 6

Source: TUD Programming Contest 2005 (Training Session), Darmstadt, Germany.

ID for online judge: POJ 2499.

Analysis
Because the root contains the pair (1, 1), and if a node contains (*a*, *b*), then its left child contains (*a* + *b*, *b*) and its right child (*a*, *a* + *b*); numbers in each pair are positive numbers, and for each pair, we can determine if it is a left child or right child by comparing the two numbers. For example, if a node contains (*a* + *b*, *b*), its parent must be (*a*, *b*) obtained by (*a* + *b*) − *b*. Therefore, the path from the root to a node can be gotten, and how often you have to go to a left child and how often to a right child can be found out. The path is unique.

For a pair (*a*, *b*), the greedy algorithm is used to calculate how often you have to go to a left child and how often to a right child.

When *a* > b, then from (*a*, *b*) it takes $\lfloor (a - 1)/b \rfloor$ steps to the left, and in each step the left parameter − *b*; otherwise, it takes $\lfloor (b - 1)/a \rfloor$ steps to the right, and in each step the right parameter − *a*. Finally, it reaches (1, 1).

Program

```
#include <iostream>
using namespace std;
int main () {
  int SC;                                    // the number of scenarios
  cin >> SC;
  for( int S=1; S<=SC; S++ ){
    cout<<"Scenario #"<< S<<":"<<endl;       // the number of the scenario
    int a, b;
    cin >> a >> b;                           //the current pair (a, b)
        int left = 0, right = 0;             // Initialize steps to
left and to rigth
    while( a > 1  ||  b > 1 ){               // return to the root or not
      if( a > b ){
        int up = (a - 1) / b;                // takes(a - 1) / b steps to
left
    left += up;                              //accumulate the number of steps to
left
    a -= up * b;
      } else {
        int up = (b - 1) / a;                //takes (b - 1) / a steps to right
        right += up;                         // accumulate the number of steps to
right
        b -= up * a;                                             }
      }
    cout << left << " " << right << endl << endl;    //Output steps to
left and right
  }
}
```

9.3 Traversal of Binary Trees

There are three ways to traverse a binary tree (Figure 9.5):

> Preorder traversal
>> Visit the tree root.
>> Traverse the left subtree by recursively calling the preorder function.
>> Traverse the right subtree by recursively calling the preorder function.
> Inorder traversal
>> Traverse the left subtree by recursively calling the inorder function.
>> Visit the tree root.
>> Traverse the right subtree by recursively calling the inorder function.

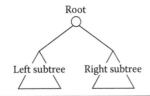

Figure 9.5   Binary tree.

Postorder traversal
    Traverse the left subtree by recursively calling the postorder function.
    Traverse the right subtree by recursively calling the postorder function.
    Visit the tree root.

In preorder traversal, the tree root is visited first. In postorder traversal, the tree root is visited finally. In inorder traversal, the substring before the tree root is the result of inorder traversal for the left subtree, and the substring after the tree root is the result of inorder traversal for the right subtree. Therefore, the results of preorder traversal and inorder traversal, and the results of postorder traversal and inorder traversal, can determine the structure of a binary tree. But the results of preorder traversal and postorder traversal can't determine the structure of a binary tree.

1. **From the results of postorder traversal and inorder traversal of a binary tree, the result of preorder traversal of a binary tree can be obtained.**

    Suppose the result of inorder traversal for a binary tree $s' = s_1' \ldots s_k' \ldots s_n'$, and the result of postorder traversal for a binary tree $s'' = s_1'' \ldots s_n''$. Obviously, from $s_1'' \ldots s_n''$ produced by postorder traversal, s_n'' is the tree root. In $s_1' \ldots s_k' \ldots s_n'$ produced by inorder traversal there is a character s_k' that is equal to s_n''.

    If $k > 1$, then the left subtree exists and the substring $s_1' \ldots s_{k-1}'$ before s_k' is the result of inorder traversal for the left subtree, and $s_1'' \ldots s_{k-1}''$ is the result of postorder traversal for the left subtree.

    If $k < n$, then the right subtree exists and the substring $s_{k+1}' \ldots s_n'$ after s_k' is the result of inorder traversal for the right subtree, and $s_k'' \ldots s_{n-1}''$ is the result of postorder traversal for the right subtree.

    If there exists the left subtree or the right subtree, the above process is called recursively.

2. **From the results of preorder traversal and inorder traversal of a binary tree, the result of postorder traversal of a binary tree can be obtained.**

    Suppose $s' = s_1' \ldots s_k' \ldots s_n'$ is the result of inorder traversal of a binary tree, and $s'' = s_1'' \ldots s_n''$ is the result of preorder traversal of a binary tree. Obviously, s_1'', the first character of the result of preorder traversal, is the tree root of the binary tree. Suppose s_k' equals s_1'' in $s_1' \ldots s_k' \ldots s_n'$.

    If $k > 1$, the left subtree exists; $s_1' \ldots s_{k-1}'$ is the result of inorder traversal of the left subtree, and $s_2'' \ldots s_k''$ is the result of preorder traversal of the left subtree.

    If $k < n$, the right subtree exists; $s_{k+1}' \ldots s_n'$ is the result of inorder traversal of the right subtree, and $s_{k+1}'' \ldots s_n''$ is the result of preorder traversal of the right subtree.

    If there exists the left subtree or the right subtree, the above process is called recursively. Finally, the root s_1'' (or s_k') is output.

9.3.1 Tree Recovery

Little Valentine liked playing with binary trees very much. Her favorite game was constructing random-looking binary trees with capital letters in the nodes. Figure 9.6 is an example of one of her creations.

To record her trees for future generations, she wrote down two strings for each tree: a preorder traversal (root, left subtree, right subtree) and an inorder traversal (left subtree, root, right subtree). For the tree drawn above, the preorder traversal is DBACEGF and the inorder traversal is ABCDEFG.

Figure 9.6   Another binary tree.

She thought that such a pair of strings would give enough information to reconstruct the tree later (but she never tried it).

Now, years later, looking again at the strings, she realized that reconstructing the trees was indeed possible, but only because she never had used the same letter twice in the same tree.

However, doing the reconstruction by hand soon turned out to be tedious. So now she asks you to write a program that does the job for her.

Input

The input will contain one or more test cases.

Each test case consists of one line containing two strings preord and inord, representing the preorder traversal and inorder traversal of a binary tree. Both strings consist of unique capital letters. (Thus, they are not longer than 26 characters.)

Input is terminated by the end of the file.

Output

For each test case, recover Valentine's binary tree and print one line containing the tree's postorder traversal (left subtree, right subtree, root).

Sample Input	Sample Output
DBACEGF ABCDEFG	ACBFGED
BCAD CBAD	CDAB

Source: Ulm Local Contest 1997.

IDs for online judges: POJ 2255, ZOJ 1944, UVA 536.

Analysis

Based on definitions of preorder traversal and inorder traversal, for a tree, the first character in preorder traversal is the root of the tree; in inorder traversal, the string before the character is inorder traversal of its left subtree, and the string after the character is inorder traversal of its right subtree.

A recursive function *recover*(*preord$_l$*, *preord$_r$*, *inord$_l$*, *inord$_r$*) is used to produce the tree's postorder traversal based on preorder traversal and inorder traversal of the tree, where preorder traversal of the tree is *preord*, and *preord$_l$* and *preord$_r$* are pointers for the front and rear, respectively. Inorder traversal of the tree is *inord*, and *inord$_l$* and *inord$_r$* are pointers for the front and rear, respectively.

1. Calculate the root's position in the inorder traversal of the tree (*inord*[*root*] == *preord*[*preord*$_l$]).
2. Calculate the size of the left subtree l_l (*root* − *inord*$_l$) and the size of the right subtree l_r (*inord*$_r$ − *root*).
3. If the left subtree isn't empty (l_l > 0), then *recover*(*preord*$_l$, *preord*$_l$ + l_l, *inord*$_l$, *root* − 1), where *preord*$_l$ and *preord*$_l$ + l_l are the front pointer and rear pointer for preorder traversal of the left subtree, and *inord*$_l$ and *root* − 1 are the front pointer and rear pointer for inorder traversal of the left subtree.
4. If the right subtree isn't empty (l_r > 0), then *recover*(*preord*$_l$ + l_l + 1, *preord*$_r$, *root* + 1, *inord*$_r$), where *preord*$_l$ + l_l + 1 and *preord*$_r$ are the front pointer and rear pointer for preorder traversal of the right subtree, and *root* + 1 and *inord*$_r$ are the front pointer and rear pointer for inorder traversal of the right subtree.
5. Output the root *inord*[*root*]).

Program by the Problemsetter

```
#include <stdio.h>
#include <string.h>
#include <assert.h>
FILE *input;
char preord[30],inord[30];   // Strings storing preorder traversal and
inorder traversal of the tree
int read_case()                  //Input preorder traversal and inorder
traversal of the tree
{
  fscanf(input,"%s %s",preord,inord);
  if (feof(input)) return 0;
  return 1;
}
void recover (int preleft, int preright, int inleft, int inright)   //
produce the tree's postorder traversal based on preorder traversal and
inorder traversal of the tree
{
  int root,leftsize,rightsize;
  assert(preleft<=preright && inleft<=inright);
  for (root=inleft; root<=inright; root++)      //Calculate the root's
position
    if (preord[preleft]==inord[root]) break;
  leftsize = root-inleft;                          //sizes of left subtree
and right subtree
  rightsize = inright-root;
  if (leftsize>0)                                  //left subtree
    recover(preleft+1,preleft+leftsize,inleft,root-1);
  if (rightsize>0)                                 //right subtree
    recover(preleft+leftsize+1,preright,root+1,inright);
  printf("%c",inord[root]);                        //Output the root
}
void solve_case()
{
  int n = strlen(preord);                          // number of nodes
  recover(0,n-1,0,n-1);                            //Output the tree's postorder
traversal
  printf("n");
```

```
}
int main()
{
  input = fopen("tree.in","r");
  assert(input!=NULL);
  while (read_case()) solve_case();
  fclose(input);
  return 0;
}
```

9.4 Problems

9.4.1 Tree Summing

LISP was one of the earliest high-level programming languages and, with FORTRAN, is one of the oldest languages currently being used. Lists, which are the fundamental data structures in LISP, can easily be adapted to represent other important data structures, such as trees.

This problem deals with determining whether binary trees represented as LISP S-expressions possess a certain property.

Given a binary tree of integers, you are to write a program that determines whether there exists a root-to-leaf path whose nodes sum to a specified integer. For example, in the tree shown in Figure 9.7 there are exactly four root-to-leaf paths. The sums of the paths are 27, 22, 26, and 18.

Binary trees are represented in the input file as LISP S-expressions having the following form:

empty tree ::= ()
tree ::= empty tree (integer tree tree)

The tree diagrammed in Figure 9.7 is represented by the expression (5 (4 (11 (7 () ()) (2 () ())) ()) (8 (13 () ()) (4 () (1 () ())))).

Note that with this formulation all leafs of a tree are of the form (integer () ()).

Since an empty tree has no root-to-leaf paths, any query as to whether a path exists whose sum is a specified integer in an empty tree must be answered negatively.

Input

The input consists of a sequence of test cases in the form of integer–tree pairs. Each test case consists of an integer, followed by one or more spaces, followed by a binary tree formatted as an

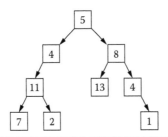

Figure 9.7 Another example of a binary tree.

S-expression, as described above. All binary tree S-expressions will be valid, but expressions may be spread over several lines and may contain spaces. There will be one or more test cases in an input file, and input is terminated by the end of the file.

Output

There should be one line of output for each test case (integer–tree pair) in the input file. For each pair I, T (I represents the integer, T represents the tree), the output is the string "yes" if there is a root-to-leaf path in T whose sum is I and "no" if there is no path in T whose sum is I.

Sample Input	Sample Output
22 (5(4(11(7()())(2()()))()) (8(13()())(4()(1()()))))	yes
20 (5(4(11(7()())(2()()))()) (8(13()())(4()(1()()))))	no
10 (3	yes
(2 (4 () ())	no
(8 () ()))	
(1 (6 () ())	
(4 () ())))	
5 ()	

Source: Duke Internet Programming Contest 1992.

IDs for online judges: POJ 1145, UVA 112.

Hint
For a test case there are two parts: an integer s, representing a sum of a root-to-leaf path, and an S-expression.

A recursive function *ParseTree(s)* is used to determine whether there is a root-to-leaf path in the current tree whose sum is s or not:

1. Omit spaces and get the first character c.
2. If c isn't '(', then report error; else, get the number v after '('.
3. For number v, $L = ParseTree(s - v)$ and $r = ParseTree(s - v)$ are used to determine whether there is a root-to-leaf path whose sum is $s - v$ or not in the left subtree and the right subtree.
4. Return "yes" or "no".

9.4.2 Trees Made to Order

We can number binary trees using the following scheme:

The empty tree is numbered 0.
The single-node tree is numbered 1.
All binary trees having m nodes have numbers less than all those having $m + 1$ nodes.
Any binary tree having m nodes with left and right subtrees L and R is numbered n such that all trees having m nodes numbered higher than n have either left subtrees numbered higher than L or a left subtree = L and a right subtree numbered higher than R.

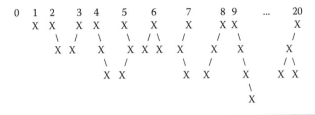

Figure 9.8   A set of binary trees.

The first 10 binary trees and tree number 20 in this sequence are shown in Figure 9.8. Your job for this problem is to output a binary tree when given its order number.

Input

Input consists of multiple problem instances. Each instance consists of a single integer n, where $1 \leq n \leq 500,000,000$. A value of $n = 0$ terminates input. (Note that this means you will never have to output the empty tree.)

Output

For each problem instance, you should output one line containing the tree corresponding to the order number for that instance. To print out the tree, use the following scheme:

A tree with no children should be output as X.
A tree with left and right subtrees L and R should be output as $(L')X(R')$, where L' and R' are the representations of L and R.
If L is empty, just output $X(R')$.
If R is empty, just output $(L')X$.

Sample Input	Sample Output
1	X
20	((X)X(X))X
31117532	(X(X(((X(X))X(X))X(X))))X(((X((X)X((X)X)))X)X)
0	

Source: ACM East Central North America 2001.

IDs for online judges: POJ 1095, ZOJ 1062, UVA 2357.

Hint
Based on the scheme numbering binary trees, the number of a binary tree with i nodes is larger than the number of a binary tree with $i - 1$ nodes. Suppose h_i is the number of binary trees with i nodes. If $i == 1$, there is only one kind of binary tree, $h_1 = 1$. If $i == 2$, one node is the root, and the other node can be the right child or the left child; that is, the node can be numbered (0, 1) or (1, 0). Therefore, $h_2 = h_0{}^*h_1 + h_1{}^*h_0 = 2$. There are two kinds of binary trees with two nodes, numbered 2 and 3. If $i == 3$, one node is the root, and the other two nodes can be numbered (2, 0), (1, 1), or (0, 2). Therefore, $h_3 = h_0{}^*h_2 + h_1{}^*h_1 + h_2{}^*h_0 = 5$. There are five kinds of binary trees, numbered 4, 5, 6,

7, and 8. By analogy, if $i \geq 2$, the number of binary trees with i nodes is $h_i = h_0{}^*h_{i-1} + h_1{}^*h_{i-2} + \ldots + h_{i-1}{}^*h_0$. It satisfies the definition of a Catalan number:

$$h_i = \begin{cases} 1 & i = 0,1 \\ h_0 * h_{i-1} + h_1 * h_{i-2} + \cdots + h_{i-1} * h_0 & i \geq 2 \end{cases}$$

There is also a recursion formula:

$$h_i = \frac{4 * i - 2}{i+1} * h_{i-1} = \frac{C_{2i}^i}{i+1} (i = 1, 2, 3, \ldots)$$

Numbers of binary trees are sorted in ascending order of the number of nodes (the numbers of binary trees with 0 node, with 1 node, with 2 nodes, …). For the binary tree numbered m, the number of nodes i satisfies

$$\sum_{k=0}^{i} h_k \leq m < \sum_{k=0}^{i+1} h_k$$

and its characteristic value

$$n = m - \sum_{k=0}^{i} h_k$$

A function for inorder traversal *work(i, n)* is used to calculate and output the binary tree based on the number of nodes i and its characteristic value n:

If the subtree has no child ($i==0$), output'X'; else
Calculate the number of nodes in the left subtree l and adjust its characteristic value n:

$$\left(\sum_{k=0}^{l} h_k * h_{i-k-1} \leq n < \sum_{k=0}^{l+1} h_k * h_{i-k-1}, n = n - \sum_{k=0}^{l} h_k * h_{i-k-1} \right).$$

If the left subtree exists ($l>0$), then output the left subtree (output '('; *work(l, n/h$_{i-l-1}$)*; output ')';);
Output the root 'X';
If the right subtree exists ($i-l-1>0$,), then output the right subtree (output'('; *work(i-l-1, n % h$_{i-l-1}$)*; output')';).

Chapter 10

Applications of Classical Trees

Binary trees are also suitable for applying dichotomy. Based on binary trees, there are some derived data structures:

1. Binary search trees (BSTs), which are used to improve efficiency for search
2. Binary heaps, which are used to store priority queues
3. Huffman trees, which are used for data encoding

10.1 Binary Search Trees

In computer science, a search algorithm is to find an item with specified properties among a collection of items. There are three kinds of search methods:

1. Sequential search: The search starts from the first element in a linear list, and elements are searched one by one to find the item with specified properties. Its time complexity is $O(n)$.
2. Binary search: In a linear list, all elements' key values are sorted in ascending order. In each step, the searched key value is compared with the key value of the middle element in the linear list. If the two keys match, then the element has been found and its index, or position, is returned. Otherwise, if the searched key value is less than the middle element's key value, then the algorithm repeats its action on the subinterval to the left of the middle element; if the searched key value is greater, on the subinterval to the right. If the remaining interval to be searched is empty, then the key cannot be found in the linear list. The time complexity is $O(\log_2 n)$.
3. BST: BSTs have the following properties. Every node has a key. The key values in the left subtree of the root are smaller than the key value in the root. The key values in the right subtree of the root are larger than the key value in the root. The left and right subtrees of the root are also BSTs. Inorder traversal for a BST produces an increasing sequence.

There are many types of BSTs, such as balanced binary trees, also called AVL search trees. A BST T is an AVL search tree if and only if its left and right subtrees are AVL search trees, and

$|h_L - h_R| \leq 1$, where h_L and h_R are the heights of the left and right subtrees, respectively. The height of AVL is $\log_2 n$.

10.1.1 BST

Consider an infinite full BST (Figure 10.1); the numbers in the nodes are 1, 2, 3, ... In a subtree whose root node is X, we can get the minimum number in this subtree by repeating going down the left node until the last level, and we can also find the maximum number by going down the right node. Now you are given some queries such as "What are the minimum and maximum numbers in the subtree whose root node is X?" Try to find answers for these queries.

Input

In the input, the first line contains an integer N, which represents the number of queries. In the next N lines, each contains a number representing a subtree with root number X ($1 \leq X \leq 2^{31} - 1$).

Output

There are N lines in total, the ith of which contains the answer for the ith query.

Sample Input	Sample Output
2	1 15
8	9 11
10	

Source: POJ Monthly, Minkerui.

ID for online judge: POJ 2309.

Analysis
Based on the problem description, if a root number X is odd, the root must be a leaf. If a root number X is even, and X divided by 2^k is odd, that is, $X \% 2^k = 0$ and $X \% 2^{k+1} \neq 0$, then the height of the subtree with root number X is k, and there are $2^{k+1} - 1$ nodes in the subtree.

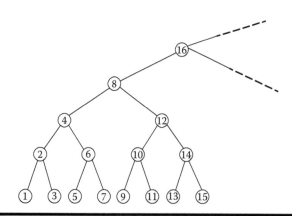

Figure 10.1   An infinite full BST.

Based on properties of the infinite full BST and the rule of node numbers, for a subtree with root number X ($X \% 2^k = 0$ and $X \% 2^{k+1} \neq 0$), the following properties hold:

- The minimum number *min* and maximum number *max* in the subtree must be odd.
- There are $2^k - 1$ nodes in its left and right subtrees, respectively.
- The interval of node numbers in the left subtree is [*min*, $X - 1$], and the interval of node numbers in the right subtree is [$X + 1$, *max*], where $min = X - 2^k + 1$ and $max = X + 2^k - 1$.

Based on the above discussion, the program is as follows:

Program

```
#include <iostream>
using namespace std;
long lowbit(long x)     // return 2^k
{
    return x & -x;
}
int main(void)
{
    long n, x;
    cin >> n;                   // the number of test cases
    for (long i = 0; i < n; i++) {
        cin >> x;               // the number of the root in the i-th test
case
        cout << x - lowbit(x) + 1 << ' ' <<  x + lowbit(x) - 1 <<
endl;   //Output the result
    }
    return 0;
}
```

10.1.2 *Falling Leaves*

Figure 10.2 shows a graphical representation of a binary tree of letters. People familiar with binary trees can skip over the definitions of a binary tree of letters, leaves of a binary tree, and a BST of letters, and go right to the problem.

A binary tree of letters may be one of two things:

1. It may be empty.
2. It may have a root node. A node has a letter as data and refers to a left and a right subtree. The left and right subtrees are also binary trees of letters.

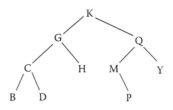

Figure 10.2   A binary tree with falling leaves.

In the graphical representation of a binary tree of letters,

1. Empty trees are omitted completely.
2. Each node is indicated by
   a. Its letter data
   b. A line segment down to the left to the left subtree, if the left subtree is nonempty.
   c. A line segment down to the right to the right subtree, if the right subtree is nonempty.

A leaf in a binary tree is a node whose subtrees are both empty. In the example in Figure 10.1, this would be the five nodes with data B, D, H, P, and Y.

The preorder traversal of a tree of letters satisfies the defining properties:

1. If the tree is empty, then the preorder traversal is empty.
2. If the tree is not empty, then the preorder traversal consists of the following, in order:
   a. The data from the root node
   b. The preorder traversal of the root's left subtree
   c. The preorder traversal of the root's right subtree

The preorder traversal of the tree in Figure 10.2 is KGCBDHQMPY.

A tree like the one in Figure 10.2 is also a BST of letters. A BST of letters is a binary tree of letters in which each node satisfies the following:

The root's data comes later in the alphabet than all the data in the nodes in the left subtree. The root's data comes earlier in the alphabet than all the data in the nodes in the right subtree.

The problem is as follows:

Consider the following sequence of operations on a BST of letters.
Remove the leaves and list the data removed.
Repeat this procedure until the tree is empty.

Starting from the tree in Figure 10.3 on the left, we produce the sequence of trees below, and then the empty tree by removing the leaves with data:

BDHPY
CM
GQ
K

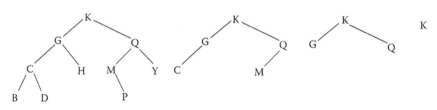

Figure 10.3 A sequence of trees.

Your problem is to start with such a sequence of lines of leaves from a BST of letters and output the preorder traversal of the tree.

Input

The input will contain one or more data sets. Each data set is a sequence of one or more lines of capital letters.

The lines contain the leaves removed from a BST in the stages described above. The letters on a line will be listed in increasing alphabetical order. Data sets are separated by a line containing only an asterisk (*).

The last data set is followed by a line containing only a dollar sign ($). There are no blanks or empty lines in the input.

Output

For each input data set, there is a unique BST that would produce the sequence of leaves. The output is a line containing only the preorder traversal of that tree, with no blanks.

Sample Input	Sample Output
BDHPY	KGCBDHQMPY
CM	BAC
GQ	
K	
*	
AC	
B	
$	

Source: ACM Mid-Central United States 2000.

IDs for online judges: POJ 1577, ZOJ 1700, UVA 2064.

Analysis

Suppose array *leaves* is the linear list for removed leaves, and the length of the list is *levels*, where *leaves*[i] are leaves removed in the *i*th times (1 ≤ *i* ≤ *levels*). For example, for the first test case in the sample input, array *leaves* is as follows:

Serial Number	String
1	BDHPY
2	CM
3	GQ
4	K

Obviously, the finally removed leaf (*leaves*[4]) 'K' is the root. Based on the alphabetical order, the first three terms for leaves are as follows:

Serial Number	Left Subtree for 'K'	Right Subtree for 'K'
1	BDH	PY
2	C	M
3	G	Q

From the third term in the above list, the left child for 'K' is 'G' and the right child for 'K' is 'Q'. That is, 'G' is the root of the left subtree and 'Q' is the root of the right subtree. Based on the alphabetical order and the first two terms in the above list, the subtrees whose root is 'G' are as follows:

Serial Number	Left Subtree for 'G'	Right Subtree for 'G'
1	BD	H
2	C	

Based on the alphabetical order, the subtrees whose root is 'Q' are as follows:

Serial Number	Left Subtree for 'Q'	Right Subtree for 'Q'
1	P	Y
2	M	

Therefore, the left child for 'Q' is 'M', and the right child is 'Y'.

By analogy, a BST can be obtained in Figure 10.4.

A recursive function *preorder* (*leaves*, *levels*) is used to compute and output the preorder traversal of the tree. The process is shown in the program.

Program

```
import java.io.*;
class leaves {                            // class leaves
   public static  void main(String[] arg) {
      String FILE = "leaves";
      ACMIO in = new ACMIO(FILE + ".in");
      PrintWriter out = null;
      try {
         out = new PrintWriter(
               new BufferedWriter(
```

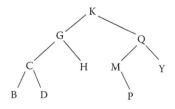

Figure 10.4   A BST.

```
                new FileWriter(FILE + ".out")));
    }catch(Exception e) {
        System.out.println("can't open output");
    }
String line;
    do {
        String[] leaves = new String[26];
        int levels = 0;                          // Initialization
        line = in.getLine();                     // a line of letters
        while (line.charAt(0) != '*' && line.charAt(0) != '$') {
            leaves[levels] = line;               // store a line of letters
            levels++;                            //next line
            line = in.getLine();
        }
        out.println(preorder(leaves, levels));   // output the preorder
traversal of the tree
    } while (line.charAt(0) == '*');
    out.close();
  }
    static String preorder(String[] leaves, int levels) {
    while (levels > 0 && leaves[levels-1].length() == 0)    // find root
        levels--;
    if (levels == 0) return "";              // If list leaves is empty,
return""; else the first character for leaves[levels-1] is as the root
        levels--;
    char root = leaves[levels].charAt(0); // last leaf is the root
    String[] left = new String[levels];   // leaves for left subtree
    String[] right = new String[levels];  // leaves for right subtree
    for (int i = 0; i < levels; i++) {    // for each String in leaves
        int past = 0;                         // the alphabet order for root
in leaves[i]
        while (past < leaves[i].length() && leaves[i].charAt(past) < root)
            past++;
        left[i] = leaves[i].substring(0,past);// The prefix whose length is
past in leaves[i] constitute the left subtree, the postfix after past
constitute the right subtree
        right[i] = leaves[i].substring(past);
    }
    return root+preorder(left, levels)+preorder(right, levels); // the
preorder traversal
  }
}
```

10.2 Binary Heaps

A binary heap is a complete binary tree for which the value in each node is greater (less) than the values of those in its children (if any). If the value in each node is greater than the values of those in its children (if any), we call it a max heap. If the value in each node is less than the values of those in its children (if any), we call it a min heap.

Obviously, in a max heap the value in the root is maximal, and in a min heap the value in the root is minimal.

A binary heap can be used to store a priority queue. In a priority queue, each element has a priority. In a min-priority queue, the search operation finds the element with minimum priority, while the deletion operation deletes this element. In a max priority queue, the search operation finds the element with maximum priority, while the deletion operation deletes this element. If a linear list is used to represent a priority queue, the time complexity finding an element with minimum (or maximum) priority is $O(n)$. Using a binary heap can improve the efficiency.

A binary heap is a complete binary tree. At the bottom level, nodes are filled from left to right. Therefore, elements in a binary heap are stored in an array $heap[0 .. n - 1]$, where the parent of $heap[i]$ is $heap[\lfloor (i - 1)/2 \rfloor]$, and its left child and right child are $heap[2*i + 1]$ and $heap[2*i + 2]$, respectively; $1 \le i \le \lfloor (n - 1)/2 \rfloor$. If $i > \lfloor (n - 1)/2 \rfloor$, node i is a leaf (Figure 10.5).

Insertion and deletion operations in a min heap are shown as follows. Insertion and deletion operations in a max heap are similar.

1. **Insertion operation in a min heap.** The process inserting an element into a min heap is as follows:

   First, the element is inserted at the end of the queue (rear). Then the position of the element is adjusted. The inserted element's value is compared with its parent's value. If its value is less than its parent's value, then the two values' positions are exchanged. The process repeats until the inserted value isn't less than its parent's value or the inserted element becomes the root. Obviously, such a complete binary tree is a min heap.

   ```
   int k = ++top;           // insert the node at rear
   heap[k] = inserted element;
   while (k > 0) {           // adjust the position
         int t = (k-1) / 2;      // parent's position t
         if (heap[t]>heap[k]) {
             heap[t] and heap[k] exchange;
             k = t;
         } else
             break;
   }
   ```

   Because there are $\lceil \log_2 n \rceil$ levels for a heap, the time complexity inserting one node is $O(\lceil \log_2 n \rceil)$.

2. **Deletion operation in a min heap.** When an element is to be deleted from a min heap, the root of the heap is deleted and the heap is adjusted. That is, the value in $heap[0]$ is deleted, $heap[0] = heap[n - 1]$, and then array $heap[0 ... n - 2]$ is adjusted to become a min heap. The process is as follows. Initially, $k = 0$.

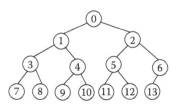

Figure 10.5 The index for a heap.

1. Suppose *heap*[*t*] is the minimal child for heap[*k*].
2. If *heap*[*k*] > *heap*[*t*], then *heap*[*k*] and *heap*[*t*] exchange, *k* = *t*, and continue to downward adjust from *k*; else, the complete binary tree is a min heap.

The process adjusting the heap after deleting *heap*[0] is as follows:

```
if (top) {          // the heap isn't empty
return the root of the heap heap[0];
int k =0;
heap[k] = heap[top--];   // the element at the rear is put the root
of the heap, the length -1
while ((k * 2+1) <= top) {    // Adjust the heap
    int t = k * 2+1;               // the minimal child t
    if (t < top && (heap[t+1] < heap[t])
        ++t;
    if (heap[k]>heap[t]) {    // Adjust
      heap[k] and heap[t] exchange;
      k = t;
    } else
        break;
}
} else
    output "the heap is empty";
```

Because there are $\lceil \log_2 n \rceil$ levels for a heap, the time complexity deleting one node is $O(\lfloor \log_2 n \rfloor)$.

10.2.1 Windows Message Queue

A message queue is the basic fundamental of a Windows system. For each process, the system maintains a message queue. If something happens to this process, such as mouse click or text change, the system will add a message to the queue. Meanwhile, the process will do a loop for getting the message from the queue according to the priority value if it is not empty. Note that less priority value means higher priority. In this problem, you are asked to simulate the message queue for putting messages to and getting messages from the message queue.

Input

There's only one test case in the input. Each line is a command, GET or PUT, which means getting message or putting message. If the command is PUT, one string means the message name and two integers means the parameter and the priority it is followed by. There will be at most 60,000 commands. Note that one message can appear twice or more, and if two messages have the same priority, the one that comes first will be processed first (i.e., first-in, first-out [FIFO] for the same priority.) Process to the end of the file (EOF).

Output

For each GET command, output the command getting from the message queue with the name and parameter in one line. If there's no message in the queue, output "EMPTY QUEUE!" There's no output for the PUT command.

Sample Input	Sample Output
GET	EMPTY QUEUE!
PUT msg1 10 5	msg2 10
PUT msg2 10 4	msg1 10
GET	EMPTY QUEUE!
GET	
GET	

Source: Zhejiang University Local Contest 2006, Preliminary.

ID for online judge: ZOJ 2724.

Analysis

A priority queue is used to store a Windows message queue. If the priority queue is not empty, the message with the highest priority is obtained. If two messages have the same highest priority, the one that comes first will be obtained first. Because the number of messages will be at most 60,000, a min heap can be used to store the priority queue.

Suppose p is a buffer area storing messages. For the ith message, $p[i].name$ is the message name, $p[i].para$ is the message parameter, $p[i].pri$ is the message priority, and $p[i].t$ is its serial number. $heap[t]$ is to store the serial number for heap node t in p; that is, $p[heap[t]]$ is to store the information for heap node t; top is the length of the heap ($0 \leq i, t \leq top$).

For each command:

If the current command is GET, the message at the top of the heap is moved out, the message at the rear is moved to the top of the heap, the length of the heap decreases 1, and then the heap property is maintained.

If the current command is PUT, the message is inserted into the heap, the new message is added at the rear, the length of the heap increases 1, and then the heap property is maintained.

Program

```
#include <cstdio>
#include <cstring>
using namespace std;
const int maxn = 60000 + 10;          // the upper limit of the number of
messages
const int maxs = 100;                  // the upper limit of the length of
message names
struct info {
    char name[maxs];                   //message's name, parameter, priority,
serial number
    int para;
    int pri, t;
} p[maxn];                             //buffer area storing message
int heap[maxn];                        //heap
int top, used;                         //length of heap, pointer for buffer
area
inline void swap(int &a, int &b)       // exchange a and b
{
```

```
    int tmp = a;
    a = b;
    b = tmp;
}
int compare(int a, int b)      // priority: 1st key, Insertion time: 2nd
key return -1, 1, or 0
{
    if (p[a].pri < p[b].pri)
        return -1;
    if (p[a].pri > p[b].pri)
        return 1;
    if (p[a].t < p[b].t)
        return -1;
    if (p[a].t > p[b].t)
        return 1;
    return 0;
}
int main(void)      // A min heap is used
{
    used = 0;
    top = 0;
    int cnt = 0;
    char s[maxs];
    while (scanf("%s", s) != EOF) {
        if (!strcmp(s, "GET")) {            // GET command
            if (top) {                      // if the heap isn't empty
                printf("%s %d\n", p[heap[1]].name, p[heap[1]].para);
                int k = 1;          // the value at the rear is moved to
the top, the length-1
                heap[k] = heap[top--];
                while (k * 2 <= top) {  // maintain the heap property
                    int t = k * 2;
                    if (t < top && compare(heap[t + 1], heap[t]) < 0)
                        ++t;
                    if (compare(heap[t], heap[k]) < 0) {
                        swap(heap[t], heap[k]);
                        k = t;
                    } else
                        break;
                }
            } else
                printf("EMPTY QUEUE!\n");
        } else {                    // PUT command
            scanf("%s%d%d", p[used].name, &p[used].para, &p[used].pri);
            p[used].t = cnt++;
            int k = ++top;          // Put the message at the rear of the
heap
            heap[k] = used++;
            while (k > 1) {         // maintain the heap property
                int t = k / 2;
                if (compare(heap[t], heap[k]) > 0) {
                    swap(heap[t], heap[k]);
                    k = t;
                } else
```

```
                            break;
                }
        }
    }
    return 0;
}
```

10.2.2 Binary Search Heap Construction

Read the statement of problem G for the definitions concerning trees. In the following we define the basic terminology of heaps. A heap is a tree whose internal nodes have each assigned a priority (a number) such that the priority of each internal node is less than the priority of its parent. As a consequence, the root has the greatest priority in the tree, which is one of the reasons why heaps can be used for the implementation of priority queues and for sorting.

A binary tree in which each internal node has both a label and a priority, and which is both a BST with respect to the labels and a heap with respect to the priorities, is called a treap. Given a set of label–priority pairs, your task is to construct a treap containing this data, with unique labels and unique priorities.

Input

The input contains several test cases. Every test case starts with an integer n. You may assume that $1 \leq n \leq 50{,}000$. Then follow n pairs of strings and numbers $l_1/p_1, \ldots, l_n/p_n$ denoting the label and priority of each node. The strings are nonempty and composed of lowercase letters, and the numbers are nonnegative integers. The last test case is followed by a zero.

Output

For each test case, output on a single line a treap that contains the specified nodes. A treap is printed as (< left subtreap >< label >/< priority >< right subtreap >). The subtreaps are printed recursively, and omitted if leaves.

Sample Input	Sample Output
7 a/7 b/6 c/5 d/4 e/3 f/2 g/1	(a/7(b/6(c/5(d/4(e/3(f/2(g/1)))))))
7 a/1 b/2 c/3 d/4 e/5 f/6 g/7	(((((((a/1)b/2)c/3)d/4)e/5)f/6)g/7)
7 a/3 b/6 c/4 d/7 e/2 f/5 g/1	(((a/3)b/6(c/4))d/7((e/2)f/5(g/1)))
0	

Source: Ulm Local Contest 2004.

IDs for online judges: POJ 1785, ZOJ 2243.

Analysis by the Problemsetter (http://www.informatik.uni-ulm.de/acm/Locals/2004/)

Since the labels and priorities are unique, there is exactly one treap for every test case. It is constructed in a straightforward manner as follows. Find the node with the greatest priority, which will become the root of the treap. Split the remaining nodes into two sets: those with labels less than that of the root and those with labels greater than that of the root. From the first or second set of nodes construct the left or right subtreap by applying this strategy recursively. If the set of nodes is empty, we have reached a leaf, and the recursion terminates.

In a straightforward implementation using lists, we need linear time to find the maximum priority node and linear time to split the set of nodes, too. Therefore, in the worst case, we get a runtime of $O(n^2)$, too slow for values of n up to 50,000. Compare this, for example, to the worst case of the quicksort algorithm.

We may, however, reduce the amount of time needed to split the set of nodes by sorting them (using an $O(n*log(n))$ complexity algorithm) according to their labels in the beginning. Then, the kind of subset of nodes needed for our procedure is represented by an interval indicating the smallest and the greatest element. In each recursive step, an interval is split into two subintervals. Still, we need linear time to find the maximum priority in such an interval, leaving us with a quadratic worst-case time.

We can find the maximum priority in logarithmic time by augmenting the list of elements with an order-statistic tree. To this end, we build a complete binary tree large enough to hold n numbers (the priorities) at its bottom level. Every internal node is marked with the maximum of the priorities below that node. Such a tree can be constructed in linear time in a bottom-up fashion. With this additional information, we no longer need to scan through a complete interval in search of the maximum priority, but can use the combined maximum numbers stored in the internal nodes. Thus, we are able to find the maximum priority in logarithmic time by a bottom-up search in the order-statistic tree from both ends of the interval. We then add a top-down search to find the element of the interval having that priority. Hence, the total runtime of our algorithm is $O(n*log(n))$.

Program by the Problemsetter (http://www.informatik.uni-ulm.de/acm/Locals/2004/)

```cpp
#include <cassert>
#include <fstream>
#include <iostream>
#include <string>
using namespace std;
ifstream in("heap.in");
const int n2 = 1<<16; // must be a power of 2
string l[n2];          // the labels
int p[n2];             // the priorities
int ost[2*n2];         // an order-statistic tree containing indexes to
labels and priorities
// for sorting by comparing labels
bool lcmp(const int ia, const int ib)
{
  return l[ia] < l[ib];
}
void recurse(int from, int to)
{
  assert(n2 <= from && from <= to && to < 2*n2);
  // r will become the index in [from..to] with maximal priority
  int r = p[ost[from]] > p[ost[to]] ? from : to;
  // move upwards in the tree until the common predecessor is reached
  for (int f=from, t=to ; f<t ; f/=2, t/=2 )
  {
    // the internal nodes between f and t store the maximal priorities
    // of the nodes in subintervals of (from..to)
    if ((f%2 == 0) && (f+1 < t) && p[ost[f+1]] > p[ost[r]]) r = f+1;
    if ((t%2 == 1) && (t-1 > f) && p[ost[t-1]] > p[ost[r]]) r = t-1;
  }
```

```
    // p[ost[r]] is already maximal in p[ost[from]]..p[ost[to]], move
downwards
    // to find the node in the bottom level from which this priority
originates
    while (r < n2)
    {
      if (ost[r] == ost[r*2])
        r = r*2;
      else if (ost[r] == ost[r*2+1])
        r = r*2+1;
      else
        assert(false);
    }
    assert(from <= r && r <= to);
    // split [from..to] at r into a left sub-treap and a right sub-treap
    cout << '(';
    if (from < r) recurse(from, r-1);
    cout << l[ost[r]] << '/' << p[ost[r]];
    if (r < to) recurse(r+1, to);
    cout << ')';
}
int main()
{
  int n;
  while (in >> n)
  {
    if (n == 0) break;
    assert(1 <= n && n <= 50000);
    // parse the labels and priorities
    for (int i=0 ; i<n ; i++)
    {
      string s;
      in >> s;
      int separator = s.find('/');
      assert(0 <= separator && separator < (int)s.size());
      l[i] = s.substr(0, separator);
      assert(sscanf(s.substr(separator+1).c_str(), "%d", &p[i]) == 1);
    }
    // the bottom level of the order-statistic tree consists of the
labels in sorted order
    for (int i=0 ; i<n ; i++)
      ost[n2+i] = i;
    sort(ost+n2, ost+n2+n, lcmp);
    // build the higher levels of the tree
    for (int a=n2, b=n2+n-1 ; a>1 ; a/=2, b/=2)
    {
      for (int i=a ; i<b ; i+=2)
        ost[i/2] = p[ost[i]] > p[ost[i+1]] ? ost[i] : ost[i+1];
      if (b%2 == 0) // the rightmost node is the single child of its
parent
        ost[b/2] = ost[b];
    }
    // recursively output the treap
    recurse(n2, n2+n-1);
```

```
        cout << endl;
    }
    return 0;
}
```

The Standard Template Library (STL) also provides a container for heap: *priority_queue*. In the header of the program, preprocessor directive "#include *<queue>*" introduces a container for class *queue*; a container *priority_queue* belongs to class *queue*.

The decode the tree problem in Section 10.2.3 shows an example for it.

10.2.3 Decode the Tree

A tree (i.e., a connected graph without cycles) with vertices numbered by the integers 1, 2, ..., n is given. The Prufer code of such a tree is built as follows: the leaf (a vertex that is incident to only one edge) with the minimal number is taken. This leaf, together with its incident edge, is removed from the graph, while the number of the vertex that was adjacent to the leaf is written down. In the obtained graph, this procedure is repeated until there is only one vertex left (which, by the way, always has number n). The written-down sequence of $n - 1$ numbers is called the Prufer code of the tree.

Your task is to reconstruct a tree, given its Prufer code. The tree should be denoted by a word of the language specified by the following grammar:

> T ::= "(N S)"
> S ::= " " T S | empty
> N ::= number

That is, trees have parentheses around them and a number denoting the identifier of the root vertex, followed by arbitrarily many (maybe none) subtrees separated by a single space character. As an example, take a look at the tree in Figure 10.6, which is denoted in the first line of the sample output. To generate further sample input, you may use your solution to Problem Code the Tree (Section 10.4.9).

Note that, according to the definition given above, the root of a tree may be a leaf as well. It is only for the ease of denotation that we designate some vertex to be the root. Usually, what we are dealing here with is called an unrooted tree.

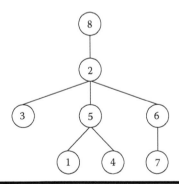

Figure 10.6 An example.

Input

The input contains several test cases. Each test case specifies the Prufer code of a tree on one line. You will find $n - 1$ numbers separated by a single space. Input is terminated by EOF. You may assume that $1 \leq n \leq 50$.

Output

For each test case generate a single line containing the corresponding tree, denoted as described above. Note that, in general, there are many ways to denote such a tree: choose your favorite one.

Sample Input	Sample Output
5 25 2 6 28	(8 (2 (3) (5 (1) (4)) (6 (7))))
2 3	(3 (2 (1)))
2 1 6 2 6	(6 (1 (4)) (2 (3) (5)))

Source: Ulm Local Contest 2001.

IDs for online judges: POJ 2568, ZOJ 1965.

Analysis by the Problemsetter (http://www.informatik.uni-ulm.de/acm/Locals/2001/)
Observe that exactly all the nodes that do not appear in the Prufer code are the leaves of the tree to be reconstructed (except maybe the node with number n). According to the definition of the Prufer code, the first number in the code is adjacent to the leaf with the smallest number. Thus, one edge is constructed. If the first number in the code appears another time somewhere in the code, it is no leaf yet. But if it does not appear one more time, it has become a leaf now. In this case, insert it into the set of leaves. Now, repeat the same procedure $n - 2$ times to get the other edges. Again, straightforward algorithms or efficient priority queue–based algorithms can be used. Obviously, the pretty printer is best implemented in a recursive-descent manner.

Program by the Problemsetter (http://www.informatik.uni-ulm.de/acm/Locals/2001/)

```
#include <cassert>
#include <fstream>
#include <queue>
#include <string>
#include <strstream>
#include <vector>
ifstream in ("decode.in");
typedef vector<int> ivec;
typedef vector<ivec> imat;
// recursive-descent tree pretty printer
// assumes numeration from 1 .. n (for root only: p=0)
void print (imat &adj, int x, int p = 0)
{
  cout << "(" << x;
  for (ivec::iterator it = adj[x].begin() ; it != adj[x].end() ; ++it)
    if (*it != p)
    {
```

```
        cout << " ";
        print (adj, *it, x);
      }
    cout << ")";
}
int main ()
{
  string line;
  while (getline (in, line))
    {
      istrstream lstr (line.c_str());
      ivec v;
      int x;
      while (lstr >> x)
        v.push_back (x);
      int n = v.size() + 1;
      // count downward degree of each node, and find leafs
      ivec deg (n+1, 0);
      for (int i=0 ; i<n-1 ; i++)
        deg[v[i]]++;
      priority_queue< int,ivec,greater<int> > leafs;
      for (int i=1 ; i<=n ; i++)
        if (deg[i] == 0)
          leafs.push (i);
      // reconstruct the adjacency list representation of the tree
      imat adj (n+1, ivec ());
      for (int i=0 ; i<n-1 ; i++)
        {
          // all the numbers that don't occur in v[i..n-2] are leafs
          // the smallest such number will be processed now
          assert (! leafs.empty());
          x = leafs.top();
          leafs.pop();
          // v[i] is the adjacent node of x
          adj[x].push_back (v[i]);
          adj[v[i]].push_back (x);
          if (--deg[v[i]] == 0)
            leafs.push (v[i]);
        }
      print (adj, n);
      cout << endl;
    }
  return 0;
}
```

10.3 Huffman Trees

In a binary tree there are n leaves with weights, where the weight of the kth leaf is w_k and the length of the path from the root to the kth leaf is p_k. Then $w_k * p_k$ is the length of the weighted path for the kth leaf, $1 <= k <= n$. The binary tree whose sum of lengths of weighted paths is minimal is called a Huffman tree. That is, if

$$L = \sum_{k=1}^{n} w_k p_k$$

is minimal, the binary tree is a Huffman tree.

Given n nodes whose weights are w_1, w_2, ..., w_n, respectively, the process for constructing a Huffman tree is as follows:

```
Firstly n nodes constitute a set of n binary trees F={T₁, T₂, ......, Tₙ},
where Tᵢ has only one node whose weight is wᵢ, 1<=i<=n.
while (F is not a tree)
{ Replace the rooted trees T and T' of least weights from F with a tree
having a new root that has T as its left subtree and T' as its right
subtree;
The weight of the new tree = the weight of T + the weight of T';
}
```

A Huffman tree is a complete binary tree. In a Huffman tree, if there are n leaves, there are $2n - 1$ nodes. Constructing a Huffman tree is a greedy method; each time two trees with least weights are selected. Therefore, a min heap is used to store roots of trees when a Huffman tree is constructed.

10.3.1 Fence Repair

Farmer John wants to repair a small length of the fence around the pasture. He measures the fence and finds that he needs N ($1 \le N \le 20,000$) planks of wood, each having some integer length L_i ($1 \le L_i \le 50,000$) units. He then purchases a single long board just long enough to saw into the N planks (i.e., whose length is the sum of the lengths L_i). Farmer John is ignoring the kerf, the extra length lost to sawdust when a saw cut is made; you should ignore it, too.

Farmer John sadly realizes that he doesn't own a saw with which to cut the wood, so he moseys over to Farmer Don's farm with this long board and politely asks if he may borrow a saw.

Farmer Don, a closet capitalist, doesn't lend Farmer John a saw but instead offers to charge Farmer John for each of the $N - 1$ cuts in the plank. The charge to cut a piece of wood is exactly equal to its length. Cutting a plank of length 21 costs 21 cents.

Farmer Don then lets Farmer John decide the order and locations to cut the plank. Help Farmer John determine the minimum amount of money he can spend to create the N planks. Farmer John knows that he can cut the board in various different orders that will result in different charges since the resulting intermediate planks are of different lengths.

Input

Line 1: One integer N, the number of planks.
Lines 2 .. N + 1: Each line contains a single integer describing the length of a needed plank.

Output

Line 1: One integer: the minimum amount of money he must spend to make $N - 1$ cuts.

Sample Input	Sample Output
3	34
8	
5	
8	

Source: USACO, November 2006, Gold.

ID for online judge: POJ 3253.

Hint
He wants to cut a board of length 21 into pieces of lengths 8, 5, and 8. The original board measures $8 + 5 + 8 = 21$. The first cut will cost 21 and should be used to cut the board into pieces measuring 13 and 8. The second cut will cost 13 and should be used to cut the 13 into 8 and 5. This would cost $21 + 13 = 34$. If the 21 was cut into 16 and 5 instead, the second cut would cost 16 for a total of 37 (which is more than 34).

Analysis
Because each cut produces two planks of wood, the process cutting planks can be represented as a binary tree. The initial single long board is the root, and the length of the board is its weight; N planks are as N leaves, where the weight of the ith leaf is the length of the ith plank w_i, and the length from the root to leaf p_i is the number of cuts producing the ith plank. Based on the problem description, the cost for cutting the ith plank is $p_i * w_i$, $1 \leq i \leq n$. Obviously, the total cost is

$$\sum_{k=1}^{n} w_k p_k$$

Therefore, calculating the minimal charge to cut a piece of wood is to calculate a Huffman tree. The process is as follows.

A min heap is constructed based on lengths of N planks. Each time, roots of the heap are deleted twice. The two deleted roots' weights are a and b, respectively. A new node whose weight is $(a + b)$ is inserted into the min heap. And the cost *ans* increases $(a + b)$. Repeat the process until there is only one node in the min heap. At that time, *ans* is the minimal cost.

Program

```
#include <iostream>
using namespace std;
const long maxn = 20000 + 10;   //size of the heap
long n, len;                    //n: the number of planks, len: the length
of the heap
long long p[maxn];              //heap
void heap_insert(long long k)
{                       //insert k into the min heap, maintain the heap
property
        long t = ++len;
        p[t] = k;
        while (t > 1)
                if (p[t/2]>p[t]) {
```

```
                              swap(p[t], p[t / 2]);
                              t /= 2;
                    } else
                              break;
}
void heap_pop(void)           // Delete the root of the min heap, maintain
the heap property
{
          long t = 1;
          p[t] = p[len--];
          while (t * 2 <= len) {
                    long k = t * 2;
                    if (k < len && p[k] > p[k + 1])
                              ++k;
                    if (p[t]>p[k]){
                              swap(p[t], p[k]);
                              t = k;
                    } else
                              break;
          }
}
int main(void)
{
          cin >> n;
          for (long i = 1; i <= n; i++)         //lengths of n planks
                    cin >> p[i];
          len = 0;
          for (long i = 1; i <= n; i++)         //a min heap is constructed
with n planks
                    heap_insert(p[i]);
          long long ans = 0;
          while (len > 1) {                     //construct a Huffman tree
                    long long a, b;
                    a = p[1];     //delete the root of heap (weight a),
maintain the heap property
                    heap_pop();
                    b = p[1];     // delete the root of heap (weight b),
maintain the heap property
                    heap_pop();
                    ans += a + b;               // the cost ans increases
(a+b)
                    heap_insert(a + b);         // A new node (a+b) is inserted
into the min heap
          }
          cout << ans << endl;                  //Output the minimal cost
}
```

10.4 Problems

10.4.1 Cartesian Tree

Let us consider a special type of a BST, called a Cartesian tree. Recall that a BST is a rooted ordered binary tree, such that for its every node *x*, the following condition is satisfied: each node in its left subtree has a key less than the key of x, and each node in its right subtree has the key greater than the key of x.

That is, if we denote the left subtree of the node x by $L(x)$, its right subtree by $R(x)$, and its key by k_x, then for each node x we have

- If $y \in L(x)$, then $k_y < k_x$
- If $z \in R(x)$, then $k_z > k_x$

The BST is called Cartesian if its every node x in addition to the main key k_x also has an auxiliary key that we will denote by a_x, and for these keys the heap condition is satisfied, that is,

- If y is the parent of x, then $a_y < a_x$

Thus, a Cartesian tree is a binary rooted ordered tree, such that each of its nodes has a pair of two keys (k, a) and the three conditions described are satisfied.

Given a set of pairs, construct a Cartesian tree out of them, or detect that it is not possible.

Input

The first line of the input file contains an integer number N—the number of pairs you should build a Cartesian tree out of $(1 \leq N \leq 50{,}000)$. The following N lines contain two numbers each—given pairs (k_i, a_i). For each pair $|k_i|, |a_i| \leq 30{,}000$. All main keys and all auxiliary keys are different, that is, $k_i != k_j$ and $a_i! = a_j$ for each $i != j$.

Output

On the first line of the output file print "YES" if it is possible to build a Cartesian tree out of the given pairs or "NO" if it is not. If the answer is positive, on the following N lines output the tree. Let nodes be numbered from 1 to N corresponding to pairs they contain as they are given in the input file. For each node output three numbers: its parent, its left child, and its right child. If the node has no parent or no corresponding child, output 0 instead.

The input ensures that there is only one possible tree.

Sample Input	Sample Output
7	YES
5 4	2 3 6
2 2	0 5 1
3 9	1 0 7
0 5	5 0 0
1 3	2 4 0
6 6	1 0 0
4 11	3 0 0

Source: ACM Northeastern Europe 2002, Northern Subregion.

ID for online judge: POJ 2201.

Hint

Because the Cartesian tree is a BST for k_i, all pairs (k_i, a_i) can be sorted as a sequence in ascending order for k_i, $1 <= i <= N$. Then, for all a_i, $1 <= i <= N$, a min heap is constructed. If this min heap is a Cartesian tree, inorder traversal for the min heap must produce the same sequence as the sequence in ascending order for k_i.

Because $k_i != k_j$ and $a_i != a_j$ for each $i != j$, the Cartesian tree must exist.

There exists a sequence that all pairs (k_i, a_i) are sorted in ascending order for k_i, $1 <= i <= N$. First, (k_1, a_1) is constructed as a Cartesian tree only with one vertex. Suppose the current Cartesian tree has contained vertices (k_1, a_1), (k_2, a_2), ..., (k_{i-1}, a_{i-1}). Then vertex (k_i, a_i) is inserted into the Cartesian tree as follows. First, vertex (k_{i-1}, a_{i-1}) is as the current vertex (k_j, a_j). Then a_i is compared with current a_j in the current vertex: for the current vertex (k_j, a_j), if $a_j > a_i$, the parent of the current vertex becomes the current vertex; if $a_i > a_j$, the right child of the current vertex j is adjusted as the left child for vertex i, and vertex i is as the right child of vertex j. If, for all vertices in the path to the root $a_j > a_i$, the root is adjusted as the left child for vertex i, and vertex i is as the root.

The algorithm is as follows:

```
Vertices are ordered in ascending order for k_i, 1 <= i <= N;
Vertex 1 is as the Cartesian tree only with one vertex;
for (i = 2; i <= n; i++)
    vertex i is inserted into the Cartesian tree by the above method;
Output the result;
```

10.4.2 Argus

A data stream is a real-time, continuous, ordered sequence of items. Some examples include sensor data, Internet traffic, financial tickers, online auctions, and transaction logs, such as web usage logs and telephone call records. Likewise, queries over streams run continuously over a period of tit1me and incrementally return new results as new data arrives. For example, a temperature detection system of a factory warehouse may run queries like the following.

Query 1: "Every 5 minutes, retrieve the maximum temperature over the past 5 minutes."
Query 2: "Return the average temperature measured on each floor over the past 10 minutes."

We have developed a data stream management system called Argus, which processes the queries over the data streams. Users can register queries to Argus. Argus will keep the queries running over the changing data and return the results to the corresponding user with the desired frequency.

For Argus, we use the following instruction to register a query:

Register *Q_num Period*

Q_num $(0 < Q\_num \leq 3000)$ is the query ID number, and *Period* $(0 < Period \leq 3000)$ is the interval between two consecutive returns of the result. After *Period* seconds of register, the result will be returned for the first time, and after that, the result will be returned every *Period* seconds.

Here we have several different queries registered in Argus at once. It is confirmed that all the queries have different *Q_num*. Your task is to tell the first *K* queries to return the results. If two or more queries are to return the results at the same time, they will return the results one by one in the ascending order of *Q_num*.

Input

The first part of the input is the register instructions to Argus, one instruction per line. You can assume the number of the instructions will not exceed 1000, and all these instructions are executed at the same time. This part ends with a line of "#".

The second part is your task. This part contains only one line, which is one positive integer K ($\leq 10,000$).

Output

You should output the Q_num of the first K queries to return the results, one number per line.

Sample Input	Sample Output
Register 2004 200	2004
Register 2005 300	2005
#	2004
5	2004
	2005

Source: ACM Beijing 2004.

IDs for online judges: POJ 2051, ZOJ 2212, UVA 3135.

Hint
The problem requires returning the first K queries. If two or more queries are to return the results at the same time, they will return the results one by one in the ascending order of Q_num. Therefore, query time is the first key, and Q_num is the second key. Tasks are stored as a min heap.

For the ith instruction, $id[i]$ is its Q_num, $per[i]$ is its *Period*, and $nt[i]$ is its query time, $1 <= i <= n$. If the instruction has been queried $k - 1$ times, then the next query time is

$$nt[i] = \sum_{p=1}^{k} pre[i]$$

For n instructions, a min heap is constructed. After instruction i is queried, instruction i is deleted from the heap, $nt[i] \mathrel{+}= per[i]$, and is added back to the heap. The algorithm is as follows:

```
For each instruction, initially nt[i]=per[i], 1<=i<=n. For each query,
delete the root of the heap, output id[root], then nt[root]+=pre[root],
and add it back to the min heap. The operation is processed K times.
```

10.4.3 Black Box

Our black box represents a primitive database. It can save an integer array and has a special i variable. At the initial moment black box is empty and i equals 0. This black box processes a sequence of commands (transactions). There are two types of transactions:

ADD (x): Put element x into black box.

GET: Increase i by 1 and give an i-minimum out of all integers contained in the black box. Keep in mind that i-minimum is a number located at the ith place after black box element sorting by nondescending order.

Let us examine a possible sequence of 11 transactions:
EXAMPLE

N	Transaction	i	Black Box Contents after Transaction Answer (Elements Are Arranged in Nondescending Order)	Output
1	ADD(3)	0	3	
2	GET	1	3	3
3	ADD(1)	1	1, 3	
4	GET	2	1, 3	3
5	ADD(−4)	2	−4, 1, 3	
6	ADD(2)	2	−4, 1, 2, 3	
7	ADD(8)	2	−4, 1, 2, 3, 8	
8	ADD(−1000)	2	−1000, −4, 1, 2, 3, 8	
9	GET	3	−1000, −4, 1, 2, 3, 8	1
10	GET	4	−1000, −4, 1, 2, 3, 8	2
11	ADD(2)	4	−1000, −4, 1, 2, 2, 3, 8	

It is required that we work out an efficient algorithm that treats a given sequence of transactions. The maximum number of ADD and GET transactions is 30,000 of each type.

Let us describe the sequence of transactions by two integer arrays:

1. $A(1), A(2), \ldots, A(M)$: A sequence of elements that are being included in the black box. A values are integers not exceeding 2 billion by their absolute value, $M \leq 30,000$. For the example we have $A = (3, 1, -4, 2, 8, -1000, 2)$.

2. $u(1), u(2), \ldots, u(N)$: A sequence setting a number of elements that are being included in the black box at the moment of first, second, ..., and N transaction GET. For the example we have $u = (1, 2, 6, 6)$.

The black box algorithm supposes that the natural number sequence $u(1), u(2), \ldots, u(N)$ is sorted in nondescending order, $N \leq M$, and for each p ($1 \leq p \leq N$), an inequality $p \leq u(p) \leq M$ is valid. It follows from the fact that for the p element of our u sequence we perform a GET transaction giving p-minimum number from our $A(1), A(2), \ldots, A(u(p))$ sequence.

Input

The input contains (in given order) $M, N, A(1), A(2), \ldots, A(M), u(1), u(2), \ldots, u(N)$. All numbers are divided by spaces and carriage return characters.

Output

Write to the output the black box answer sequence for a given sequence of transactions, one number per line.

Sample Input	Sample Output
7 4	3
3 1 –4 2 8 –1000 2	3
1 2 6 6	1
	2

Source: ACM Northeastern Europe 1996.

IDs for online judges: POJ 1442, ZOJ 1319, UVA 501.

Hint

There are two kinds of commands: ADD(x) and GET. In this problem two heaps are used to represent the black box, a min heap and a max heap. The current first i-minimum integers are in the max heap. Therefore, the root of the min heap is the current $(i + 1)$-minimum integer.

For an ADD transaction ADD (x), first, element x is inserted into the min heap. Second, the root of the min heap is inserted into the max heap and is deleted from the min heap. Finally, the root of the max heap is inserted into the min heap and is deleted from the max heap. After it, the current first i-minimum integers are in the max heap. The root of the min heap is larger than or equal to any element in the max heap. That is, the root of the min heap is the current $(i + 1)$-minimum out of all integers contained in the black box.

For a GET transaction, first, i is increased by 1. That is, the current i-minimum out of all integers is the root of the min heap. Therefore, the GET transaction returns the root of the min heap. Then the root of the min heap is deleted from the min heap and is inserted into the max heap.

10.4.4 Heap

A (binary) heap is an array that can be viewed as a nearly complete binary tree. In this problem, we are talking about max heaps.

A max heap holds the property that for each node other than the root, its key is no greater than its parent's. Upon this, we further require that for every node that has two children, the key of any node in the subtree rooted at its left child should be less than that of any node in the subtree rooted at its right child.

Any array can be transformed into a max heap satisfying the above requirement by modifying some of its keys. Your task is to find the minimum number of keys that have to be modified.

Input

The input contains a single test case. The test case consists of nonnegative integers distributed on multiple lines. The first integer is the height of the heap. It will be at least 1 and at most 20. Then follow the elements of the array to be transformed into the heap described above, which do not exceed 10^9. Modified elements should remain integral though not necessarily nonnegative.

Output

Output only the minimum number of elements (or keys) that have to be modified.

Sample Input	Sample Output
3	4
1	
3 6	
1 4 3 8	

Source: POJ Monthly, April 1, 2007.

ID for online judge: POJ 3214.

Hint

The max heap is a nearly complete binary tree. For each node other than the root, its key is no greater than its parent's. And for every node that has two children, the key of any node in the subtree rooted at its left child should be less than that of any node in the subtree rooted at its right child.

Therefore, postorder traversal for such a max heap will produce an increasing sequence.

Suppose $a[1 .. n]$ is an array for the max heap, where $a[1]$ is the root. If $2*i \leq n$, then $a[2*i]$ is the left child for $a[i]$, and if $2*i + 1 \leq n$, then $a[2*i + 1]$ is the right child for $a[i]$.

During postorder traversal for such a max heap, the number of vertices in each vertex's right subtree x can be obtained and array b is constructed, where $b[i] = a[n - i + 1] - x$ $(1 \leq i \leq n)$. Therefore, if array a is a max heap, array b must be in ascending order. The solution is to find the minimum number of elements that should be modified. The longest increasing subsequence is used to solve the problem.

10.4.5 How Many Trees?

The balanced binary tree is defined recursively as follows:

1. The difference in the depth of its left child tree and right child tree is at most 1.
2. Its left child tree is a balanced binary tree.
3. Its right child tree is also a balanced binary tree.

Now it is your job to calculate the number of balanced binary trees with a given number of nodes and leafs.

Input

The input consists of multiple tests. Each test consists of two numbers n and m in a single line, the number of the nodes and the number of leafs $(0 < m \leq n \leq 20)$.

Output

The number of balanced binary trees that have exactly n nodes and m leafs is shown:

Sample Input	Sample Output
5 2	4
15 9	0

Source: ZOJ Monthly, December 2002.

ID for online judge: ZOJ 1470.

Hint

The problem shows the numbers of nodes and leafs. Based on the definition of balanced binary tree, the difference in the depth of the left child tree and right child tree is at most 1. Suppose $f[i][j][k]$ is the number of balanced binary trees with i nodes and j leafs, and its depth is k.

There are two special cases:

If the number of nodes or the depth is 0, then the number of nodes i, the number of leafs j, and the depth k are all 0.

If the number of nodes or the depth is 1, then the number of nodes i, the number of leafs j, and the depth k are all 1.

$$f[i][j][k]=\begin{cases}(i==0)\,\&\,\&(j==0)\,\&\,\&(k==0) & (i==0)\|(k==0)\\(i==1)\,\&\,\&(j==1)\,\&\,\&(k==1) & (i==1)\|(k==1)\end{cases}$$

Cases in which the number of nodes i and the depth k are larger than 1 are discussed as follows. Suppose the number of nodes in the left subtree is l, $0 \le l \le i-1$, where the number of leafs in the left subtree is l_y, $0 \le l_y \le \min\{l, j\}$, and the number of nodes in the right subtree is r, $r = i-l-1$, where the number of leafs in the right subtree is r_y, $r_y = j - l_y$.

Based on the definition of the balanced binary tree, there are three cases for the depths of its left subtree and right subtree:

Depths of its left subtree and right subtree are the same, that is, the depths of its left subtree and right subtree are $k-1$; or The difference in the depth of its left child tree and right child tree is 1, that is, the depth of its left child tree is $k-2$, that of its right child tree is $k-1$; or The depth of its left child tree is $k-1$, and right child tree is $k-2$.

Based on the addition principle and the multiplication principle,

$$f[i][j][k]=\sum_{l=0}^{i-1}\sum_{l_y=0}^{\min\{l,j\}}(f[l][l_y][k-1]*f[i-l-1][j-l_y][k-1]+f[l][l_y][k-2]*$$

$$f[i-l-1][j-l_y][k-1]+f[l][l_y][k-1]*f[i-l-1][j-l_y][k-2])$$

The problem shows the numbers of nodes and leafs. The upper limit of the depth is 6. Therefore,

$$ans=\sum_{k=1}^{6}f[n][m][k]$$

Program

```
#include<iostream>
#include<cstring>
using namespace std;
//f[i][j][k]  : the number of balanced binary trees with i nodes and j
leaves, whose height is k
int f[21][21][7];
// f is calculated by memory search
int calculate(int nodes,int leaves,int height){
      //result and f[nodes][leaves][height] share common address for
convenience
         int &result=f[nodes][leaves][height];
         if(result!=-1) return result;
         result=0;
         //boundary case
         if(nodes==0||height==0)return
result=(nodes==0&&leaves==0&&height==0);
         if(nodes==1||height==1)return
result=(nodes==1&&leaves==1&&height==1);
         // enumerate numbers of nodes and leaves in the left subtree
         for(int left=0;left<=nodes-1;left++){
                int right=nodes-1-left;
                for(int left_leaves=0;left_leaves<=left&&left_
leaves<=leaves;left_leaves++){
                       int right_leaves=leaves-left_leaves;
         // Enumerate heights of the left subtree and the right subtree
         result+=calculate(left,left_leaves,height-1)*calculate(right,ri
ght_leaves,height-1);
         result+=calculate(left,left_leaves,height-2)*calculate(right,ri
ght_leaves,height-1);
         result+=calculate(left,left_leaves,height-1)*calculate(right,ri
ght_leaves,height-2);
                }
         }
         return result;
}
int n, m;
int main( ){
         memset(f,-1,sizeof(f));
      cin>>n>>m;
        while(!cin.eof( )){
                int result=0;
                // heights are enumerated
                for(int h=1;h<=6;h++) result+=calculate(n,m,h);
                cout<<result<<endl;
                cin>>n>>m;
        }
}
```

10.4.6 The Number of the Same BST

Many people know BST. The keys in a BST are always stored in such a way as to satisfy the BST property.

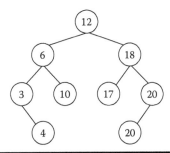

Figure 10.7 A BST.

Let x be a node in a BST. If y is a node in the left subtree of x, then $key[y] \le key[x]$. If y is a node in the right subtree of x, then $key[y] > key[x]$. For example, see Figure 10.7. It is a BST, and it can be built by inserting the elements of vector A <12, 6, 3, 18, 20, 10, 4, 17, 20> sequentially. But it can also be built by vector B <12, 18, 17, 6, 20, 3, 10, 4, 20>.

Now given a vector X, you may get a BST tree from X. Your job is to calculate how many different vectors can build the same BST. To make it easy, you should just output the number of different vectors mod 9901.

Input

The input consists of several cases. Each case starts with a line containing one positive integer n, which is the length of the test vector. The integer n is less than 100. Following this, there will be n positive integers, which are less than 10,000, on the next line. The input will end with a case starting with $n = 0$. This case should not be processed.

Output

For each test case, print a line with a single integer, which is the number of different vectors mod 9901.

Sample Input	Sample Output
3	
2 1 3	
9	
5 6 3 18 20 10 4 17 20	
0	

Source: POJ Monthly, March 26, 2006.

ID for online judge: POJ 2775.

Hint

There are two steps to solve the problems:

> Step 1: A BST a is constructed based on the input vector. The length of a is *tot*. For node i, its key is $a[i].key$; pointers pointing to its left child and right child are $a[i].l$ and $a[i].r$, respectively; and the size of the subtree whose root is node i is $a[i].s$, $0 \le i \le tot$. The root of the tree is $a[0]$, that is, *root* = 0.

Step 2: Calculate how many different vectors can build the same BST *a*.

Suppose *calculate*(*t*) returns the number of vectors building the same BST whose root is *t*. It is implemented by postorder traversal.

1. If its subtrees are empty (*t* == 0), then *calculate*(0) = 1.
2. If its subtrees aren't empty (*t* ≠ 0), then
   The number of its left subtree is *calculate*(*a*[*t*].*l*).
   The number of its right subtree is *calculate*(*a*[*t*].*r*).
   The number of combinations for merging vectors is

$$c_{a[t].s-1}^{a[a[t].l].s}$$

where the total number of nodes for the left subtree and the right subtree is *a*[*t*].*s* − 1, and the number of nodes for the left subtree is *a*[*a*[*t*].*l*].*s*.

Based on the multiplication principle,

$$calculate(t) = \left(calculate\left(a[t].l\right) * calculate\left(a[t].r\right) * c_{a[t].s-1}^{a[a[t].l].s} \right) \%9901$$

Therefore,

$$calculate(t) = \begin{cases} 1 & t = 0 \\ (calculate(a[t].l) * calculate(a[t].r) * c_{a[t].s-1}^{a[a[t].l].s})\%9901 & t \neq 0 \end{cases}$$

Obviously, the recursive function *calculate*(*root*) returns the solution to the problem.

Program

```
#include<iostream>
using namespace std;
const long long md=9901;
long long c[1100][1100],p[1100];
long long n, tot, root;
struct tree{
      long long key, l, r, s;
};
tree a[1100];
// Initialization for the number of combinations
void make(long long n){
      for (long long i=0; i<=n; i++) c[i][0]=1;
      for (long long i=1; i<=n; i++)
      for (long long j=1; j<=i; j++) c[i][j]=(c[i-1][j]+c[i-1][j-1])%md;
}
// adding nodes, that is, constructing a tree
void ins(long long &t,long long key){
```

```
      if (t==0){
            t=++tot;
            a[t].key=key;
            a[t].l=0;
            a[t].r=0;
            a[t].s=1;
            return;
      }
      a[t].s++;
      if (key<=a[t].key) ins(a[t].l,key); else ins(a[t].r,key);
}
// counting process
long long calculate(long long t){
      // If its subtrees are empty
      if (t==0) return 1; else
      // If its subtrees aren't empty
      return (calculate(a[t].l)*calculate(a[t].r)*c[a[t].s-1]
[a[a[t].l].s])%md;
}
int main( ){
      make(1001);
      cin>>n;
      while (n){
            root=0; tot=0;
            for (long long i=1; i<=n; i++) cin>>p[i];
            //Constructing trees
            for (long long i=1; i<=n; i++) ins(root,p[i]);
            cout<<calculate(root)<<endl;
            cin>>n;
      }
}
```

10.4.7 The Kth BST

Definition: A binary tree is a finite set of nodes that is either empty or consists of a root and two disjoint binary trees called the left subtree and the right subtree.

Definition: A BST is a binary tree. It may be empty. If it is not empty, it satisfies the following properties:

1. Every element has a key, and no two elements have the same key; that is, the keys are unique.
2. The keys in a nonempty left subtree must be smaller than the key in the root of the subtree.
3. The keys in a nonempty right subtree must be larger than the key in the root of the subtree.
4. The left and right subtrees are also BSTs.

In this problem, we just care about the preorder traversal of a BST. Here is the pseudocode for preorder traversal:

```
void preorder(tree_pointer ptr) /* preorder tree traversal */
{
      if (ptr)
      { printf("%d", ptr->data);
```

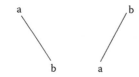

Figure 10.8 Two BSTs.

```
        preorder(ptr->left_child);
        preorder(ptr->right_child);
    }
}
```

Now, you are given *n*, the number of nodes in a BST, and the nodes of the BST, which consist of the first *n* lowercase letters. Of course, more than one BST can be constructed except when *n* is 1. You task is to sort the BSTs according to their preorder representations and give the *K*th BST.

For example, when *n* is 2, there are two BSTs that can be constructed, as shown in Figure 10.8. Their preorder representations are *ab* and *ba*, so the first one is *ab* and the second one is *ba*.

Input

There are multiple test cases in this problem. The input is terminated by EOF.

For each test case, there are two inputs: *n* and *K*, representing the number of nodes in the BST and the index of the BST you need to output.

Note:

- *n* is between 1 and 19,
- *K* is between 1 and the number of ways to construct the BST.

Output

For each input, you should first output the *K*th preorder representation of the BST. Next, for each node (in the order *a*, *b*, *c*, ...), output it first, and then output the left subnode (output * if it does not exist) and the right subnode (output * if it does not exist), separated by a single blank space. *K* will not be greater than the number of representations of BST given *n* nodes. Output a blank line between two test cases.

Sample Input	Sample Output
2 2	ba
4 9	a * *
	b a *
	cbad
	a * *

Sample Input	Sample Output
	b a *
	c b d
	d * *

Source: Zhejiang Provincial Programming Contest 2006, Preliminary.

ID for online judge: ZOJ 2738.

Hint
Because the problem requires outputting each node's left subnode and right subnode, the BST is stored in a multiple linked list. The data structure is as follows.

Suppose s is the multiple linked list storing every node and its children, where the letter whose serial number is i and its left child and right child are stored in $s[i]$ (the serial number for 'a' is 1, ..., the serial number for 'z' is 26, and if the node doesn't exist, the corresponding character is '*'); a is the BST, where $a[t].key$ is the serial number for node t's letter; $a[t].l$ and $a[t].r$ are pointers pointing to its left child and right child, respectively; and $f[i][j]$ is the number of BSTs that j is the root and in which there are i nodes, where its left child is l, and its left subtree contains $j-1$ nodes (i.e., the number of BSTs for the left subtree is $f[j-1][l]$ [$0 \le l \le j-1$]), and its right child is r, and its right subtree contains $i-j$ nodes (i.e., the number of BSTs for the right subtree is $f[i-j][r]$ [$0 \le r \le i-j$]). Obviously,

$$f[i][j]=\begin{cases} 1 & (i=1)\,\&\&(j=0) \\ \sum_{l=0}^{j-1} f[j-1][l] * \sum_{r=0}^{i-j} f[i-j][r] & (i\ge1)\,\&\&(j\ge0) \end{cases}$$

The number of BSTs containing i nodes is stored in $catalan[i]$. Obviously,

$$catalan[i]=\sum_{j=0}^{i} f[i][j]$$

For each BST, there is a preorder representation. And $catalan[i]$ preorder representations are sorted in alphabetical order. For a BST subtree with n nodes, whose index for preorder representation is k, root's letter serial number key satisfies

$$\sum_{i=1}^{key} f[n][i] \le k < \sum_{i=1}^{key+1} f[n][i]$$

The serial number for the letter at the root key can be calculated. Based on the definition of BST, there are $key-1$ nodes in the left subtree, and the index for the left subtree's preorder representation is $k_l= (k-1)/catalan[n - key] + 1$; there are $n - key$ nodes in the right subtree, and the index for the right subtree's preorder representation is $k_r = (k - 1)\% \, catalan[n - key] + 1$.

A recursive function *build(&t, n, k, plus)* is used to calculate the *k*th preorder representation of the BST with *n* nodes and record the multiple linked list *s* storing every node and its children, where *t* is the root. In a BST, keys in a left subtree must be smaller than the key in the root of the subtree, and keys in a right subtree must be larger than the key in the root of the subtree. Therefore, *plus* is the increment for the serial number for *t*. When *t* is a right child, and there are *key* nodes for the left subtree and the root, the serial number is *t* plus *key*, that is, *plus = plus + key*. The process for *build(&t, n, k, plus)* is as follows:

```
build(lolo &t, lolo n, lolo k, lolo plus){
        if (n==0){ // If the number of nodes is 0, the root is 0 and
backtracking
                t=0;
                return;
        }
        t=++tot;          //Calculate the serial number for the root
        Calculate the serial number for node t's letter a[t].key=key+plus; (
```

$\sum_{i=1}^{key} f[n][i] \le k < \sum_{i=1}^{key+1} f[n][i]$, plus is used to distinguish whether t is a right child or not.);

```
Output the letter for node t in preorder traversal (char(a[t].key+'a'-1);
Calculate indexes for preorder representations for left subtree and
right subtree k₁ and k_r (k₁=(k-1)/catalan[n-key]+1, k_r=(k-1)%
catalan[n-key]+1);
   build(a[t].l, key-1, k₁, plus);
   build(a[t].r, n-key, k_r, plus+key);
```

Record information for node *t* and its left child and right child *s[a[t].key]*:

1. Record node *t* (*s[a[t].key]+=char(a[t].key +'a'-1)*);
2. Record the left child (if (*a[t].l*) *s[a[t].key]+=char(a[a[t].l].key+'a'-1)*; else *s[a[t].key]+='*'*);
3. Record the right child (if (*a[t].r*) *s[a[t].key]+=char(a[a[t].r].key+'a'-1)*; else *s[a[t].key]+='*'*);

In the main program, initially *tot=0*; *root=0*; then *build(root,n,k,0)* outputs the *k*th preorder representation of the BST; finally output *n* nodes and their children *s[1]…s[n]*.

Program

```
#include<iostream>
#include<string>
#include<cstring>
#define lolo long long
using namespace std;
struct tree{
      lolo l, r, key;
};
// f[i][j] is the number of BSTs that j is the root and in which there
are i nodes
//sum[i] is the number of BSTs with i nodes (catalan number)
lolo n, k, f[30][30], sum[30], tot, root;
```

```
bool v[30];
tree a[2000];
string s[30];
// Initialization
void make(int n){
      memset(f, 0, sizeof(f));
      f[0][0]=1;
      for (int i=1; i<=n; i++)
      for (int j=1; j<=i; j++){
            for (int l=0; l<j; l++)
            for (int r=j; r<=i; r++) f[i][j]+=f[j-1][l]*f[i-j][r-j];
      }
      memset(sum, 0, sizeof(sum));
      for (int i=0; i<=n; i++)
      for (int j=0; j<=i; j++) sum[i]+=f[i][j];
}
// Constructing tree and output preorder traversal
void build(lolo &t, lolo n, lolo k, lolo plus){
      if (n==0){
            t=0;
            return;
      }
      t=++tot;
      lolo key=1;
      //Enumerate letter for the current node
      while (k>f[n][key]){
            k-=f[n][key];
            key++;
      }
      a[t].key=key+plus;
      cout<<char(a[t].key+'a'-1);
      // Recursion
      build(a[t].l,key-1,(k-1)/sum[n-key]+1,plus);
      build(a[t].r,n-key,(k-1)%sum[n-key]+1,plus+key);
      s[a[t].key].clear();
      s[a[t].key]+=char(a[t].key+'a'-1);
      s[a[t].key]+=" ";
      if (a[t].l) s[a[t].key]+=char(a[a[t].l].key+'a'-1); else s[a[t].
key]+='*';
      s[a[t].key]+=" ";
      if (a[t].r) s[a[t].key]+=char(a[a[t].r].key+'a'-1); else s[a[t].
key]+='*';
}
int main(){
      make(19);
      bool flag=0;
      cin>>n>>k;
      while (!cin.eof()){
            if (flag) cout<<endl; else flag=1;
            // Initialization
            tot=0;
            root=0;
            build(root,n,k,0);
            cout<<endl;
```

```
            for (int i=1;i<=n;i++) cout<<s[i]<<endl;
            cin>>n>>k;
        }
}
```

10.4.8 The Prufer Code

A tree (i.e., a connected graph without cycles) with vertices is given ($N \geq 2$). Vertices of the tree are numbered by the integers 1, ..., N. A Prufer code for the tree is built as follows: a leaf (a vertex that is incident to the only edge) with a minimal number is taken. Then this vertex and the incident edge are removed from the graph, and the number of the vertex that was adjacent to the leaf is written down. In the obtained graph, once again a leaf with a minimal number is taken and removed, and this procedure is repeated until the only vertex is left. It is clear that the only vertex left is the vertex with the number N. The written-down set of integers ($N - 1$ numbers, each in a range from 1 to N) is called *a Prufer code* of the graph.

Given a Prufer code, your task is to reconstruct a tree, that is, find out the adjacency lists for every vertex in the graph.

You may assume that $2 \leq N \leq 7500$.

Input

The input is a set of numbers corresponding to a Prufer code of some tree. The numbers are separated with spaces or line breaks.

Output

The output is adjacency lists for each vertex. It is in the format of vertex number, colon, and numbers of adjacent vertices separated with a space. The vertices inside lists and lists themselves should be sorted by vertex number in ascending order (see sample output).

Sample Input	Sample Output
2 1 6 2 6	1: 4 6
	2: 3 5 6
	3: 2
	4: 1
	5: 2
	6: 1 2

Source: Ural State University Personal Contest Online February 2001, Student Session.

ID for online judge: Ural 1069.

Hint

This problem and that in Section 10.2.3 for decoding the tree are similar. The inputs for the two problems are all Prufer codes. The output for decoding the tree problem is a tree represented by brackets, and the output for this problem is a tree represented by an adjacency list.

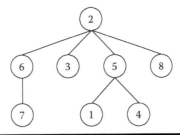

Figure 10.9 Another example.

10.4.9 Code the Tree

A tree (i.e., a connected graph without cycles) with vertices numbered by the integers 1, 2, ..., n is given. The Prufer code of such a tree is built as follows: the leaf (a vertex that is incident to only one edge) with the minimal number is taken. This leaf, together with its incident edge, is removed from the graph, while the number of the vertex that was adjacent to the leaf is written down. In the obtained graph, this procedure is repeated until there is only one vertex left (which, by the way, always has number n). The written-down sequence of $n - 1$ numbers is called the Prufer code of the tree.

Your task is, given a tree, to compute its Prufer code. The tree is denoted by a word of the language specified by the following grammar:

```
T ::= "(" N S ")"
S ::= " " T S | empty
N ::= number
```

That is, trees have parentheses around them and a number denoting the identifier of the root vertex, followed by arbitrarily many (maybe none) subtrees separated by a single space character. As an example, take a look at the tree in Figure 10.9, which is denoted in the first line of the sample input. To generate further sample input, you may use your solution to Problem Decode the Tree (Section 10.2.3).

Note that according to the definition given above, the root of a tree may be a leaf as well. It is only for the ease of denotation that we designate some vertex to be the root. Usually, what we are dealing with here is called an unrooted tree.

Input

The input contains several test cases. Each test case specifies a tree as described above on one line of the input file. Input is terminated by EOF. You may assume that $1 \leq n \leq 50$.

Output

For each test case generate a single line containing the Prufer code of the specified tree. Separate numbers by a single space. Do not print any spaces at the end of the line.

Sample Input	Sample Output
(2 (6 (7)) (3) (5 (1) (4)) (8))	5 25 2 6 28
(1 (2 (3)))	2 3
(6 (1 (4)) (2 (3) (5)))	2 16 2 6

Source: Ulm Local Contest 2001.

ID for online judge: POJ 2567, ZOJ 1097.

Hint by the Problemsetter (http://www.informatik.uni-ulm.de/acm/Locals/2001/)
The parser can be implemented in a recursive-descent manner. It constructs an adjacency set representation of the tree. For the generation of the Prufer code, any straightforward algorithm can be applied, since runtime is no problem, when $n \leq 50$. An $O(n^*\log(n))$ algorithm, which keeps track of the current leaf nodes in a priority queue, is of course more efficient.

SUMMARY OF SECTION III

A tree is a data structure representing hierarchical data. Tree traversal can transform a tree into a linear structure.

In this section, experiments were organized in two fields: trees and binary trees.

In experiments for trees, trees are used to solve not only hierarchical problems, but also other problems, such as union–find sets.

In experiments for binary trees, experiments for applications of binary trees and some classical binary trees, such as BST, Heap, and Huffman tree, are shown.

EXPERIMENTS FOR GRAPHS

Graphs are discrete structures consisting of vertices and edges connecting these vertices. Therefore, graphs are used to represent objects and relationships between these objects in the real world.

This section mainly focuses on four parts of graph theory to organize experiments:

1. Graph traversal: First, two kinds of graph traversals, breadth-first search (BFS) and depth-first search (DFS), are introduced. Many graph algorithms are based on BFS and DFS. Then based on BFS and DFS, topological sort and connectivity of undirected graphs are introduced.
2. Algorithms for minimal spanning trees, the Kruskal and Prim algorithms, are introduced.
3. Algorithms for the best path, the Warshall, Floyd–Warshall, Dijkstra, Bellman–Ford, and shortest path faster (SPFA) algorithms, are introduced.
4. Some algorithms of graphs, such as maximum matching on bipartite graphs and network flows, are introduced.

There are two kinds of storage methods for graphs:

1. An adjacency matrix is used to store relationships between vertices.
2. An adjacency list is used to store information for edges.

Which kind of storage method is selected for graphs is based on problems.

Chapter 11

Applications of Graph Traversal

In some applications, all vertices in a graph need to be visited exactly once. Such a process is called graph traversal. Tree traversal is a special case of graph traversal. Because the structure of a graph is more complex than the structure of a tree, algorithms for graph traversal are also more complex than algorithms for tree traversal. In the process of graph traversal, each visited vertex should be marked to avoid a vertex being visited more than once.

In this chapter, first, two kinds of graph traversals, breadth-first search (BFS) and depth-first search (DFS), are introduced. Given a vertex in a graph, all vertices in the connected component containing the vertex can be visited by BFS or DFS. Then, based on BFS and DFS, topological sort and connectivity of undirected graphs are introduced.

11.1 BFS Algorithm

Given a graph $G(V, E)$ and a source vertex s in G, BFS visits all vertices that can be reached from s layer by layer; and calculates distances from s to all vertices (i.e., numbers of edges from s to these vertices). The distance from s to vertex v $d[v]$, $v \in V$, is as follows:

$$d[v] = \begin{cases} -1 & \text{if } s \text{ and } v \text{ are not connected} \\ \text{the length of the shortest path from } s \text{ to } v & \text{otherwise} \end{cases}$$

Initially, $d[s] = 0$, and for $v \in V - \{s\}$, $d[v] = -1$. The process for BFS is as follows.

Every visited vertex u is processed in order: for every vertex v that is adjacent to u and is not visited, that is, $(u, v) \in E$ and $d[v] = -1$, v will be visited. Because vertex u is the parent or the precursor for vertex v, $d[v] = d[u] + 1$.

Because the traversal order is based on hierarchy and the traversal is implemented through the first-in, first-out (FIFO) access rule, a queue Q is used to store visited vertices: Initially, source vertex s is added into queue Q, and $d[s] = 0$. Then, vertex u, which is the front, is deleted from queue Q; vertices that aren't visited and are adjacent to u, that is, for such a vertex v, $(u, v) \in E$ and $d[v] = -1$, are visited in order: $d[v] = d[u] + 1$, and vertex v is added into queue Q. The process repeats until queue Q is empty. That is, BFS traversal starts from source s, visits all connected vertices, and forms a BFS traversal tree whose root is s.

The BFS algorithm starting from source s visits all vertices that can be reached from s top down and layer by layer. The BFS algorithm is as follows:

```
void BFS(VLink G[ ], int s) // BFS algorithm starting from source s in G
{ int v;
  visit s;
  d[s]=0;                    // distance d[s]
  ADDQ(Q, s);                // s is added into queue Q
  while (!EMPTYQ(Q))         // while queue Q is not empty, visit other
vertices
  { u=DELQ(Q);               // the front is deleted from queue Q
    Get the first adjacent vertex v for vertex u ( if there is no
adjacent vertex for u, v=-1);
    while (v != -1)
    { if (d[v] == -1)        // if vertex v hasn't been visited
      { visit v;
        ADDQ(Q, v);          // adjacent vertex v is added into queue Q
        d[v] =d[u]+1;        // distance d[v]
      }
      Get the next adjacent vertex v for vertex u;
    }
  }
}
```

$BFS(G, s)$ can visit all vertices that can be reached from s in G, that is, vertices in the connected component containing s. The algorithm of graph traversal based on BFS is as follows:

```
void TRAVEL_BFS (VLink G[ ], int d[ ], int n)
{ int i;
  for (i = 0; i < n; i ++)           // Initialization
    d[i] =-1;
  for (i = 0; i < n; i ++)           // BFS for all unvisited vertices
    if (d[i] == -1)
      BFS(G, i);
}
```

Suppose there are n vertices and e edges for graph G. Through the BFS algorithm, every visited vertex can be added into the queue exactly once. Therefore, the *while* repetition statement runs at most n times in BFS. If an adjacency list is used, its time complexity is O(e), and if an adjacency matrix is used, its time complexity is O(n^2).

11.1.1 Prime Path

The ministers of the cabinet were quite upset by the message from the chief of security stating that they would all have to change the four-digit room numbers on their offices.

— It is a matter of security to change such things every now and then, to keep the enemy in the dark.
— But look, I have chosen my number 1033 for good reasons. I am the prime minister, you know!
— I know, so therefore your new number, 8179, is also a prime. You will just have to paste four new digits over the four old ones on your office door.

— No, it's not that simple. Suppose that I change the first digit to an 8, then the number will read
8033, which is not a prime!

— I see, being the prime minister, you cannot stand having a non–prime number on your door
even for a few seconds.

— Correct! So I must invent a scheme for going from 1033 to 8179 by a path of prime numbers
where only one digit is changed from one prime to the next prime.

Now, the minister of finance, who had been eavesdropping, intervened.

— No unnecessary expenditure, please. I happen to know that the price of a digit is 1 pound.

— Hmm, in that case I need a computer program to minimize the cost. You don't know some very
cheap software gurus, do you?

— In fact, I do. You see, there is this programming contest going on....

Help the prime minister find the cheapest prime path between any two given four-digit primes.
The first digit must be nonzero, of course. Here is a solution for the case above:

1033
1733
3733
3739
3779
8779
8179

The cost of this solution is 6 pounds. Note that the digit 1, which got pasted over in step 2,
cannot be reused in the last step—a new 1 must be purchased.

Input

The input is one line with a positive number: the number of test cases (at most 100). Then for each
test case, there is one line with two numbers separated by a blank. Both numbers are four-digit
primes (without leading zeros).

Output

The output is one line for each case, either with a number stating the minimal cost or containing
the word *impossible*.

Sample Input	Sample Output
3	6
1033 8179	7
1373 8017	0
1033 1033	

Source: ACM Northwestern Europe 2006.

ID for online judge: POJ 3126.

Analysis

Every number is a four-digit number. There are 10 possible values for each digit ([0 .. 9]), and the first digit must be nonzero. The problem is represented by a graph: the initial prime and all primes obtained by changing a digit are vertices. If prime *a* can be changed into prime *b* by changing a digit, there is an edge (*a*, *b*) whose length is 1 connecting two vertices corresponding to *a* and *b*, respectively. Obviously, if there is a path from initial prime *x* to goal prime *y*, then the number of edges in the path is the cost; else, there is no solution. Therefore, solving the problem is to calculate the shortest path from initial prime *x* to goal prime *y*, and BFS is used to find the shortest path.

Suppose array *s*[ ] is used to store lengths of the shortest paths for all obtained primes; the type for elements in queue *h*[ ] is *struct*, where *h*[ ].*k* and *h*[ ].*step* are used to store primes and lengths of paths, respectively, and pointers for the front and the rear of *h* are *l* and *r*, respectively.

First, the sieve method is used to calculate all primes between 2 and 9999, and all primes are put into array *p*. Only the minimal cost is required to calculate the problem. Therefore, the graph need not be stored, and we only need focus on calculating the shortest paths.

The algorithm is as follows:

Step 1: Initialization. The initial prime *x* is added into queue *h*. Its path length is 0 (*h*[1].*k*=*x*; *h*[1].*step*=0;). The minimal cost *ans* is initialized –1;

Step 2: Front *h*[l] is operated as follows:

If the front is the goal prime (*h*[l].*k*==*y*), then note the length of the path (*ans*=*h*[l].*step*) and exceed the loop;

Enumerate all possibilities for the front: enumerate the number of digit *i* from 1 to 4, enumerate value *j* for digit *i* from 0 to 9, and the first digit must be nonzero (!((*j*==0)&&(*i*==4))):

- Get the number *tk* by changing the front *h*[l].*k*'s digit *i* into *j*.
- If *tk* is a composite number (*p*[*tk*]==true), then continue to enumerate.
- Get the length of the path *ts* for prime number *tk* (*ts*=*h*[l].*step*+1).
- If *ts* is not the shortest (*ts*≥*s*[*tk*]), then continue to enumerate.
- If *tk* is the goal prime (*tk*==*y*), then note the length of the path (*ans*=*ts*) and exceed the loop.
- Note the length of the path for prime *tk* (*s*[*tk*]=*ts*).
- Add prime *tk* and its length of the path (*r*++; *h*[*r*].*k*=*tk*; *h*[*r*].*step*=*ts*;) into the queue.

If the queue is empty (*l*==*r*) or the goal prime has been gotten (*ans*≥0), then exceed the loop.

The front is deleted from queue (*l*++).

Step 3: Output the result: if the goal prime is gotten (*ans*≥0), then output the length of the shortest path *ans*; else, output "Impossible."

Program

```
#include<iostream>
#include<string>
using namespace std;
struct node{
        int k, step;        // current prime number k, the length of the path
(the number of changed digits ) step
};
node h[100000];              //Queue
bool p[11000];               // Sieve
```

```
int x, y, tot, s[11000];              // Initial prime x, goal prime y,
the number of remainder test cases tot, the current shortest path s[x]
for prime x
void make(int n){                     // Get primes in [2..n] by sieve method
      memset(p, 0, sizeof(p));
      p[0]=1;
      p[1]=1;
      for (int i=2; i<=n; i++) if (!p[i])
      for (int j=i*i; j<=n; j+=i) p[j]=1;
}
int change(int x, int i, int j){      // change the i-th digit of x into j
      if (i==1) return (x/10)*10+j; else
      if (i==2) return (x/100)*100+x%10+j*10; else
      if (i==3) return (x/1000)*1000+x%100+j*100; else
      if (i==4) return (x%1000)+j*1000;
}
int main(){
      make(9999);                     // Get primes in [2..9999]
      cin>>tot;                       // the number of test cases
      while (tot--){
            cin>>x>>y;                // initial prime x and goal prime y
            h[1].k=x;                 // initial prime x is pushed into
the queue
            h[1].step=0;
            int l=1,r=1;              // Initialize pointers of the queue
            memset(s, 100, sizeof(s));   //Initialize the length of the
path
            int ans=-1;               // Initialize the minimal cost
while (1){
                  if (h[l].k==y) {    // goal prime y is gotten
                        ans=h[l].step;
                        break;
                  }
                  int tk,ts;
                  for (int i=1; i<=4; i++)   // every digit of the front
for the queue is changed
                        for (int j=0; j<=9; j++) if (!((j==0)&&(i==4))){//
Enumerate
                              tk=change(h[l].k, i, j);
                              if (p[tk]) continue;    // If tk isn't a prime
                              ts=h[l].step+1;    // the length of the path to tk
                              if (ts>=s[tk]) continue;
                              if (tk==y){    // If tk is the goal prime
                                    ans=ts;
                                    break;
                              }
                              s[tk]=ts;             // the length of the path to tk
                              r++;
                              h[r].k=tk;            // Prime tk and its length of
the path is pushed
                              h[r].step=ts;
                        }
                  if (l==r||ans>=0) break; // If the queue is empty or
the goal prime is arrived
```

```
                    l++;
            }
            if (ans>=0) cout<<ans<<endl; else cout<<"Impossible"<<endl;
// Output the result
            }
}
```

11.2 DFS Algorithm

The DFS for a graph is similar to preorder traversal for a tree or a binary tree.

The DFS algorithm starts from a vertex *u*. First, vertex *u* is visited. Then unvisited vertices adjacent from *u* are selected one by one, and for each vertex DFS is initiated. The algorithm is as follows:

```
void DFS(VLink G[ ], int u)      // DFS starts from a vertex u
{ int w;
  visited[u] = 1;                // Vertex u is visited.
  Get a vertex w adjacent from u (If there is no such a vertex w, w=-1.);
  while (w != -1)                // adjacent vertices are selected one by one
  { if (visited[w] == 0)         //If vertex w hasn't been visited
    { visited[w]=1;
      DFS(G, w) ;                //Recursion
    }
    Get the next vertex w adjacent from u (If there is no such a vertex
w, w=-1.);
  }
}
```

DFS(G, u) visits the connected component containing vertex *u*. The DFS for a graph is as follows:

```
void TRAVEL_DFS(VLink G[ ], int visited[ ], int n)
{ int i;
  for (i = 0; i < n; i ++)        //Initialization
    visited[i] = 0;
  for (i = 0; i < n; i ++)        // DFS for every unvisited vertex
    if (visited[i] == 0)
      DFS(G, i);
}
```

For a graph with *n* vertices and *e* edges, the time complexity for DFS that initializes all vertices' marks is $O(n)$, and the time complexity for DFS is $O(e)$. Therefore, if $n \le e$, the time complexity for DFS is $O(e)$.

11.2.1 House of Santa Claus

In your childhood you most likely had to solve the riddle of the house of Santa Claus. Do you remember that the importance was on drawing the house in a stretch without lifting the pencil and not drawing a line twice? As a reminder it has to look like what is shown in Figure 11.1.

Well, a couple of years later, like now, you have to draw the house again, but on the computer. As one possibility is not enough, we require *all* the possibilities when starting in the lower left corner. Follow the example in Figure 11.2 while defining your sketch.

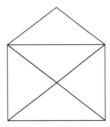

Figure 11.1   House of Santa Claus.

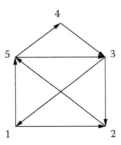

Figure 11.2   This sequence would give the output line 153125432.

All the possibilities have to be listed in the output file by increasing order, meaning that 1234... is listed before 1235....

Output

    12435123
    13245123
    ...
    15123421

Source: ACM Scholastic Programming Contest ETH Regional Contest 1994.
IDs for online judge: UVA 291.

Analysis
The house of Santa Claus is an undirected graph with eight edges (Figure 11.3). A symmetrical adjacency matrix *map*[ ][ ] is used to represent the graph. In the diagonal of the matrix, *map*[1][4], *map*[4][1], *map*[2][4], and *map*[4][2] are 0, and other elements are 1. Because the graph is a connected graph, DFS for the graph starting from any vertex can visit all vertices and edges.

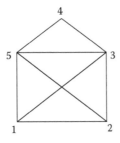

Figure 11.3   House of Santa Claus.

The problem requires you to implement drawing the house in a stretch without lifting the pencil and not drawing a line twice. That is, the drawing must cover all eight edges exactly once. And the problem requires listing all possibilities by increasing order. Therefore, DFS must visit all vertices starting from vertex 1.

Program

```cpp
#include<iostream>
#include<string>
#include<cstring>
using namespace std;
int map[6][6];                          // adjacency matrix
void makemap(){                         //Generating the adjacency matrix
      memset(map,0,sizeof(map));
      for (int i=1;i<=5;i++)
      for (int j=1;j<=5;j++) if(i!=j) map[i][j]=1;
      map[4][1]=map[1][4]=0;
      map[4][2]=map[2][4]=0;
}
void dfs(int x,int k,string s){         //DFS Traversal
      s+=char(x+'0');                   // Vertex x is pushed into the
sequence
      if (k==8) {                       // Drawing the house is finished.
            cout<<s<<endl;
            return;
      }
      for (int y=1;y<=5;y++)            // In increasing order visit edges
      if (map[x][y]){
            map[x][y]=map[y][x]=0;
            dfs(y,k+1,s);
            map[x][y]=map[y][x]=1;
      }
}
int main(){
      makemap();                        //Generating the adjacency matrix
      dfs(1,0,"");        // All the possibilities are outputted by
increasing order
}
```

11.3 Topological Sort

Sort for a linear list is to sort elements based on keys' ascending or descending order. Topological sort is different from sort for a linear list. Topological sort is to sort all vertices in a directed acyclic graph (DAG) into a linear sequence. If there is an arc (*u*, *v*) in DAG, *u* appears before *v* in the sequence.

There are two methods to implement topological sort: deleting arcs and topological sort implemented by DFS.

Deleting Arcs

Step 1: Select a vertex whose in-degree is 0 and output the vertex;

Step 2: Delete the vertex and arcs that start at the vertex, that is, in-degrees for vertices at which arcs' end decrease 1.

Repeat the above steps. If all vertices are output, the process of topological sort ends; else, there exists cycles in the graph and there is no topological sort in the graph.

The time complexity for the algorithm is $O(E)$.

Using the algorithm for deleting arcs once, we can get one topological sort. Using the recursive method, this algorithm is applied for all vertices whose in-degree is 0 successively, and all topological sorts can be obtained.

11.3.1 Following Orders

Order is an important concept in mathematics and computer science. For example, Zorn's lemma states: "a partially ordered set in which every chain has an upper bound contains a maximal element." Order is also important in reasoning about the fixed-point semantics of programs.

This problem involves neither Zorn's lemma nor fixed-point semantics, but does involve order.

Given a list of variable constraints of the form $x < y$, you are to write a program that prints all orderings of the variables that are consistent with the constraints.

For example, given the constraints $x < y$ and $x < z$, there are two orderings of the variables x, y, and z that are consistent with these constraints: $x\,y\,z$ and $x\,z\,y$.

Input

The input consists of a sequence of constraint specifications. A specification consists of two lines: a list of variables on one line followed by a list of contraints on the next line. A constraint is given by a pair of variables, where $x\,y$ indicates that $x < y$.

All variables are single-character, lowercase letters. There will be at least 2 variables and no more than 20 in a specification. There will be at least 1 constraint and no more than 50 in a specification. There will be at least 1 ordering and no more than 300 consistent with the contraints in a specification.

Input is terminated by the end of the file.

Output

For each constraint specification, all orderings consistent with the constraints should be printed. Orderings are printed in lexicographical (alphabetical) order, one per line.

The outputs for different constraint specifications are separated by a blank line.

Sample Input	Sample Output
a b f g	abfg
a b b f	abgf
v w x y z	agbf
v y x v z v w v	gabf

(Continued)

Sample Input	Sample Output
	wxzvy
	wzxvy
	xwzvy
	xzwvy
	zwxvy

Source: Duke Internet Programming Contest 1993.

ID for online judge: POJ 1270, UVA 124.

Analysis

Every variable (letter) is represented as a vertex, and a constraint $x < y$ is represented as an arc (x, y). Therefore, a list of contraints is represented as a directed graph:

1. A directed graph is constructed based on the input.

   Suppose *var* is the string for a list of variables. Because there are spaces in the string, $var[0]$, $var[2]$, $var[4]$, ... , are vertices, and the number of vertices is $\lfloor length(\text{var})/2 \rfloor + 1$.

   Suppose v is the string for a list of contraints. Array *pre* is used to store the sequence for vertices' in-degrees, where $pre[ch]$ is the in-degree for vertex *ch*. Array *pre* is calculated as follows:

   for (int i=0; i<the length of v; i+= 4) ++pre[the i+2-th letter in v];

2. Get all Topological Sorts through DFS

   All Topological Sorts in a directed graph can be obtained through DFS. Initial state is a subsequence *res* whose length is *dep*-1:

```
dfs(dep, res) {
If a Topological Sort is gotten (dep==n+1), then output res and
backtrack (return);
   Search vertex i whose in-degree is 0 (has[i]&& pre[i]==0,
'a'<=i<='z'):
   { Delete vertex i (has[i]=false);
     Delete all arcs which start form vertex i ( for(int k=0;k< the
length of v; k+=4) if (the kth character in v ==i)--pre[the k+2-th
character in v] );
       dfs(dep+1, res+i);
       return to the state before the recursion ( for(int k=0; k< the
length of v; k+=4) if ( the k-th character of v==i) ++pre[the k+2-th
character in v]; has[i]=true );
   }
}
```

Obviously, *dfs*(1, "") is called recursively and all topological sorts can be obtained.

Program

```
import java.util.*;
import java.io.Reader;
```

```
import java.io.Writer;
import java.math.*;
public class Main {
   public static void print(String x) { // print(x): output Topological
Sort x
      System.out.print(x);
   }
   static int N;                         //The number of vertices N
   static int[ ] pre;
   static boolean[ ] has;
   static String var, v;
  static void dfs(int dep, String res) { // dfs(dep, res) used to get
  topological sorts
      if (dep == N + 1) {                // a topological sort is gotten
         print(res + "\n");
         return;
      }
      for (int i = 'a'; i <= 'z'; i++)// Search a vertex whose in-degree
is 0
         if (has[i] && pre[i] == 0) {
            has[i] = false;
            for (int k = 0; k < v.length(); k += 4) //Delete arcs
starting from the vertex
                  if (v.charAt(k)==i)--pre[v.charAt(k + 2)];
               dfs(dep+1, res+(char)i);// the dep+1-th vertex
               for (int k = 0; k < v.length(); k += 4) //Recovery
                  if (v.charAt(k)==i)++pre[v.charAt(k + 2)];
               has[i] = true;
         }
   }
   public static void main(String[ ] args) {
      Scanner input = new Scanner(System.in);     //java standard input
      while (input.hasNextLine()) {
         var = input.nextLine();      // Input a list of variables
         v = input.nextLine();        // Input a list of contraints
         has = new boolean[1 << 8];
         for (int i = 0; i<var.length();i+= 2)
            has[var.charAt(i)]=true;
         N = var.length() / 2 + 1;    // The number of vertices
         pre = new int[1 << 8];
       for (int i = 0; i < v.length(); i += 4)   //Calculate every
vertices' in-degrees
               ++pre[v.charAt(i+2)];
         dfs(1, "");
         print("/n");
      }
   }
}
```

Topological Sort Implemented by DFS

Suppose *x* and *y* are vertices in a directed graph and (*x*, *y*) is an arc. If *x* is in the set of vertices obtained by DFS(*y*), then arc (*x*, *y*) is a back edge, and its time complexity is O(*E*).

There is no cycle in a directed graph if and only if there is no back edge in the graph.

Based on it, the algorithm of topological sort implemented by DFS is as follows:

Suppose it takes one time unit to visit a vertex, the end time when vertex u and its descendants are all visited is $f[u]$. And $f[u]$ can be calculated by DFS algorithm as follows. Obviously, if there exists a topological sort in the graph, there is no back edge in DFS traversal for the graph. That is, for any arc (u, v) in the graph, $f[v] < f[u]$.

The topological sequence is stored in a stack *topo*. In *topo*, array $f[\ ]$ for vertices are in descending order from top to bottom.

```
void DFS-visit (u);              //DFS traversal for the subtree whose
root is u
  { Set a visited mark for u;
    time=time+1;
    for each arc (u, v)
    if (v hasn't been visited)
        DFS-visit (v);
    f[u]=time;
    add u into stack topo;
};
```

Initially, *time* = 0 and set unvisited marks to all vertices. For every unvisited vertex v, *DFS-visit* (v) is called. Then stack *topo* and $f[\ ]$ can be obtained. If there exists an arc (u, v) in the graph such that $f[v] > f[u]$, then (u, v) is a back edge and topological sort fails; else, all vertices from top to bottom in stack *topo* constitute a topological sequence.

The time complexity for DFS is $O(E)$, and the time complexity for adding all vertices into stack *topo* is $O(1)$. Therefore, the time complexity for topological sort is $O(E)$.

11.3.2 Sorting It All Out

An ascending sorted sequence of distinct values is one in which some form of a less than operator is used to order the elements from smallest to largest. For example, the sorted sequence A, B, C, D implies that $A < B$, $B < C$, and $C < D$. In this problem, we will give you a set of relations of the form $A < B$ and ask you to determine whether a sorted order has been specified or not.

Input

The input consists of multiple problem instances. Each instance starts with a line containing two positive integers n and m. The first value indicates the number of objects to sort, where $2 \leq n \leq 26$. The objects to be sorted will be the first n characters of the uppercase alphabet. The second value, m, indicates the number of relations of the form $A < B$, which will be given in this problem instance. Next will be m lines, each containing one such relation consisting of three characters: an uppercase letter, the character '<', and a second uppercase letter. No letter will be outside the range of the first n letters of the alphabet. Values of $n = m = 0$ indicate the end of the input.

Output

For each problem instance, the output consists of one line. This line should be one of the following three:

Sorted sequence determined after *xxx* relations: *yyy* ... *y*
Sorted sequence cannot be determined
Inconsistency found after *xxx* relations

where *xxx* is the number of relations processed at the time either a sorted sequence is determined or an inconsistency is found, whichever comes first, and *yyy* ... *y* is the sorted, ascending sequence.

Sample Input	Sample Output
4 6	Sorted sequence determined after 4 relations: ABCD.
A<B	Inconsistency found after 2 relations.
A<C	Sorted sequence cannot be determined.
B<C	
C<D	
B<D	
A<B	
3 2	
A<B	
B<A	
26 1	
A<Z	
0 0	

Source: ACM East Central North America 2001.

IDs for online judges: POJ 1094, ZOJ 1060, UVA 2355.

Analysis

The purpose of the problem is to determine whether a sorted sequence exists or not through topological sort. When a relation is input, an arc is added into a directed graph. There are three probabilities:

1. If arc (*x*, *y*) is added, and *x* can be reached from *y*, then (*x*, *y*) is a back edge and an inconsistency is found.
2. If arc (*x*, *y*) is added, and there exists a sorted sequence for *n* vertices, then a sorted sequence is determined. It can be determined by deleting arcs for the current graph.
3. After *m* arcs have been added and there is no sorted sequence for *n* vertices, output "Sorted sequence cannot be determined."

The set of relations can be represented as a directed graph $G(V, E)$, where objects to be sorted are represented as vertices, and relations are represented as arcs. If there is a relation $x < y$, and *x* can't be reached from *y* through DFS, then arc (*x*, *y*) is added. The adjacency matrix *g* for *G* is defined as follows:

$$g[i,j] = \begin{cases} 1 & \text{There is an arc } (i, j) \\ 0 & \text{There is no arc } (i, j) \end{cases} \quad (0 \le i, j \le n-1)$$

Array *go* is used to mark visited vertices, where *go*[*i*] == true shows vertex *i* has been visited; array *f* is used to show vertices' in-degrees, where *f*[*i*] is the in-degree for vertex *i*; $0 \leq i \leq n - 1$. Sequence *Q* is used to store vertices whose in-degrees are 0, and the length of *Q* is *tot*. If there is an inconsistency found, then *doit* = false. *finish* is the mark for topological sort. If *finish* == true, there is a topological sort.

The process is as follows:

```
Input the number of objects n and the number of relations m;
    Adjacency matrix g falls to zero;
    doit=true;
    for m relations:
    { Input the current relation x<y;
      if (doit==true)
        { DFS starts from y;
          if (x is reachable from y)
              Inconsistency found;
          Determine whether a topological sort exists or not by
deleting arcs;
              if (finish ==true)
                  there is a Topological Sort;
          }
        }
      if (doit==true) Sorted sequence cannot be determined
```

Program

```
import java.util.*;
import java.math.*;
public class Main {
    static boolean[ ] go;
    static int[ ][ ] g;                       // adjacency matrix
    static int N,K;                           //Numbers of vertices and arcs
    public static void find(int x){// find all vertices reached from x
        go[x] = true;
        for (int i=0; i<N; i++)
            if (g[x][i]==1&&!go[i]) find(i);
    }
    public static void main(String[ ] args){
        Scanner input = new Scanner(System.in);
        while (true){
            N = input.nextInt();      //number of vertices n and number of arcs
k
            K = input.nextInt();
            if (N==0) break;
            g = new int [N][N];                     //Initialize the adjacency
matrix
            for (int i=0; i<N; i++)
                for (int j=0; j<N; j++)
                    g[i][j] = 0;
            boolean doit = true;           //The mark of topological sort
            int[ ] f = new int [N+1];
            for (int i=1; i<=K; i++){      //Input and deal with k
relations
```

```
              String p = input.next(); //the k-th relation and vertices x
and y
              int x = p.charAt(0)-'A',c = p.charAt(1),y =
p.charAt(2)-'A';
                if (c=='>'){
                    c = x;
                    x = y;
                    y = c;
                }
                go = new boolean[N];
                for (int j=0;j<N;j++) go[j] = false;
                if (doit){                    //DFS start from y
                    find(y);
                    if (go[x]){         // If(x, y) is a back edge
                        System.out.println("Inconsistency found after " + i
+" relations.");
                        doit = false; continue;
                    }
                    g[x][y] = 1;
                    f[y]++;                    //the in-degree of vertex y ++
                    int[ ] Q = new int[N+1];
                    int tot = 0;
                    for (int k=0; k<N&&tot<=1; k++)
                        if (f[k]==0) Q[++tot]=k;
                    if (tot==1) {        //Only one vertex whose in-degree is
0
                        boolean finish = true;
                        while (tot<N){
                            int xx = Q[tot], tmp = 0;// Deleting arcs
                            for (int k=0; k<N; k++)
                                if (g[xx][k]==1&&0==(f[k]-=g[xx][k])){
                                    Q[++tot] = k;
                                    ++tmp;
                                }
                            if (tmp>1){
                                finish = false;
                                break;
                            }
                        }
                        if (finish&&tot==N){ // There is a sorted sequence
                            System.out.print("Sorted sequence determined
after "+ i + " relations: ");
                            for (int k=1; k<=N; k++)
                                System.out.print((char)('A'+Q[k]));
                            System.out.println(".");
                            doit = false;
                        }
                        for(int k=0; k<N; k++) f[k]=0;
                        for (int j=0; j<N; j++)
                            for (int k=0; k<N; k++)
                                f[k] += g[j][k];
                    }
                }
            }
```

```
        if (doit)
          System.out.println("Sorted sequence cannot be determined.");
      }
    }
}
```

11.4 Connectivity of Undirected Graphs

Let $G(V, E)$ be a connected graph. A cut vertex of G is a vertex whose removal disconnects G. A bridge (or a cut edge) of G is an edge whose removal disconnects G. The vertex connectivity of a graph is the minimum number k of vertices that must be removed to disconnect the graph. And the edge connectivity of a graph is the minimum number k of edges that must be removed to disconnect the graph. The vertex connectivity and edge connectivity of a graph show the connectivity of a graph.

A connected component of a graph G is a connected subgraph of G that is not a proper subgraph of another connected subgraph of G. In an unconnected graph, how many connected components without a cut vertex can be obtained? Such connected components are called biconnected components. A connected subgraph without a cut vertex is also called a block.

The function *low* is used to get cut vertices and bridges of a connected graph and biconnected components of a graph. Suppose $pre[v]$ is the sequence number of vertex v in DFS traversal. That is, $pre[v]$ is the time that vertex v is visited. Function $low[u]$ is the $pre[v]$ of vertex v, which is the earliest visited ancestor of u and u's descendants. That is,

$$low[u] = \min_{(u,s),(u,w)\in E}\{pre[u], low[s], pre[w]\}$$

where s is a child of u, and (u, w) is a back edge.

A vertex itself is considered one of its ancestors. Therefore, $low[u] = pre[u]$ or $low[u] = pre[w]$ can hold. $low[u]$ is calculated as follows:

$$low[u] = \begin{cases} pre[u] & u \text{ is visited for the first time in DFS} \\ \min\{low[u], pre[w]\} & (u,w) \text{ is a back edge} \\ \min\{low[u], low[s]\} & \text{all edges related to } u\text{'s children are inspected} \end{cases}$$

In the algorithm, $low[u]$ is changed until the DFS subtree whose root is u and array *low* and array *pre* for u and its descendants are produced.

In DFS, edges can be classified into four types:

Branch edge T: Edge (u, v) is a branch edge if it is the first time that v is visited in DFS.

Back edge B: Edge (u, v) is a back edge if u is a descendant of v and v has been visited, but all descendants of v haven't been visited.

Forward edge F: Edge (u, v) is a forward edge if v is a descendant of u, all descendants of v have been visited, and $pre[u] < pre[v]$.

Cross-edge C: All other edges (u, v). That is, u and v has no ancestor–descendant relationship in a DFS tree, or u and v are in different DFS trees. All descendants of v have been visited and $pre[u] > pre[v]$.

1. **Function *low* is used to get cut vertices in a connected graph.** We determine whether a vertex is a cut vertex or not based on the two following properties:

   Property 1: If vertex U isn't a root, U is a cut vertex if and only if there exists a child s of U, $low[s] \geq pre[U]$. That is, there is no back edge from s and its descendants to U's ancestors.

   In Figure 11.4a, although in the subtree whose root is $s1$ there is a back edge to U's ancestor, there is no back edge to U's ancestor from s_2 or s_2's descendants. If U is removed, the graph is not connected.

   In an undirected graph, there are only branch edges and back edges. We can calculate *low* and *pre* through DFS, and find whether property 1 holds or not. The process is as follows:

   If (v, w) is a branch edge T ($pre[w]==-1$), and if there is no back edge from w or w's descendants to v's ancestors ($low[w] \geq pre[v]$), then vertex v is a cut vertex, and $low[v]=min\{low[v], low[w]\}$.

```
    If (v, w) is a back edge B (pre[w]!=-1), then
low[v]=min{low[v], pre[w]}.
void fund_cut_point(int v)    // DFS starts from v to calculate a
cut vertex in an undirected graph
    { int w;
    low[v]=pre[v]= ++d;               // Initialization
    for ( w∈the set of adjacent vertices for v) &&(w!=v)// Search
edge (v, w) for vertex v
    { if (pre[w]==-1)                 //If (v, w) is branch edge T,
w is called recursively. If w and its descendants can't return
to v's ancestors, v is a cut vertex, calculate low[v]
        { fund_cut_point(w);        //w's all children's related
edges
        if (low[w]>=pre[v])   // v is a cut vertex
            v is a cut vertex;
        low[v]=min{ low[v], low[w]};
        };
    else low[v]=min{ low[v], pre[w]};// If (v, w) is a back
edge, calculate low[v]
    }
}
```

   Property 2: If U is selected as the root, then U is a cut vertex if and only if it has more than one child (Figure 11.4b).

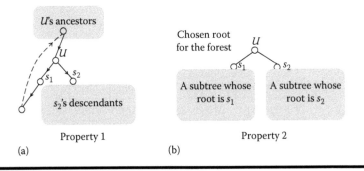

<table>
<tr><td>Property 1</td><td>Property 2</td></tr>
<tr><td>(a)</td><td>(b)</td></tr>
</table>

Figure 11.4   Properties 1 (a) and 2 (b).

In Figure 11.4b, root U has two subtrees whose roots are s_1 and s_2, respectively, and there is no cross-edge C between the two trees (in an undirected graph, there is no cross-edge C). Therefore, the graph isn't connected after vertex U is deleted, and vertex U is a cut vertex.

Based on the above two properties, the algorithm calculating cut vertices is as follows:

```
for(i = 0; i < n; i ++)          //Initialization
pre[i] =-1;
low[s]=pre[s]=d=0;   // vertex s: start vertex
p=0;       // the number of children for vertex s
for (each w∈adj[s]) p++;
if (p>1)
  s is a cut vertex and exit;       //Property 2
  fund_cut_point(s);        // Property 1
```

2. **Function *low* is used to get the bridge in a connected graph.** In an undirected graph, edge (u, v) is a bridge if and only if (u, v) is not in any simple circuit.

The method determining whether an edge is a bridge or not is as follows. Edge (u, v) is a branch edge discovered by DFS. If there is no back edge connecting v and its descendants to u's ancestors, that is, $low[v] > pre[u]$ or $low[v] == pre[v]$, then deleting (u, v) leads to u and v unconnected. Therefore, edge (u, v) is a bridge.

In Figure 11.5a, DFS is used, a DFS tree is obtained as Figure 11.5b, and *pre* and *low* for all vertices are showed in Figure 11.5c. Obviously, for v_5, v_7, and v_{12}, $low[v] == pre[v]$, and (v_0, v_5), (v_6, v_7), and (v_{11}, v_{12}) satisfy $low[v] > pre[u]$ for edge (u, v). These edges are bridges in Figure 11.5a.

In an undirected graph there are only branch edges and back edges. DFS can be used to calculate *low* and *pre* for vertices (initial values for *pre*[ ] are −1) and calculate bridges in the undirected graph. The method is as follows:

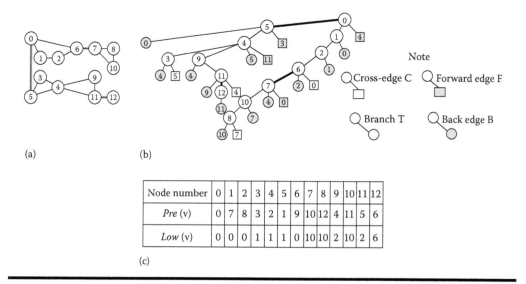

(a) (b)

Node number	0	1	2	3	4	5	6	7	8	9	10	11	12
Pre (v)	0	7	8	3	2	1	9	10	12	4	11	5	6
Low (v)	0	0	0	1	1	1	0	10	10	2	10	2	6

(c)

Figure 11.5 DFS. (a) Undirected graph. (b) DFS tree. (c) The nodes of the *pre* value and *low* value.

```
    If (v, w) is a branch edge (pre[w]==-1), and if there is no back
edge from w or w's descendants to u's ancestors, ((low[w]==pre[w])
||(low[w]>pre[v])), then (v, w) is a bridge, and
low[v]=min{low[v], low[w]}.
    If (v, w) is a back edge (pre[w]!=-1), then low[v]=min{low[v],
pre[w]}.
void fund_bridge (v);                // DFS to find bridges from
vertex v
{ int w;
  low[v]= pre[v]=++d;
  for (each w∈ the set of adjacent vertices for v) &(w!=v) //
Search edge(v, w)
    { if (pre[w]==-1)                 // if (v, w) is a branch edge
      { fund_bridge (w);
          if ((low[w]== pre[w])||( low[w]> pre[v]))
              (v, w) is a branch edge;
          low[v]=min{ low[v], low[w]};
      };
      else low[v]=min{ low[v], pre[w]}; // if (v, w) is a back edge
    }
}
```

3. **Function *low* is used to get biconnected components.** A biconnected component is a connected component without a cut vertex. Biconnected components of a graph are partitions of edges of the graph; that is, every edge must be in a block, and two different blocks don't contain common edges. In Figure 11.6, vertex *b* is a common vertex for blocks 3 and 4, vertex *c* is a common vertex for blocks 3 and 1, and vertex *e* is a common vertex for blocks 2 and 4. The three vertices are cut vertices for the graph. The graph isn't connected when one of the three vertices is deleted.

The key to finding a block in an undirected graph is to find a cut vertex. DFS is used to get *low* and *pre* (initial values for *pre*[ ] are –1) and calculate blocks in the undirected graph. The process is as follows.

For vertex *v*, *u* is the parent for *v*: if *u* is the root, (*u*, *v*) is the first edge for the block; else, suppose *f* is *u*'s parent. If *u* is deleted, *v* and *f* aren't connected, and then {*f*, *u*, *v*} isn't biconnected, and (*u*, *v*) is the first edge for the new block; else, (*u*, *v*) and (*f*, *u*) are in a same block. A stack is used to store vertices in the current block. Suppose

```
    st is a stack, sp is the pointer pointing to the top of the
    stack;
```

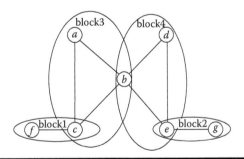

Figure 11.6 Three cut vertices for the graph. Cut vertices *b,c,e* are common vertices for two blocks.

```
    r is the number of blocks in the graph;
    ans is used to store blocks, where all vertices for the t-th
block are stored in ans[t][0]...ans[t][k], and ans[t][k+1]=-1 (end
mark for block t, 1<=t<=r);
void dfs(v)                   //calculate block ans containing vertex v
{ st[sp++] = v;                          //v is pushed into the stack
  pre[v]=low[v] =++d;                    // set pre and low for v
  for (each w∈ the set of adjacent vertices for v) &(w!=v)
//search adjacent edge (v, w) for v
        { if (pre[w]==-1) {                    //(v, w) is a branch edge T
            dfs(w);
            if (low[w]< low[v])         // all children's related
edges for w have been checked, low[v]=min{ low[w], low[v]}
                low[v]=low[w];
            if (low[w]>=pre[v]) {             //w and its descendants
can't return to an ancestor earlier than v, then v is a cut
vertex, the block is sent to ans[r]
                k = 0;
                st[sp] = -1;
                ans[r][0] = v; // vertex v enters ans[r]
                while (st[sp] != w) // vertices above w enter
ans[r]
                    ans[r][++k] = st[--sp];
                ans[r][++k] = -1; // end mark for ans[r]
                if (k>2) //if number of vertices in the block >
2, accumulation
                    r++;
                }
        } else if (pre[w]< low[v])//(v, w) is back edge B,
low[v]=min{ pre[w], low[v]}
                low[v]= pre[w];
        }
}
```

11.4.1 Knights of the Round Table

Being a knight is a very attractive career: searching for the Holy Grail, saving damsels in distress, and drinking with the other knights are fun things to do. Therefore, it is not very surprising that in recent years, the kingdom of King Arthur has experienced an unprecedented increase in the number of knights. There are so many knights now that it is very rare that every knight of the Round Table can come at the same time to Camelot and sit around the Round Table; usually only a small group of knights is there, while the rest are busy doing heroic deeds around the country.

Knights can easily get overexcited during discussions—especially after a couple of drinks. After some unfortunate accidents, King Arthur asked the famous wizard Merlin to make sure that in the future no fights break out between the knights. After studying the problem carefully, Merlin realized that the fights can only be prevented if the knights are seated according to the following two rules:

■ The knights should be seated such that two knights who hate each other should not be neighbors at the table. (Merlin has a list that says who hates whom.) The knights are sitting around a roundtable; thus, every knight has exactly two neighbors.

■ An odd number of knights should sit around the table. This ensures that if the knights cannot agree on something, then they can settle the issue by voting. (If the number of knights is even, then it can happen that yes and no have the same number of votes, and the argument goes on.)

Merlin will let the knights sit down only if these two rules are satisfied; otherwise, he cancels the meeting. (If only one knight shows up, then the meeting is canceled as well, as one person cannot sit around a table.) Merlin realized that this means that there can be knights who cannot be part of any seating arrangements that respect these rules, and these knights will never be able to sit at the Round Table (one such case is if a knight hates every other knight, but there are many other possible reasons). If a knight cannot sit at the Round Table, then he cannot be a member of the knights of the Round Table and must be expelled from the order. These knights have to be transferred to a less prestigious order, such as the knights of the square table, the knights of the octagonal table, or the knights of the banana-shaped table. To help Merlin, you have to write a program that will determine the number of knights that must be expelled.

Input

The input contains several blocks of test cases. Each case begins with a line containing two integers $1 \le n \le 1000$ and $1 \le m \le 1,000,000$. The number n is the number of knights. The next m lines describe which knight hates which knight. Each of these m lines contains two integers, k_1 and k_2, which means that knight number k_1 and knight number k_2 hate each other (the numbers k_1 and k_2 are between 1 and n).

The input is terminated by a block with $n = m = 0$.

Output

For each test case you have to output a single integer on a separate line: the number of knights that have to be expelled.

Sample Input	Sample Output
5 5	2
1 4	
1 5	
2 5	
3 4	
4 5	
0 0	

Source: ACM Central Europe 2005.

IDs for online judges: POJ 2942, UVA 3523.

Analysis

Knights are represented as vertices. If two knights don't hate each other, there is an edge connecting the two corresponding vertices. Such a graph is called a friendship graph. A friendship graph is constructed as follows. Initially, a complete graph is given, and then edges connecting knights who hate each other are deleted in the graph.

Then the friendship graph is partitioned such that each edge is contained in a subgraph, and there is no common edge in two different subgraphs. There is at most one common vertex for two different subgraphs, and such a vertex is a cut vertex for the friendship graph. If a partitioned subgraph contains an odd cycle, then the odd cycle corresponds to a roundtable, and vertices in the odd cycle are knights of the Round Table. Finally, the number of vertices that aren't in cycles are calculated. These vertices are knights that have to be expelled.

1. A friendship graph is partitioned as follows.
   Suppose G is the adjacency matrix for the friendship graph, where

   $$G[i,j] = \begin{cases} true & \text{knight } i \text{ and knight } j \text{ are friends} \\ false & \text{knight } i \text{ and knight } j \text{ hate each other;} \end{cases}$$

   pre and *low* are defined as above; for vertex i, $pre[i]$ is the time that vertex i is visited, and $low[i]$ is the $pre[v]$ of vertex v, which is the earliest visited ancestor of i and i's descendants;
   st is a stack, and *sp* is the pointer pointing to the top of stack;
   r is the number of blocks for the friendship graph;
   ans is used to store blocks for the friendship graph; where all vertices in the tth block are stored in $ans[t][0]…ans[t][k]$, and $ans[t][k+1]=-1$ (end mark, $1 \leq t \leq r$).
   A subprogram *dfs(c)* is used to calculate the block *ans* in which there is vertex c. For unvisited vertices DFS is processed to calculate all blocks in the graph (*for*(int $i = 0$; $i < N$; i++) if ($pre[i] == 0$)).

2. Determine whether the current block contains an odd cycle.
   If the current block contains an odd cycle, that is, a cycle with odd vertices, then the cycle corresponds to a roundtable. DFS is used to determine whether the current block contains an odd cycle or not. Suppose marks for vertices have been visited:

   $$color[i] = \begin{cases} 0 & \text{Vertex } i \text{ isn't in the current block} \\ 1 & \text{Vertex } i \text{ is an unvisited vertex in the current block} \\ -2 & \text{Vertex } i \text{ is a vertex visited in even times in the current block} \\ 2 & \text{Vertex } i \text{ is a vertex visited in odd times in the current block} \end{cases}$$

and *flag* is the mark whether the current block contains an odd cycle or not.
Initially, *flag* = false (block c hasn't an odd cycle); for the first visited vertex, its *color* is 2, and other vertices' *color* are 1:

```
now = 0;
while (ans[c][now] != -1) {color[ans[c][now]] = 1; ++now };
    color[ans[c][0]] = 2;
```

Then the subprogram *dfs*(-1, *ans*[c][0], -2) is called to determine whether block c contains an odd cycle or not:

```
void dfs(pnt, c, col) { //edge (pnt, c) has been visited, the mark
    for unvisited vertices which are adjacent to c is col.
```

```
          if (flag)  // block has an odd cycle
              return;
          for (int i = 0; i < N; ++i) {// search vertex i (i is in the
  block, and isn't pnt and c)
              if (G[c][i] && color[i] != 0 && i != pnt && i != c) {
                  if (color[i] == 1){ // i isn't in the path
                      color[i] = col;
                      dfs(c, i, -col);
                  } else if (color[i] == color[c]) {
                      // i is in the path, and the cycle is an odd
  cycle
                      flag = true;
                      return;
                  }
              }
          }
      }
  }
```

3. Determine whether vertices in block *c* are expelled or not.

   If block *c* contains an odd cycle, then all vertices in the cycle are knights of a roundtable. Suppose *ok*[*i*] is the mark for knights of a roundtable. Function *solve*(*c*) is used to calculate *ok*[ ] in block *c*:

```
static void solve(int c) {
int now = 0;
while (ans[c][now] != -1) {
  color[ans[c][now]] = 1;
  ++now;
 }
flag = false;
color[ans[c][0]] = 2;           // The first vertex in the block has
been visited
  dfs(-1, ans[c][0], -2);        // Determine whether the block has
an odd cycle
  now = 0;
  while (ans[c][now] != -1) {
    color[ans[c][now]] = 0;
    if (flag)
      ok[ans[c][now]] = true;
    ++now;
   }
}
```

4. Main program

```
   Input a test case and construct a friendship graph G;
   Calculate all blocks in the friendship graph (for(i = 0; i < N;
   i++) if (pre[i] == 0) dfs(i));
   Determine whether vertices are expelled in r blocks (for(int
   i = 0; i < r; i++) solve(i));
   Calculate and output the number of expelled knights
```

$$kick = \sum_{i=0}^{n-1} (ok[i] == false);$$

Program

```java
import java.lang.*;
import java.math.*;
import java.util.*;
public class Main {
    public static void print(String x) {
        System.out.print(x);
    }
    static int N, M;                            // Numbers of vertices and edges
    static boolean G[ ][ ], ok[ ], flag;       //G: an adjacency matrix; ok:
a mark sequence for expelled knights; flag: a mark for an odd cycle
    static int[ ] low, pre, color, st;
    static int[ ][ ] ans = new int[1 << 10][1 << 10];   // ans: current
block;
    static int r, sp, cnt;
    static void dfs(int c) {
        st[sp++] = c;                          //c is pushed into the stack
        pre[c] = low[c] = ++cnt;               //Initialization
        for (int i = 0; i < N; i++)            // Search vertices
adjacent to c
            if (G[i][c]) {                     // If (c, i)∈G,
                if (pre[i] == 0) {             //If it is the first time
that t is visited
                    dfs(i);
                    if (low[i]<low[c])
                        low[c] = low[i];
                    if (low[i] >= pre[c]) {
                        int k = 0;
                        st[sp] = -1;
                        ans[r][0] = c;         // vertex c into ans[r]
                        while (st[sp] != i)
                            ans[r][++k] = st[--sp];
                        ans[r][++k] = -1;
                        if (k > 2)             // if number of vertices > 2,
it is a block
                            r++;
                    }
                } else if (pre[i] < low[c])   //(c, i) is a back edge,
low[c]=min{pre[i], low[c]}
                    low[c] = pre[i];
            }
    }
    static void dfs(int pnt, int c, int col) {
        if (flag)                              // if the block has an odd cycle
            return;
        for (int i = 0; i < N; ++i) {
            if (G[c][i] && color[i] != 0 && i != pnt && i != c) {
                if (color[i] == 1){            // if vertex i isn't in the path
                    color[i] = col;
                    dfs(c, i, -col);
                } else if (color[i] == color[c]) { // the block has an odd
cycle
                    flag = true;
                    return;
```

```
                        }
                    }
                }
            }
    static void solve(int c) {         // determine vertices whether should
be expelled
        int now = 0;
        while (ans[c][now] != -1) {
            color[ans[c][now]] = 1;
            ++now;
        }
        flag = false;
        color[ans[c][0]] = 2;
        dfs(-1, ans[c][0], -2);
        now = 0;
        while (ans[c][now] != -1) {
            color[ans[c][now]] = 0;
            if (flag)
                ok[ans[c][now]] = true;
            ++now;
        }
    }
    public static void main(String args[ ]) {
        Scanner input = new Scanner(System.in);
        st = new int[1 << 10];
        while ((N = input.nextInt()) != 0) {   // Input numbers of vertices
and edges N and M
            M = input.nextInt();
            G = new boolean[N][N];
            ok = new boolean[N];
            low = new int[N];
            pre = new int[N];
            color = new int[N];
            sp = r = 0;
            for (int i = 0; i < N; i++)         // Initialization: a complete
graph
                for (int j = 0; j < N; j++)
                    G[i][j] = true;
            for (int i = 0; i < M; i++) {       //delete edges who hate each
other
                int x = input.nextInt() - 1, y = input.nextInt() - 1;
                G[x][y] = false;
                G[y][x] = false;
            }
            pre = new int[N];
            for (int i = 0; i < N; i++)
                if (pre[i] == 0)                // vertex i isn't visited
                    dfs(i);
            for (int i = 0; i < r; i++)         // Calculate expelled knights
                solve(i);
            int kick = 0;                       // Calculate the number of expelled
knights
            for (int i = 0; i < N; ++i)
                if (ok[i] == false)
```

```
                    ++kick;
         print(kick + "\n");              // output the number of expelled
knights
         }
    }
}
```

11.5 Problems

11.5.1 Ordering Tasks

John has *n* tasks to do. Unfortunately, the tasks are not independent and the execution of one task is only possible if other tasks have already been executed.

Input

The input will consist of several instances of the problem. Each instance begins with a line containing two integers, $1 \leq n \leq 100$ and *m*. *n* is the number of tasks (numbered from 1 to *n*), and *m* is the number of direct precedence relations between tasks. After this, there will be *m* lines with two integers, *i* and *j*, representing the fact that task *i* must be executed before task *j*. An instance with $n = m = 0$ will finish the input.

Output

For each instance, print a line with *n* integers representing the tasks in a possible order of execution.

Sample Input	Sample Output
5 4	1 4 2 5 3
1 2	
2 3	
1 3	
1 5	
0 0	

Source: GWCF Contest 2, Golden Wedding Contest Festival.

ID for online judge: UVA 10305.

Hint

Tasks are represented as vertices. Relationships between tasks are represented as arcs, where if task *i* is executed before task *j*, there is an arc $(i - 1, j - 1)$. Therefore, tasks in a possible order of execution constitute a topological sort in a directed graph. Suppose in-degrees for vertices are stored in *ind*[ ], where the in-degree for vertex *i* is *ind*[*i*] $(0 \leq i \leq n - 1)$; the adjacency list is *lis*[ ], and queue *q* is used to store vertices whose in-degree are 0.

The adjacency list *lis*[ ] and array *ind*[ ] are set up when a test case is input. All vertices whose in-degrees are 0 are added into *q*.

Repeat the following process until *q* is empty.

Delete the front *x*.

For every vertex in *lis*[*x*], its in-degree decreases 1.

Add vertices whose in-degrees are 0 into *q*.

The sequence of vertices deleted from *q* is a topological sort.

11.5.2 Spreadsheet

In 1979, Dan Bricklin and Bob Frankston wrote VisiCalc, the first spreadsheet application. It became a huge success and, at that time, was the killer application for the Apple II computers. Today, spreadsheets are found on most desktop computers.

The idea behind spreadsheets is very simple, though powerful. A spreadsheet consists of a table where each cell contains either a number or a formula. A formula can compute an expression that depends on the values of other cells. Text and graphics can be added for presentation purposes.

You are to write a very simple spreadsheet application. Your program should accept several spreadsheets. Each cell of the spreadsheet contains either a numeric value (integers only) or a formula, which only support sums. After having computed the values of all formulas, your program should output the resulting spreadsheet where all formulas have been replaced by their value.

Input

The first line of the input file contains the number of spreadsheets to follow. A spreadsheet starts with a line consisting of two integer numbers, separated by a space, giving the number of columns and rows. The following lines of the spreadsheet each contain a row. A row consists of the cells of that row, separated by a single space.

A cell consists of either a numeric integer value or a formula. A formula starts with an equal sign (=). After that, one or more cell names follow, separated by plus signs (+). The value of such a formula is the sum of all values found in the referenced cells. These cells may again contain a formula. There are no spaces within a formula.

You may safely assume that there are no cyclic dependencies between cells. So each spreadsheet can be fully computed.

The name of a cell consists of one to three letters for the column, followed by a number between 1 and 999 (including) for the row. The letters for the column form the following series: A, B, C, ..., Z, AA, AB, AC, ..., AZ, BA, ..., BZ, CA, ..., ZZ, AAA, AAB, ..., AAZ, ABA, ..., ABZ, ACA, ..., ZZZ. These letters correspond to numbers from 1 to 18278. The top left cell has the name A1. See Figure 11.7.

A1	B1	C1	D1	E1	F1	---
A2	B2	C2	D2	E2	F2	---
A3	B3	C3	D3	E3	F3	---
A4	B4	C4	D4	E4	F4	---
A5	B5	C5	D5	E5	F5	---
A6	B6	C6	D6	E6	F6	---
---	---	---	---	---	---	---

Figure 11.7 Naming of the top left cells.

Output

The output of your program should have the same format as the input, except that the number of spreadsheets and the number of columns and rows are not repeated. Furthermore, all formulas should be replaced by their value.

Sample Input	Sample Output
1	10 34 37 81
4 3	40 17 34 91
10 34 37 =A1+B1+C1	50 51 71 172
40 17 34 =A2+B2+C2	
=A1+A2 =B1+B2 =C1+C2 =D1+D2	

Source: ACM Southwestern European Regional Contest 1995.

ID for online judge: UVA 196.

Hint

The name of a cell consists of one to three letters for the column, followed by a number between 1 and 999 (including) for the row. Therefore, letters and numbers are transferred into sequence numbers for column and row, respectively, as follows.

The sequence number for the column of letter sequence $c_k \,..\, c_1$ is

$$y = \sum_{i=1}^{k} (c_i - 64) * 26^{i-1}$$

and the sequence number for the row of number sequence $b_p \,..\, b_1$ is

$$x = \sum_{i=1}^{p} (b_i - 48) * 10^{i-1}$$

That is, a cell in a formula $c_k \,..\, c_1 \ b_p \,..\, b_1$ corresponds to a position (x, y) in a table. Suppose the table is array $w[\ ][\ ]$, and a cell's position (i, j) corresponds to a number $d = j*1000 + i$, that is, the row number is $d \% 1000$, and the column number is $\lfloor d/1000 \rfloor$.

A table can be represented as a directed graph: cells are represented as vertices, and relationships between cells and formulas are represented as arcs, where if cell (x, y) is in cell (i, j), there is an arc from (x, y) to (i, j). An adjacency matrix g is used to store the directed graph, where $g[x][y]$ is used to store the numeric integer value.

The algorithm is as follows:

A directed graph is constructed;
Deleting arcs is used to find a Topological Sort and calculate numeric integer values.

11.5.3 Genealogical Tree

The system of Martians' blood relations is confusing enough. Actually, Martians bud when they want and where they want. They gather together in different groups, so that a Martian can have 1 parent as well as 10. Nobody will be surprised by a hundred children. Martians have become used to this, and their style of life seems natural to them.

In the Planetary Council, the confusing genealogical system leads to some embarrassment. There meet the worthiest of Martians, and therefore, in order to offend nobody in all of the discussions, it is used first to give the floor to the old Martians, then to the younger ones, and only then to the youngest, childless assessors. However, the maintenance of this order really is not a trivial task. A Martian does not always know all of his parents (and there's nothing to tell about his grandparents). But if by a mistake a grandson speaks first and only then his young-appearing great-grandfather, this is a real scandal.

Your task is to write a program that would define, once and for all, an order that would guarantee that every member of the council takes the floor earlier than each of his descendants.

Input

The first line of the standard input contains an only number N, $1 \leq N \leq 100$—a number of members of the Martian Planetary Council. According to the centuries-old tradition, members of the council are enumerated with integers from 1 up to N. Further, there are exactly N lines; moreover, the ith line contains a list of ith member's children. The list of children is a sequence of serial numbers of children in an arbitrary order separated by spaces. The list of children may be empty. The list (even if it is empty) ends with 0.

Output

The standard output should contain in its only line a sequence of speakers' numbers, separated by spaces. If several sequences satisfy the conditions of the problem, you are to write to the standard output any of them. At least one such sequence always exists.

Sample Input	Sample Output
5	2 4 5 3 1
0	
4 5 1 0	
1 0	
5 3 0	
3 0	

Source: Ural State University Internal Contest, October 2000 Junior Session.

ID for online judge: Ural 1022.

Hint
A directed graph is used to represent the system of Martians' blood relations, where Martians are represented as vertices, and arcs are from parents to children. The sequence of speakers is a topological sort for the directed graph.

An adjacency list *g* is constructed when a test case is input, where *g*[*x*] is used to store children of vertex *x*. Array *ind* is used to store vertices' in-degrees, where *ind*[*x*] is the in-degree of vertex *x*.

All vertices whose in-degrees are 0 are added into queue *q*. Then repeat the following steps until *q* is empty:

Delete the front *x* from *q*.
In-degrees for *x*'s children subtract 1.
If the in-degree for a vertex is 0, then the vertex is added into *q*.
The sequence deleted from *q* is the topological sort.

11.5.4 Rare Order

A rare book collector recently discovered a book written in an unfamiliar language that used the same characters as the English language. The book contained a short index, but the ordering of the items in the index was different from what one would expect if the characters were ordered the same way as in the English alphabet. The collector tried to use the index to determine the ordering of characters (i.e., the collating sequence) of the strange alphabet, but gave up with frustration at the tedium of the task.

You are to write a program to complete the collector's work. In particular, your program will take a set of strings that has been sorted according to a particular collating sequence and determine what that sequence is.

Input

The input consists of an ordered list of strings of uppercase letters, one string per line. Each string contains at most 20 characters. The end of the list is signaled by a line containing a single character #. Not all letters are necessarily used, but the list will imply a complete ordering among those letters that are used.

Output

Your output should be a single line containing uppercase letters in the order that specifies the collating sequence used to produce the input data file.

Sample Input	Sample Output
XWY	XZYW
ZX	
ZXY	
ZXW	
YWWX	
#	

Source: ACM-ICPC World Finals 1990.

ID for online judge: UVA 200.

Hint

Array $T[\ ]$ is used to store the ordered list of strings (the length of T is *tot*). Each uppercase letter is represented as a vertex, and a vertex is labeled with a number that the letter corresponds to, where 'A' is represented with 1, 'B' is represented with 2, and so on. Then T is transferred into an adjacency matrix v as follows:

```
for (int i = 0; i < tot; i++)
 for (int j = i + 1; j < tot; j++) {
  len = min(the length of T[i], the length of T[j]);
  for (int k=0; k<len; k++)
   if (the k-th letter in T[i] != the k-th letter in T[j]) {
    v[the number of the k-th letter in T[i], the number of the k-th
letter in T[j]]=true;
    break;
   }
 }
```

A topological sort is found based on the adjacency matrix v, and the topological sort is the letters' sequence.

Initialization: Calculate the vertices' in-degrees (if $v[i][j]$ == true, then $inq[i] = inq[j] = $ true, and $++ind[j]$; $1 \le i, j \le 26$), and add vertices whose in-degrees are 0 ($inq[i]$ && $ind[i]$ == 0) into q.

Repeat the following process until q is empty.

Delete the front x.

In-degrees for all vertices WHICH are adjacent to x substract 1.

If the in-degree of vetrice i is 0 ($v[x][i]$ && $--ind[i]$ == 0), then vertex i is added into q.

The sequence of vertices deleted from q is the letters' sequence.

11.5.5 Pushing Boxes

Imagine you are standing inside a two-dimensional maze composed of square cells that may or may not be filled with rock. You can move north, south, east, or west one cell at a step. These moves are called walks.

One of the empty cells contains a box that can be moved to an adjacent free cell by standing next to the box and then moving in the direction of the box. Such a move is called a push. The box cannot be moved in any other way than by pushing, which means that if you push it into a corner, you can never get it out of the corner again.

One of the empty cells is marked as the target cell. Your job is to bring the box to the target cell by a sequence of walks and pushes. As the box is very heavy, you would like to minimize the number of pushes (Figure 11.8). Can you write a program that will work out the best such sequence?

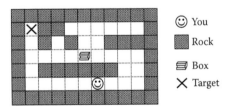

Figure 11.8 Two-dimensional maze.

Input

The input contains the descriptions of several mazes. Each maze description starts with a line containing two integers, r and c (both <=20), representing the number of rows and columns of the maze.

Following this are r lines, each containing c characters. Each character describes one cell of the maze. A cell full of rock is indicated by a "#", and an empty cell is represented by a ".". Your starting position is symbolized by "S", the starting position of the box by "B", and the target cell by "T".

Input is terminated by two zeros for r and c.

Output

For each maze in the input, first print the number of the maze, as shown in the sample output. Then, if it is impossible to bring the box to the target cell, print "Impossible"; otherwise, output a sequence that minimizes the number of pushes. If there is more than one such sequence, choose the one that minimizes the number of total moves (walks and pushes). If there is still more than one such sequence, any one is acceptable.

Print the sequence as a string of the characters N, S, E, W, n, s, e, and w, where uppercase letters stand for pushes, lowercase letters stand for walks, and the letters themselves stand for the directions north, south, east, and west.

Output a single blank line after each test case.

Sample Input	Sample Output
1 7	Maze #1
SB....T	EEEEE
1 7	
SB..#.T	Maze #2
7 11	Impossible.
###########	
#T##.....#	Maze #3
#.#.#..####	eennwwWWWWWeeeeeesswwwwwwwwnNN
#....B....#	
#.######..#	Maze #4
#.....S...#	swwwnnnnnneeessSSS
###########	
8 4	
....	
.##.	
.#..	

Sample Input	Sample Output
.#..	
.#.B	
.##S	
....	
###T	
0 0	

Source: ACM Southwestern European Regional Programming Contest 1997.

IDs for online judges: UVA 589, ZOJ 1249, POJ 1475.

Hint

The two-dimensional maze composed of square cells that may or may not be filled with rock can be represented as an undirected graph. Therefore, the problem requires calculating the shortest path in the graph. Double BFS is used to calculate the shortest path. There are two cases for the problem.

1. Your pushing the box: The cell where the box is in will be the cell where you will move into.
2. Your walking to the cell where you can push the box.

Therefore, the main stem for the program is moving the box, and for each move, your walking to the cell where you can push the box should be considered. That is, a BFS is nested in another BFS.

There are two problems that we should solve.

Problem 1: How Do You Walk to the Cell Where You Can Push the Box?

Suppose your current position is (sr, sc), and the current position for the box is (br, bc). In order to push the box from (br, bc) to its adjacent cell $(nextr, nextc)$, you must walk from (sr, sc) to (er, ec), where (er, ec) and (br, bc) are also adjacent, and (er, ec), (br, bc), and $(nextr, nextc)$ are in the same row or same column. That is, (er, ec) and $(nextr, nextc)$ are across from (br, bc).

BFS is used to solve the problem whether you can walk to (er, ec) or from (sr, sc) or not, and if you can, BFS calculates the sequence with the minimal number of walks.

States are defined based on positions for you and the box, and the sequence of cells that you walk through.

A matrix *visPerson*[ ][ ] is used to represent the path that you walk, to avoid repetition and walking to the box's position, where

$$visPerson[x][y] = \begin{cases} \text{true} & \text{the box is at } (x, y) \text{ or you have walked through } (x, y) \\ \text{false} & \text{otherwise} \end{cases}$$

Initially, *visPerson*[br][bc] = true, and other values for *visPerson*[ ][ ] are false. Each time a state is taken out from a queue and analyzed as follows.

If you walk into (er, ec), return a successful mark and the sequence of walks.

If you have walked through the current position, that is, the value for *visPerson*[ ][ ] has been true, then the next state is taken out from the queue.

Otherwise, for this cell, the value for *visPerson*[ ][ ] is set to true and four directions are enumerated.

If the adjacent cell in direction i ($0 \leq i \leq 3$) is in the maze, isn't filled with rock, and its value for *visPerson*[ ][ ] is false, then a new state is produced. In the new state, your current position is the adjacent cell, and the character in direction i is added into the sequence of walks. The new state is added into the queue. Then the next new direction is considered.

The process repeats until you walk into (*er*, *ec*) or the queue is empty. If the queue is empty, there isn't a path from (*sr*, *sc*) to (*er*, *ec*), and a failed mark is returned.

Problem 2: How Is the Box Pushed to the Target Cell?

Similarly, states are defined based on positions for you and the box, and the sequence of cells that the box is pushed through.

A matrix *visBox*[ ][ ] is used to represent the path on which the box is pushed, where

$$
visBox[x][y] = \begin{cases} \text{true} & \text{the box has been pushed through } (x, y) \\ \text{false} & \text{otherwise} \end{cases}
$$

Initially, all values in *visBox*[ ][ ] are false. The initial state, consisting of the initial position for you and the box, and empty sequence are pushed into the queue.

A state is taken out from the queue and analyzed as follows.

If the box has been pushed through the position in the state, that is, the value for *visBox*[ ][ ] at the position has been true, then the next state is taken out from the queue; else, the value for *visBox*[ ][ ] at the position is set as true.

If the box is at the target cell in the state, then the sequence of minimal moves is output; else, four directions are enumerated.

Its adjacent cell in direction i ($0 \leq i \leq 3$) (*nextr*, *nextc*) and its adjacent cell in the opposite direction (*backR*, *backC*) are calculated. Obviously, you can push the box into (*nextr*, *nextc*) if and only if you can walk to (*backR*, *backC*). If the two cells are in the maze and aren't filled with rock, and *visBox*[*nextr*][*nextc*] == false, then BFS is used to determine whether you can walk to (*backR*, *backC*) or not. If you can, the new state is produced. In the new state, your position is the position of the box, the position of the box becomes (*nextr*, *nextc*), and the sequence of moves = the sequence of moves in the old state + the sequence that you walk from the original position to (*backR*, *backC*) + the character for direction i. Then the new state is added into the queue.

The above process repeats until the sequence of moves with minimal numbers is obtained or the queue is empty. If the queue is empty, then output "Impossible."

Program

```
#include <iostream>
#include <queue>
#include <string>
using namespace std;
const int MAX = 20 + 5;              //the upper limit of the maze
size
char map[MAX][MAX];                  //maze
```

```
bool visPerson[MAX][MAX];              // the path that you walk in the maze
bool visBox[MAX][MAX];                 // the path that the box is pushed
int R, C;                              // the size of the maze R*C
int dir[4][2] = {{0,1}, {0,-1}, {1,0}, {-1,0}}; // displacement in 4
directions
char pushes[4] = {'E', 'W', 'S', 'N'};          //Characters for pushing
the box
char walks[4] = {'e', 'w', 's', 'n'};           // Characters for your
walking
string path;                              // the sequence of minimal moves
struct NODE                               // the structure for state
{
    int br, bc;                           // the position of the box
    int pr, pc;                           //your position
    string ans;                           // the sequence of moves
};
bool InMap(int r, int c)                  //whether (r, c) is in the maze
{
    return (r >= 1 && r <= R && c >= 1 && c <= C);
}
bool Bfs2(int sr, int sc, int er, int ec, int br, int bc, string & ans)
// BFS is used to solve the problem whether you can walk to (er, ec) from
(sr, sc) or not.
//(er, ec) must be adjacent to (br, bc).
{
    memset(visPerson, false, sizeof(visPerson));  //Initialization
    queue<NODE> q;                     //Queue q is used to store states
    NODE node, tmpNode;                // node: the front of q, tmpNode:
new extended state
    node.pr = sr; node.pc = sc; node.ans = "";
    // Initial state (your current position(sr, sc), the sequence of
moves is empty) is added into q
    q.push(node);
    visPerson[br][bc] = true;                     //the current position of
box (br, bc)
    while (!q.empty())               //while q isn't empty, the front is
taken out
    {
        node = q.front();
        q.pop();
        if (node.pr==er && node.pc==ec) { ans = node.ans; return true; }
        // if you walk to (er, ec), return successful mark and the
sequence of moves
        if (visPerson[node.pr][node.pc]) continue;
        visPerson[node.pr][node.pc] = true;
        for (int i=0; i<4; i++)                    //4 directions are
enumerated
        {  //the adjacent cell in direction i (nr, nc): if the cell is in
the maze, isn't filled with rock and you didn't walk to it, a new state
tmpNode is produced (the current position (nr, nc)), the character of
direction i is added to the sequence of moves, and tmpNode is added
into q
            int nr = node.pr + dir[i][0]; int nc = node.pc + dir[i][1];
            if (InMap(nr, nc) && !visPerson[nr][nc] && map[nr][nc] !=
'#')
```

```
                {
                    tmpNode.pr = nr; tmpNode.pc = nc; tmpNode.ans = node.ans
        + walks[i];
                    q.push(tmpNode);
                }
            }
        }
        return false;
}
bool Bfs1(int sr, int sc, int br, int bc)              //your position (sr,
sc), the position of the box is (br, bc). BFS is used to determine
whether the box can be pushed to the target cell
{
        memset(visBox, false, sizeof(visBox));       // Initialization
        queue<NODE> q;                                //queue q is used to store states
        NODE node, tmpNode;           // node: the front of q, tmpNode: new
extended state
        // Initial state (your current position(sr, sc), the position of the
box is (br, bc), the sequence of moves is empty) is added into q
        node.pr = sr; node.pc = sc; node.br = br; node.bc = bc; node.ans =
"";
        q.push(node);
        while (!q.empty())                            // while q isn't empty, the
front is taken out
        {
            node = q.front();
            q.pop();
            if (visBox[node.br][node.bc]) continue;
            visBox[node.br][node.bc] = true;
            if (map[node.br][node.bc] == 'T')      // Target cell
            {
                path = node.ans; return true;
            }
            for (int i=0; i<4; i++)                  // 4 directions are
enumerated
            {
// the adjacent cell in direction i (nextr, nextc) for the box, and its
opposite adjacent cell (backR, backC). The box can be pushed to (nextr,
nextc) if and only if you can walk to (backR, backC).
                int nextr = node.br + dir[i][0]; int nextc = node.bc + dir[i]
[1];
                int backR = node.br - dir[i][0]; int backC = node.bc - dir[i]
[1];
                string ans = "";                    //Initialize the sequence of
moves
//If (backR, backC) and (nextr, nextc) are in the maze, aren't filled
with rock, and the box didn't come to (nextr, nextc), then BFS is used to
determine whether you can walk to (backR, backC). If successful, the new
state tmpNode is produced and pushed into q.
                if (InMap(backR, backC) && InMap(nextr, nextc) && map[nextr]
[nextc] != '#'
                    && map[backR][backC] != '#' && !visBox[nextr][nextc])
                    {
                        if (Bfs2(node.pr, node.pc, backR, backC, node.br, node.
bc, ans))
```

```
                    {
                        tmpNode.pr = node.br;  tmpNode.pc = node.bc;
                        tmpNode.br = nextr;  tmpNode.bc = nextc;
                        tmpNode.ans = node.ans + ans + pushes[i];
                        q.push(tmpNode);
                    }
                }
            }
        }
    }
    return false;
}
int main()
{
    int sr, sc;                            // your starting position
    int br, bc;                            // the position of box
    int cases = 1;                          //the number of test
case
    while (scanf("%d%d", &R, &C) && R && C)    //input test cases
    {
        for (int r=1; r<=R; r++)                    //input maze
        {
            for (int c=1; c<=C; c++)
            {
                cin >> map[r][c];
                if (map[r][c] == 'S'){ sr = r; sc = c; }          //
your starting position
                    else if (map[r][c] == 'B') { br = r; bc = c; }    //
the position of box
            }
        }
                path = "";                        //Initialize the sequence
of moves
//If the box can be pushed to the target, then output the minimal
sequence of moves; else output "Impossible."
        (Bfs1(sr, sc, br, bc)) ? cout << "Maze #" << cases << endl <<
path << endl :
                            cout << "Maze #" << cases << endl <<
"Impossible." << endl;
        cases++;
        cout << endl;
    }
    return 0;
}
```

11.5.6 Basic Wall Maze

In this problem you have to solve a very simple maze consisting of

- A 6 × 6 grid of unit squares
- Walls of length between 1 and 6 that are placed either horizontally or vertically to separate squares
- One start and one end marker

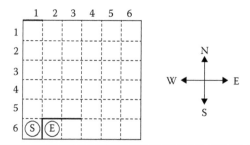

Figure 11.9   Maze.

A maze may look like that shown in Figure 11.9.

You have to find a shortest path between the square with the start marker and the square with the end marker. Only moves between adjacent grid squares are allowed; adjacent means that the grid squares share an edge and are not separated by a wall. It is not allowed to leave the grid.

Input

The input consists of several test cases. Each test case consists of five lines: The first line contains the column and row number of the square with the start marker, and the second line the column and row number of the square with the end marker. The third, fourth, and fifth lines specify the locations of the three walls. The location of a wall is specified by either the position of its left end point followed by the position of its right end point (in the case of a horizontal wall) or the position of its upper end point followed by the position of its lower end point (in the case of a vertical wall). The position of a wall end point is given as the distance from the left side of the grid, followed by the distance from the upper side of the grid.

You may assume that the three walls don't intersect with each other, although they may touch at some grid corner, and that the wall end points are on the grid. Moreover, there will always be a valid path from the start marker to the end marker. Note that the sample input specifies the maze from Figure 11.9.

The last test case is followed by a line containing two zeros.

Output

For each test case print a description of a shortest path from the start marker to the end marker. The description should specify the direction of every move ('N' for up, 'E' for right, 'S' for down, and 'W' for left).

There can be more than one shortest path; in this case you can print any of them.

Sample Input	Sample Output
1 6	NEEESWW
2 6	
0 0 1 0	

Sample Input	Sample Output
1 5 1 6	
1 5 3 5	
0 0	

Source: Ulm Local Contest 2006.

ID for online judge: POJ 2935.

Hint

Because walls are placed either horizontally or vertically to separate squares, squares and lines are all represented as vertices. A maze can be represented as a 13*13 matrix (Figure 11.10).

Vertices for walls are labeled with "visit." Obviously, for walls around the maze,

$visit[0][0] = visit[1][0] = \ldots = visit[12][0] = true$
$visit[0][0] = visit[0][1] = \ldots = visit[0][12] = true$
$visit[0][12] = visit[1][12] = \ldots = visit[12][12] = true$
$visit[12][0] = visit[12][1] = \ldots = visit[12][12] = true$

If a wall from $(x1, y1)$ to $(x2, y2)$ is a vertical wall $(x1 == x2)$,
$visit[2*x1][2*y1] = visit[2*x1][2*y1 + 1] = \ldots = visit[2*x1][2*y2] = true$

If a wall from $(x1, y1)$ to $(x2, y2)$ is a horizontal wall $(y1 == y2)$,
$visit[2*x1][2*y1] = visit[2*x1 + 1][2*y1] = \ldots = visit[2*x2][2*y1] = true$

The start marker (sx, sy) is the vertex whose coordinate is $(sx*2 - 1, sy*2 - 1)$, and the end marker (ex, ey) is the vertex whose coordinate is $(ex*2 - 1, ey*2 - 1)$.

Based on the above graph, BFS can be used to calculate a shortest path from the start marker to the end marker, where queue Q is used to store visited vertices' coordinates (except vertices for walls) and $prev[\ ][\ ]$ is used to store directions for visited vertices.

When a shortest path from the start marker to the end marker is found, based on $prev[\ ][\ ]$ the path from the start marker to the end marker can be output.

11.5.7 Firetruck

The Center City Fire Department collaborates with the transportation department to maintain maps of the city that reflect the current status of the city streets. On any given day, several streets are closed for repairs or construction. Firefighters need to be able to select routes from the fire stations to fires that do not use closed streets.

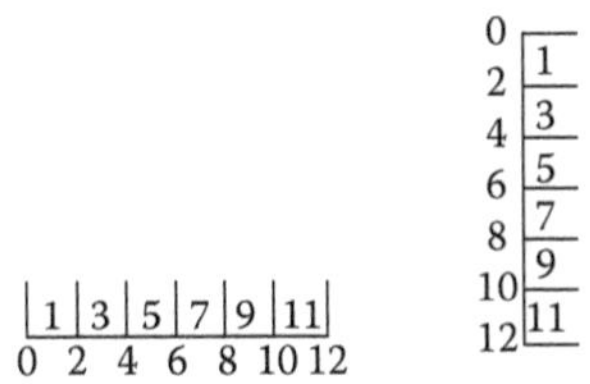

Figure 11.10 13*13 matrix.

Central City is divided into nonoverlapping fire districts, each containing a single fire station. When a fire is reported, a central dispatcher alerts the fire station of the district where the fire is located and gives a list of possible routes from the fire station to the fire. You must write a program that the central dispatcher can use to generate routes from the district fire stations to the fires.

Input

The city has a separate map for each fire district. Street corners of each map are identified by positive integers less than 21, with the fire station always on corner 1. The input file contains several test cases representing different fires in different districts.

- The first line of a test case consists of a single integer that is the number of the street corner closest to the fire.
- The next several lines consist of pairs of positive integers separated by blanks that are the adjacent street corners of open streets. (E.g. if the pair 4 7 is on a line in the file, then the street between street corners 4 and 7 is open. There are no other street corners between 4 and 7 on that section of the street.)
- The final line of each test case consists of a pair of zeros.

Output

For each test case, your output must identify the case by number (CASE #1, CASE #2, etc.). It must list each route on a separate line, with the street corners written in the order in which they appear on the route. And it must give the total number of routes from the fire station to the fire. Include only routes that do not pass through any street corner more than once. (For obvious reasons, the fire department doesn't want its trucks driving around in circles.)

Output from separate cases must appear on separate lines.

The following sample input and corresponding correct output represent two test cases.

Sample Input	Sample Output
6	CASE 1:
1 2	1 2 3 4 6
1 3	1 2 3 5 6
3 4	1 2 4 3 5 6
3 5	1 2 4 6
4 6	1 3 2 4 6
5 6	1 3 4 6
2 3	1 3 5 6
2 4	There are 7 routes from the fire station to street corner 6.
0 0	CASE 2:
4	1 3 2 5 7 8 9 6 4
2 3	1 3 4

Sample Input	Sample Output
3 4	1 5 2 3 4
5 1	1 5 7 8 9 6 4
1 6	1 6 4
7 8	1 6 9 8 7 5 2 3 4
8 9	1 8 7 5 2 3 4
2 5	1 8 9 6 4
5 7	There are 8 routes from the fire station to street corner 4.
3 1	
1 8	
4 6	
6 9	
0 0	

Source: ACM World Finals 1991.

ID for online judge: UVA 208.

Hint

An undirected graph is modeled as follows. Street corners are represented as vertices, and unclosed streets are represented as edges. Routes from the fire station (street corner 1) to the fire (the street corner closest to the fire, numbered *en*) are required.

Backtracking is used to find all possible routes. There is a restriction for a new added vertex. The new added vertex *y* must be connected with the fire vertex *en*. The Warshall algorithm is used to find connectivity between vertex 2 .. *n* and vertex *en*. Suppose *p* is the transitive closure, where *p*[*i*][*j*] shows the connectivity between vertex *i* and vertex *j*.

```
p is initialized as the adjacency matrix;
for (int i=1; i<=n; i++) p[i][i]=1;
for (int k=2; k<=n; k++)
  for (int i=2; i<=n; i++)
    if (p[i][k]) for (int j=2; j<=n; j++) p[i][j]|=p[k][j];
  for (int i=2; i<=n; i++) if (!p[i][en]) cut[i]=1;  // vertex i and
vertex en aren't connected
```

Therefore, *cut*[*y*] is the pruning condition: if (vertex *y* is adjacent to vertex *x*, which is the last vertex in the route) && (vertex *y* hasn't been visited) && (!*cut*[*y*]), then vertex *y* can be added into the route.

11.5.8 Dungeon Master

You are trapped in a three-dimensional (3D) dungeon and need to find the quickest way out. The dungeon is composed of unit cubes that may or may not be filled with rock. It takes 1 minute to move one unit north, south, east, west, up, or down. You cannot move diagonally, and the maze is surrounded by solid rock on all sides.

Is an escape possible? If yes, how long will it take?

Input

The input consists of a number of dungeons. Each dungeon description starts with a line containing three integers, L, R, and C (all limited to 30 in size).

L is the number of levels making up the dungeon.

R and C are the number of rows and columns making up the plan of each level.

Then there will follow L blocks of R lines each containing C characters. Each character describes one cell of the dungeon. A cell full of rock is indicated by a '#', and empty cells are represented by a '.'. Your starting position is indicated by 'S', and the exit by the letter 'E'. There's a single blank line after each level. Input is terminated by three zeros for L, R, and C.

Output

Each maze generates one line of output. If it is possible to reach the exit, print a line of the form

Escaped in x minute(s).

where x is replaced by the shortest time it takes to escape.

If it is not possible to escape, print the line
Trapped!

Sample Input	Sample Output
3 4 5	Escaped in 11 minutes.
S....	Trapped!
.###.	
.##..	
###.#	
#####	
#####	
##.##	
##...	
#####	
#####	
#.###	
####E	

Sample Input	Sample Output
1 3 3 S## #E# ### 0 0 0	

Source: Ulm Local Contest 1997.

ID for online judge: POJ 2251.

Hint

The 3D dungeon is represented as a graph, where unit cubes are represented as vertices, and two adjacent unit cubes are connected with an edge. Given a starting position (*sx, sy, sz*) and an exit (*ex, ey, ez*), if there are paths from the starting position to the exit without rocks, you are required to calculate the shortest path. The backtracking method can be used to solve the problem.

Suppose $d[x][y][z]$ is the shortest distance from the starting position to the unit cube (*x, y, z*). Initially, elements in *d* are set to 100.

The recursive function *dfs*(*x, y, z, k*) is used to calculate it. Suppose the current position (*x, y, z*) is the *k*th position in the shortest path.

```
dfs( x, y, z, k) {
    d[x][y][z]=k;
    if ((x, y, z) is exit (ex, ey, ez)) return;
    if (|x - ex| + |y - ey| + |z - ez|>=d[ex][ey][ez]) return;   //Pruning
    Analyze six adjacent cubes for (x, y, z):
      if (the current cube (x', y', z') satisfies the condition that ((x',
y', z') is within bounds)&&(k+1<d[x'][y'][z']) &&((x', y', z') isn't a
rock))
        { dfs(x', y', z', k+1); }
}
```

Obviously, *dfs*(*sx, sy, sz*, 0) can calculate the shortest paths from the starting position (*sx, sy, sz*) to all cubes. If $d[ex][ey][ez] \geq 1{,}600{,}000{,}000$, then there is no possibility to escape; else, $d[ex][ey][ez]$ is the shortest time it takes to escape.

11.5.9 A Knight's Journey

Background

The knight is getting bored of seeing the same black and white squares again and again and has decided to make a journey around the world. Whenever a knight moves, it is two squares in one direction and one square perpendicular to this. The world of a knight is the chessboard he is living on. Our knight lives on a chessboard that has a smaller area than a regular 8*8 board, but it is still rectangular (Figure 11.11). Can you help this adventurous knight make travel plans?

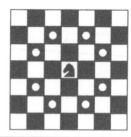

Figure 11.11 Eight possible moves of a knight.

Problem

Find a path such that the knight visits every square once. The knight can start and end on any square of the board.

Input

The input begins with a positive integer n in the first line. The following lines contain n test cases. Each test case consists of a single line with two positive integers, p and q, such that $1 \leq p^*q \leq 26$. This represents a p^*q chessboard, where p describes how many different square numbers 1, ..., p exist, and q describes how many different square letters exist. These are the first q letters of the Latin alphabet: A, ...

Output

The output for every scenario begins with a line containing "Scenario #i:," where i is the number of the scenario starting at 1. Then print a single line containing the lexicographically first path that visits all squares of the chessboard with knight moves, followed by an empty line. The path should be given on a single line by concatenating the names of the visited squares. Each square name consists of a capital letter followed by a number.

If no such path exists, you should output "Impossible" on a single line.

Sample Input	Sample Output
3	Scenario #1:
1 1	A1
2 3	
4 3	Scenario #2:
	Impossible
	Scenario #3:
	A1B3C1A2B4C2A3B1C3A4B2C4

Source: TUD Programming Contest 2005, Darmstadt, Germany.

ID for online judge: POJ 2488.

Hint

The backtracking method can be used to find a path such that the knight visits every square once. Suppose $v[x][y]$ is the number of moves arriving at (x, y).

The recursive function $dfs(x, y, step)$ means in move $step$ the knight visits (x, y). The process is as follows:

- `v[x][y]= step;`
- If all squares have been visited (`step==n*m`), then output the path and exit;
- 8 adjacent squares are enumerated; if the current square (x', y') is within bounds and isn't visited (`v[x'][y']==0`), then `dfs(x', y', step+1);`

In the main program, every possible starting position (i, j) $(1 \le i \le p, 1 \le j \le q)$ is enumerated. If $dfs(i, j, 1)$ can visit all squares, then output the path and exit; else, output "Impossible."

Because of the lexicographically first path, eight adjacent squares should be sorted. Suppose $dx[\ ]$ and $dy[\ ]$ are incremental arrays for the x direction and y direction, respectively. They are sorted as follows:

int $dx[\ ] = \{-1, 1, -2, 2, -2, 2, -1, 1\}$
int $dy[\ ] = \{-2, -2, -1, -1, 1, 1, 2, 2\}$

11.5.10 *Children of the Candy Corn*

The cornfield maze is a popular Halloween treat. Visitors are shown the entrance and must wander through the maze facing zombies, chainsaw-wielding psychopaths, hippies, and other terrors on their quest to find the exit.

One popular maze-walking strategy guarantees that the visitor will eventually find the exit. Simply choose either the right or left wall and follow it. Of course, there's no guarantee which strategy (left or right) will be better, and the path taken is seldom the most efficient. (It also doesn't work on mazes with exits that are not on the edge; those types of mazes are not represented in this problem.)

As the proprieter of a cornfield that is about to be converted into a maze, you'd like to have a computer program that can determine the left- and right-hand paths along with the shortest path so that you can figure out which layout has the best chance of confounding visitors.

Input

The input to this problem will begin with a line containing a single integer n indicating the number of mazes. Each maze will consist of one line with a width, w, and height, h ($3 \le w, h \le 40$), followed by h lines of w characters each that represent the maze layout. Walls are represented by hash marks ('#'), empty space by periods ('.'), the start by an 'S', and the exit by an 'E'.

Exactly one 'S' and one 'E' will be present in the maze, and they will always be located along one of the maze edges and never in a corner. The maze will be fully enclosed by walls ('#'), with the only openings being the 'S' and 'E'. The 'S' and 'E' will also be separated by at least one wall ('#').

You may assume that the maze exit is always reachable from the start point.

Output

For each maze in the input, output on a single line the number of (not necessarily unique) squares that a person would visit (including the 'S' and 'E') for (in order) the left, right, and shortest paths, separated by a single space each. Movement from one square to another is only allowed in the horizontal or vertical direction; movement along the diagonals is not allowed.

Sample Input	Sample Output
2	37 5 5
8 8	17 17 9
########	
#......#	
#.####.#	
#.####.#	
#.####.#	
#.####.#	
#...#..#	
#S#E####	
9 5	
#########	
#.#.#.#	
S.......E	
#.#.#.#	
#########	

Source: ACM South Central United States 2006.

ID for online judge: POJ 3083.

Hint

The key to the problem is to calculate the path following either the right or left wall. Suppose four directions are shown as in Figure 11.12.

If the direction t turns counterclockwise 90°, the new direction $t_1 = (t + 3)\%4$. And if the direction t turns clockwise 90°, the new direction $t_2 = (t + 1)\%4$.

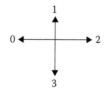

Figure 11.12   Four directions.

A recursive function *dfs_left(x, y, t)* means that the visitor enters *(x, y)* from direction *t*; calculate the number of steps *step* to the exit following the left wall.

```
dfs_left(x, y, t) {
    if (((x, y) is '#')||((x, y) is wall)) return 2;
    the number of steps step ++;
    if ((x, y) is the exit) return 1;
    The successful mark flag is initialized 0;
    direction t turns counter clockwise 90°, and the new direction
tt=(t+3)%4;
    for (int i=0; i<3; i++){        // tt turns clockwise 90° 4 times
        Calculate the value r for dfs_left(x', y', tt) ((x', y') is
(x, y)'s adjacent square in direction tt);
        if (r==1(it is an exit)) { flag=1; break;}
        else if (r==0) {step ++; tt turns clockwise 90° (tt=(tt+1)%4); }
    }
    return flag;
}
```

We can also simulate the visitor following the right wall in the same way. A recursive function *dfs_right (x, y, t)* is used to implement it. In this function, direction *t* turns clockwise 90°, and direction *tt* turns counterclockwise 90°.

It is easier to calculate the shortest path. Suppose *d[x][y]* is the number of steps for the shortest path from the entrance to *(x, y)* and is called the distance value for *(x, y)*.

A recursive function *dfs(x, y, k)* means the visitor leaves *(x, y)*, where he enters it in step *k*. Calculate the shortest path *step*.

```
dfs( x, y, k){
    if ((x, y) is wall or(k>=d[x][y])) exit;
    d[x][y] =k;
    if ((x, y) is the exit), then the shortest path step=k and exit;
    4 adjacent squares (x', y') for (x, y) are enumerated: dfs(x', y',
k+1);
}
```

Suppose (x_s, y_s) is the entrance. In the main program, for four adjacent squares for (x_s, y_s), first, *dfs_left(x_s, y_s, t)* runs and returns the number of steps following the left wall. In the same way, the number of steps following the right wall can be calculated. The number of steps in the shortest path can be calculated by *dfs $(x_s, y_s, 1)$*.

11.5.11 Curling 2.0

On Planet MM-21, after their Olympic games this year, curling is getting popular. But the rules are somewhat different from ours. The game is played on an ice game board on which a square mesh is marked. They use only a single stone. The purpose of the game is to lead the stone from the start to the goal with the minimum number of moves.

Figure 11.13 shows an example of a game board. Some squares may be occupied with blocks. There are two special squares, the start and the goal, which are not occupied with blocks. (These two squares are distinct.) Once the stone begins to move, it will proceed until it hits a block. In

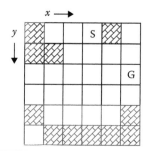

Figure 11.13 Example of board (S, start; G, goal).

order to bring the stone to the goal, you may have to stop the stone by hitting it against a block and throw again.

The movement of the stone obeys the following rules:

- At the beginning, the stone stands still at the start square.
- The movements of the stone are restricted to the *x* and *y* directions. Diagonal moves are prohibited.
- When the stone stands still, you can make it move by throwing it. You may throw it to any direction unless it is blocked immediately (Figure 11.14a).
- Once thrown, the stone keeps moving in the same direction until one of the following occurs:
  - The stone hits a block (Figure 11.14b and c).
    - The stone stops at the square next to the block it hit.
    - The block disappears.
  - The stone goes off the board.
    - The game ends in failure.
  - The stone reaches the goal square.
    - The stone stops there and the game ends in success.
- You cannot throw the stone more than 10 times in a game. If the stone does not reach the goal in 10 moves, the game ends in failure.

Under the rules, we would like to know whether the stone at the start can reach the goal and, if yes, the minimum number of moves required.

With the initial configuration shown in Figure 11.13, four moves are required to bring the stone from the start to the goal. The route is shown in Figure 11.15a. Notice when the stone reaches the goal, the board configuration has changed, as in Figure 11.15b.

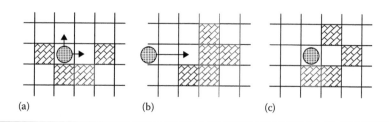

(a)   (b)   (c)

Figure 11.14 Stone movements.

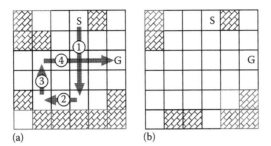

(a)   (b)

Figure 11.15   Solution for Figure 11.14 and the final board configuration.

Input

The input is a sequence of data sets. The end of the input is indicated by a line containing two zeros separated by a space. The number of data sets never exceeds 100.

Each data set is formatted as follows:

The width (w) and the height (h) of the board
First row of the board

...

hth row of the board

The width and the height of the board satisfy $2 \le w \le 20$, $1 \le h \le 20$.

Each line consists of w decimal numbers delimited by a space. The number describes the status of the corresponding square:

0 vacant square
1 block
2 start position
3 goal position

The data set for Figure 11.13 is as follows:

```
6 6
1 0 0 2 1 0
1 1 0 0 0 0
0 0 0 0 0 3
0 0 0 0 0 0
1 0 0 0 0 1
0 1 1 1 1 1
```

Output

For each data set, print a line having a decimal integer indicating the minimum number of moves along a route from the start to the goal. If there are no such routes, print –1 instead. Each line should not have any character other than this number.

Sample Input	Sample Output
2 1	1
3 2	4
6 6	−1
1 0 0 2 1 0	4
1 1 0 0 0 0	10
0 0 0 0 0 3	−1
0 0 0 0 0 0	
1 0 0 0 0 1	
0 1 1 1 1 1	
6 1	
1 1 2 1 1 3	
6 1	
1 0 2 1 1 3	
12 1	
2 0 1 1 1 1 1 1 1 1 1 3	
13 1	
2 0 1 1 1 1 1 1 1 1 1 3	
0 0	

Source: ACM Japan 2006, Domestic.

ID for online judge: POJ 3009.

Hint

Suppose *ans* is the minimum number of moves along a route from the start to the goal. Initially *ans* is 11. A recursive function $dfs(x, y, k)$ is to calculate *ans*, where k means the stone moves into (x, y) in step k. The process is as follows:

```
dfs( x, y, k){
if (k>=10|| k>=ans) exit;
 4 adjacent squares (x', y') for (x, y) are enumerated:
 { if ((x', y') is occupied with a block)
     { The block disappears; dfs(x', y', k+1); recover the
block; }
    else if ((x', y') is the goal)
        { ans=min{ans, k+1}; exit;}
    }
 }
```

The start is (x_s, y_s). The main program runs $dfs(x_s, y_s, 0)$. If *ans* is still 11, then there is no solution; else, *ans* is the minimum number of moves.

11.5.12 Shredding Company

You have just been put in charge of developing a new shredder for the Shredding Company. Although a normal shredder would just shred sheets of paper into little pieces so that the contents would become unreadable, this new shredder needs to have the following unusual basic characteristics:

1. The shredder takes as input a target number and a sheet of paper with a number written on it.
2. It shreds (or cuts) the sheet into pieces, each of which has one or more digits on it.
3. The sum of the numbers written on each piece is the closest possible number to the target number, without going over it.

For example, suppose that the target number is 50 and the sheet of paper has the number 12,346. The shredder would cut the sheet into four pieces, where one piece has 1, another has 2, the third has 34, and the fourth has 6. This is because their sum, 43 (1 + 2 + 34 + 6), is closest to the target number 50 of all possible combinations without going over 50. For example, a combination where the pieces are 1, 23, 4, and 6 is not valid because the sum of this combination, 34 (1 + 23 + 4 + 6), is less than the above combination's 43. The combination of 12, 34, and 6 is not valid either because the sum, 52 (12 + 34 + 6), is greater than the target number of 50 (Figure 11.16).

There are also three special rules:

1. If the target number is the same as the number on the sheet of paper, then the paper is not cut. For example, if the target number is 100 and the number on the sheet of paper is also 100, then the paper is not cut.
2. If it is not possible to make any combination whose sum is less than or equal to the target number, then "error" is printed on a display. For example, if the target number is 1 and the number on the sheet of paper is 123, it is not possible to make any valid combination, as the combination with the smallest possible sum is 1, 2, 3. The sum for this combination is 6, which is greater than the target number, and thus "error" is printed.

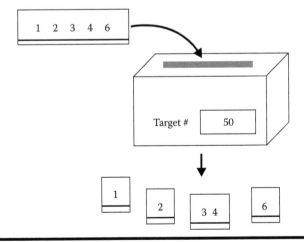

Figure 11.16 Shredding a sheet of paper having the number 12,346 when the target number is 50.

3. If there is more than one possible combination where the sum is closest to the target number without going over it, then "rejected" is printed on a display. For example, if the target number is 15 and the number on the sheet of paper is 111, then there are two possible combinations with the highest possible sum of 12: (a) 1 and 11 and (b) 11 and 1; thus, "rejected" is printed. In order to develop such a shredder, you have decided to first make a simple program that would simulate the above characteristics and rules. Given two numbers, where the first is the target number and the second is the number on the sheet of paper to be shredded, you need to figure out how the shredder should cut up the second number.

Input

The input consists of several test cases, each on one line, as follows:

t_1 *num$_1$*
t_2 *num$_2$*
...
t_n *num$_n$*
0 0

Each test case consists of the following two positive integers, which are separated by one space: (1) the first integer (t_i above) is the target numer and (2) the second integer (*num$_i$* above) is the number that is on the paper to be shredded.

Neither integers may have a 0 as the first digit; for example, 123 is allowed, but 0123 is not. You may assume that both integers are at most six digits in length. A line consisting of two zeros signals the end of the input.

Output

For each test case in the input, the corresponding output takes one of the following three types:

■ *sum part$_1$ part$_2$* ...
■ *rejected*
■ *error*

In the first type, *part$_j$* and *sum* have the following meanings:

1. Each *part$_j$* is a number on one piece of shredded paper. The order of *part$_j$* corresponds to the order of the original digits on the sheet of paper.
2. *sum* is the sum of the numbers after being shredded, that is, *sum* = *part$_1$* + *part$_2$* + ...

Each number should be separated by one space.

The message "error" is printed if it is not possible to make any combination, and "rejected" if there is more than one possible combination.

No extra characters, including spaces, are allowed at the beginning of each line or at the end of each line.

Sample Input	Sample Output
50 12346	43 1 2 34 6
376 144139	283 144 139
927438 927438	927438 927438
18 3312	18 3 3 12
9 3142	error
25 1299	21 1 2 99
111 33333	rejected
103 862150	103 86 2 15 0
6 1104	rejected
0 0	

Source: ACM Japan 2002, Kanazawa.

IDs for online judges: POJ 1416, ZOJ 1694, UVA 2570.

Hint

Suppose *limit* is the target number, *n* is the number on the sheet of paper, and *Max* is the best sum of the numbers written on each piece. The purpose of the problem is to shred the target number into several numbers such that the sum of the numbers written on each piece *Max* is the closest possible number to the target number, without going over it, and we should consider whether it is not possible to make any combination (the number of combinations $r == 0$), or whether there is more than one possible combination (the number of combinations $r > 1$).

The target number *n* is shredded from right to left. If *i* digits for *n* is shredded, that is, $n\%10^i$ is the number on the piece, then

$$\left\lfloor \frac{n}{10^i} \right\rfloor$$

is to be shredded. There are two shredded situations:

1. The number on the piece need not be shredded.
2. The number on the piece needs to be shredded.

A recursive function *dfs*(*n*, *sum*, *now*, *k*, *p*) is used to solve the problem, where *n* is the target number, *sum* is the sum of shredded numbers, *now* is the number to be shredded and is intended to put into the *k*th piece, and *p* is the weight for the next decimal digit.

```
dfs(n, sum, now, k, p){
    if (n==0){// Shred is finished; and now is put into the k-th piece.
        t[k]=now;
        if (sum+now >limit) return;    // The sum of shredded numbers
> Max; exit;
        if (sum+now==Max) r++;    // The sum of shredded numbers =
Max; r++;
```

```
                else if (sum+now >Max){       // adjust Max
                Max=sum+ now;
                   r=1;                    //the number of combinations that the sum is
Max is 1
                   ansk=k;                 //The number of pieces
                   for (int i=1; i<=k; i++) ans[i]=t[i];
                   return;                 // Backtracking
            }
            int m=n%10;
            dfs(n/10, sum, now+p*m, k, p*10);     // Situation [1]
            t[k]=now; //now is put into the kth piece.
            dfs(n/10, sum+now, m, k+1, 10);       // Situation [2]
     }
}
```

In the main program, initially *Max* = 0, *r* = 0; then *dfs*(n/10, 0, n%10, 1, 10) runs. If *r* > 1, then there are more than one possible combination where the sum is closest to the target number without going over it; if *r* == 1, then *Max* is the sum, and shredded numbers are *ans*[*ansk*] .. *ans*[1].

11.5.13 Be Wary of Roses

You've always been proud of your prize rose garden. However, some jealous fellow gardeners will stop at nothing to gain an edge over you. They have kidnapped, blindfolded, and handcuffed you, and dumped you right in the middle of your treasured roses. You need to get out, but you're not sure you can do it without trampling any precious flowers.

Fortunately, you have the layout of your garden committed to memory. It is an $N \times N$ collection of square plots (*N* odd), some containing roses. You are standing on a square marble plinth in the exact center. Unfortunately, you are quite disoriented and have no idea which direction you're facing! Thanks to the plinth, you can orient yourself facing one of north, east, south, or west, but you have no way to know which.

You must come up with an escape path that tramples the fewest possible roses, no matter which direction you're initially facing. Your path must start in the center, consist only of horizontal and vertical moves, and end by leaving the grid.

Input

Every case begins with *N*, the size of grid (1 ≤ *N* ≤ 21), on a line. *N* lines with *N* characters each follow, describing the garden: '.' indicates a plot without any roses, 'R' indicates the location of a rose, and 'P' stands for the plinth in the center.

Input will end on a case where *N* = 0. This case should not be processed.

Output

For each case, output a line containing the minimum guaranteed number of roses you can step on while escaping.

Sample Input	Sample Output
5	At most 2 rose(s) trampled.
.RRR.	
R.R.R	

Sample Input	Sample Output
R.P.R	
R.R.R	
.RRR.	
0	

Source: Bangladesh National Computer Programming Contest 2004.

ID for online judge: UVA 10798.

Hint

Based on the problem description, you are blindfolded and standing on a square marble plinth in the exact center. Suppose you are standing on (x, y), then (x, y), $(n-1-y, x)$, $(y, n-1-x)$ and $(n-1-x, n-1-y)$ constitute a rhombus. If you move into an adjacent square (x', y'), then (x', y'), $(n-1-y', x')$, $(y', n-1-x')$ and $(n-1-x', n-1-y')$ constitute a new rhombus, and either $|x-x'|==1$ or $|y-y'|==1$ holds.

Therefore the successor function is as follows.

Suppose you move from (x, y) into an adjacent square (x', y').

If there is a rose in (x', y'), then the number of roses for (x', y') in direction 1 = the number of roses for (x, y) in direction 1 + 1;

If there is a rose in $(y', n-1-x')$, then the number of roses for (x', y') in direction 2 = the number of roses for (x, y) in direction 2 + 1;

If there is a rose in $(n-1-x', n-1-y')$, then the number of roses for (x', y') in direction 3 = the number of roses for (x, y) in direction 3 + 1;

If there is a rose in $(n-1-y', x')$, then the number of roses for (x', y') in direction 4 = the number of roses for (x, y) in direction 4 + 1;

Obviously, the maximum number of trampled roses in the 4 directions *val* is the upper limit for the number of trampled roses that you move into (x', y').

A memorized BFS is used to calculate the minimum number of trampled roses. Greedy method is also used to solve the problem. A min heap is used to store a priority queue using *val* as priority.

A memorization technology is used to encapsulate the current state. Elements for an encapsulated state are as follow. The current position for the square marble plinth (x, y), the number of trampled roses in 4 directions (up, left, down, right) and its maximal value *val*. A boolean array $vis[x][y][d_1][d_2][d_3][d_4]$ is used to represent whether the state that the number of trampled roses in 4 directions (up, left, down, right) are d_1, d_2, d_3 and d_4 for (x, y) respectively has been searched or not. If a state has been produced, a pruning technology is used and the state isn't into the priority queue.

Program

```cpp
#include <cstdio>
#include <cstring>
#include <algorithm>
#include <queue>
using namespace std;
const int N = 21;                      //The upper limit size of the garden
const int d[4][2]={{1, 0}, {-1, 0}, {0, -1}, {0, 1}};   //4 directions
```

```
int n, vis[N][N][11][11][11][11];];                    // A memorized list,
where vis[x][y][d1][d2][d3][d4] is used to represent whether the state
that the number of trampled roses in 4 directions (up, left, down, right)
are d₁, d₂, d₃ and d₄ for (x, y) respectively, when you arrive at (x, y).
char g[N][N];                            // N×N collection of square plots
struct State {                       // Definition of State
     int x, y, val;                    //(x,y): the position for square
marble plinth; val is the maximal the number of trampled roses in 4
directions
     int up, left, down, right;          // numbers of trampled roses in 4
directions
     State() {x= y=up=left=down=right=0;} //  Initial state (starting
position (0, 0), numbers of trampled roses in 4 directions are all 0)
     State(int x, int y, int up, int left, int down, int right) { //
encapsulate the current state
          this->x = x;
          this->y = y;
          this->up = up;
          this->left = left;
          this->down = down;
          this->right = right;
          val = max(max(max(up,left), down), right);
     }
     bool operator<(const State& c)const {   //priority: the number of
trampled roses
          return val > c.val;
     }
} s;                                //State s
void init(){                               //Input rose garden, and record the
position of for square marble plinth
     for (int i = 0; i < n; i++) {
        scanf("%s", g[i]);
        for (int j = 0; j < n; j++)
          if (g[i][j] == 'P') s.x = i, s.y = j;
     }
}
int bfs(){// A memorized BFS is used to calculate the minimum number of
trampled roses
     memset(vis, 0, sizeof(vis));
     priority_queue<State> Q; //Q : a priority queue storing states
     Q.push(s);
     vis[s.x][s.y][0][0][0][0]=1; // Initial state is pushed into the
     memorized list
     while (!Q.empty()) {
          State u = Q.top();
          Q.pop();
          if (u.x==0||u.x==n-1||u.y==0||u.y==n-1)return u.val; //return
the minimum number of roses
          for (int i = 0; i < 4; i++) {                       // 4 directions
are enumerated
               int xx = u.x + d[i][0];                       //the adjacent
position (xx, yy) in direction i
               int yy = u.y + d[i][1];
               int up = u.up;
```

```
              int left = u.left;
              int down = u.down;
              int right = u.right;
              if (g[xx][yy] == 'R') up++;                    //Accumulation
              if (g[n - 1 - yy][xx] == 'R') left++;
              if (g[n - 1 - xx][n - 1 - yy] == 'R') down++;
              if (g[yy][n - 1 - xx] == 'R') right++;
              if (!vis[xx][yy][up][left][down][right]) { //new state is
pushed into the memorized list and queue
                     vis[xx][yy][up][left][down][right] = 1;
                     Q.push(State(xx, yy, up, left, down, right));
              }
         }
    }
}
int main() {
    while (~scanf("%d", &n) && n) {
         init();
         printf("At most %d rose(s) trampled.n",bfs());
    }
    return 0;
}
```

11.5.14 Monitoring the Amazon

A network of autonomous, battery-powered data acquisition stations has been installed to moni-
tor the climate in the region of the Amazon. An order dispatch station can initiate transmission
of instructions to the control stations so that they change their current parameters. To avoid
overloading the battery, each station (including the order dispatch station) can only transmit to
two other stations. The destinations of a station are the two closest stations. In the case of a draw,
the first criterion is to chose the westernmost (leftmost on the map), and the second criterion is to
chose the southernmost (lowest on the map).

   You are commissioned by Amazon state government to write a program that decides if, given
the localization of each station, messages can reach all stations.

Input

The input consists of an integer N, followed by N pairs of integers X_i, Y_i, indicating the localization
coordinates of each station. The first pair of coordinates determines the position of the order dispatch
station, while the remaining $N - 1$ pairs are the coordinates of the other stations. The following
constraints are imposed: $-20 \leq X_i$, $Y_i \leq 20$, and $1 \leq N \leq 1000$. The input is terminated with $N = 0$.

Output

For each given expression, the output will echo a line indicating if all stations can be reached or
not (see sample output for the exact format).

Sample Input	Sample Output
4	All stations are reachable.
1 0 0 1 –1 0 0 –1	All stations are reachable.

(Continued)

Sample Input	Sample Output
8	There are stations that are unreachable.
1 0 1 1 0 1 −1 1 −1 0 −1 −1 0 −1 1 −1	
6	
0 3 0 4 1 3 −1 3 −1 −4 −2 −5	
0	

Source: 2004 Federal University of Rio Grande do Norte Classifying Contest— Round 1.

ID for online judge: UVA 10687.

Hint

Control stations are represented as vertices. Based on criterions that the first criterion is to choose the westernmost (leftmost on the map), the second criterion is to choose the southern-most (lowest on the map), and out-degree for each vertex k ($0 \le k \le n - 1$) of 2 (If there is only one control station, the out-degree is 1.), the method calculating arcs whose tail is vertex k is as follows.

Coordinate sequence w is constructed and sorted in ascending order as follows. The distance from vertex k is the first key, the X axis is the second key, and the Y axis is the third key. There are arcs from vertex k to vertex $w[1]$ and vertex $w[2]$. (If there is only one control station, there is one arc $(k, w[1])$.)

After the directed graph is constructed, backtracking is used to find the set of reachable vertices from vertex 1. If the set can contain all vertices, then all stations are reached; else, there are stations that are unreachable.

11.5.15 Graph Connectivity

Let us consider an undirected graph $G = <V, E>$. At first there is no edge in the graph. You are to write a program to calculate the connectivity of two different vertices. Your program should maintain the functions inserting or deleting an edge.

Input

The first line of the input contains an integer number N ($2 \le N \le 1000$)—the number of vertices in G. The second line contains the number of commands Q ($1 \le Q \le 20,000$). Then the following Q lines describe each command, and there are three kinds of commands:

I *u v*: Insert an edge (*u, v*). We guarantee that there is no edge between nodes *u* and *v* when you face this command.

D *u v*: Delete an existing edge (*u, v*). We guarantee that there is an edge between nodes *u* and *v* when you face this command.

Q *u v*: A querying command to ask the connectivity between nodes *u* and *v*.

You should notice that the nodes are numbered from 1 to N.

Output

Output one line for each querying command. Print Y if two vertices are connected, or print N otherwise.

Sample Input	Sample Output
3	N
7	Y
Q 1 2	N
I 1 2	Y
I 2 3	
Q 1 3	
D 1 2	
Q 1 3	
Q 1 1	

Source: POJ Monthly June 25, 2006, Zheng Zhao.

ID for online judge: POJ 2838.

Hint
In this problem, the undirected graph G is dynamic. Therefore, an adjacency list is used to store G. Each command is processed as follows:

I x y: Insert y into the adjacency list for x, and the length of the adjacency list for x increases 1; insert x into the adjacency list for y, and the length of the adjacency list for y increases 1.

D x y: Delete y in the adjacency list for x, and the length of the adjacency list for x decreases 1; delete x in the adjacency list for y, and the length of the adjacency list for y decreases 1.

Q x y: If $x == y$, then x and y are connected; else, BFS is used to find all reachable vertices from x. If y is a reachable vertex, then x and y are connected; else, x and y aren't connected.

11.5.16 The Net

Taking into account the present interest in the Internet, smart information routing becomes a must. This job is done by routers situated in the nodes of the network. Each router has its own list of routers that are visible for it (the so-called routing table). It is obvious that the information should be directed in the way that minimizes the number of routers it has to pass (the so-called hop count).

Your task is to find an optimal route (minimal hop count) for the given network from the source of the information to its destination.

Input

The first line contains the number of routers in the network (n). The next n lines contain a description of the network. Each line contains the router *ID*, followed by a hyphen and a comma separated list of *ID*s of visible routers. The list is sorted in ascending order. The next line contains a number of routes (m) you should determine. The consecutive m lines contain starting and ending routers for the route separated by a single space.

Input data may contain descriptions of many networks.

Output

For each network you should output a line with five hyphens and then, for each route, a list of routers passed by information sent from starting to the destination routers.

In case passing of information is impossible (no connection exists), you should output a string "Connection impossible." In the case of multiple routes with the same hop count, the one with lower *ID*s should be output (in the case of a route from router 1 to router 2 as 1 3 2 and 1 4 2, 1 3 2 should be output).

Assumptions: A number of routers is not greater than 300, and there are at least 2 routers in the network. Each routers "sees" no more than 50 routers.

Sample Input	Sample Output
6	-----
1-2,3,4	1 3 6
2-1,3	1 3 5
3-1,2,5,6	2 1 4
4-1,5	2 3 5
5-3,4,6	3 6
6-3,5	2 1
6	-----
1 6	9 7 3 4 8 10
1 5	Connection impossible
2 4	9 6 2
2 5	
3 6	
2 1	
10	
1-2	
2-	

Sample Input	Sample Output
3-4	
4-8	
5-1	
6-2	
7-3,9	
8-10	
9-5,6,7	
10-8	
3	
9 10	
5 9	
9 2	

ID for online judge: UVA 627.

Hint

Routers are represented as vertices in a graph, and the list of routers is the adjacency list for the graph. The Floyd–Warshall algorithm is used to find the lengths of the shortest paths between all pairs of vertices *dist*[ ][ ] (initial values for *dist*[ ][ ] are ∞).

Computing the hop count is to find the shortest path between vertex x and vertex y. Based on the matrix of the shortest paths *dist*[ ][ ], it can be calculated easily.

If *dist*[x][y] == ∞, then router x and router y aren't connected; else, router x and router y are connected. The algorithm is as follows.

```
Output the first vertex x in the route;
while (dist[x][y] != 1) // vertex y is not the last vertex in the
shortest path
    for (int k=1; k<=N; k++)
        if (dist[x][k]==1 && dist[x][k]+dist[k][y]== dist[x][y]) {
            Output vertex k;
            x=k;
            break;
            }
Output the last vertex y in the shortest path;
```

11.5.17 The Warehouse

Secret Agent Θ-7 has found the secret weapon warehouse of the mad scientist Dr. Matroid. The warehouse is full of large boxes (possibly with deadly weapons inside them). While inspecting the warehouse, Θ-7 accidentally triggered the alarm system. The warehouse has a very effective protection against intruders: if the alarm is triggered, then the floor is filled with deadly acid. Therefore, the only way Θ-7 can escape is to climb onto the boxes and somehow reach the exit on top of them. The exit is a hole in the ceiling; if Θ-7 can climb through this hole, then he can escape using the helicopter parked on the roof. There is a ladder and a box below the hole; thus, the goal is to reach this box.

The floor of the warehouse can be divided into a grid containing $n \times n$ cells, the size of each cell is 1×1 meter. Each cell is either fully occupied by one box or unoccupied. Each box is rectangular: the size of the base is 1×1 meter, and the height is either 2, 3, or 4 meters. In Figure 11.17a, you can see an example warehouse, where the numbers show the height of the boxes, E shows the exit, and the circle shows that Secret Agent Θ-7 is currently on the top of that box.

Θ-7 can do two things. If he is standing on top of a box, and in an adjacent cell there is another box, then he can move to the top of this other box. For example, in the situation depicted in Figure 11.17a, he can move to either north or east, but not west or south. Note that only these four directions are allowed; diagonal moves are not possible. The height difference between the two boxes does not matter.

The second thing Θ-7 can do is topple the box he is standing on in one of the four directions. The effect of toppling is best shown by an example: in the situation shown in Figure 11.17b, he can topple the box west (Figure 11.17c) or north (Figure 11.17d). If a box of height h is toppled north (west, south, etc.), then it will occupy h consecutive cells to the north (west, south, etc.) of its original position. The original position will be unoccupied (but can be later occupied again by toppling another box). A box can only be toppled if the cells where it will fall are unoccupied. For example, in Figure 11.17a, the box where Θ-7 is standing cannot be toppled in any of the four directions.

By toppling a box, Θ-7 jumps one step in the direction that the box is toppled (see Figure 11.17c and d). If a box is toppled, then it cannot be toppled again later. Recall that there is a box below the exit (at the cell marked with E in the figure); thus, it is not possible to topple a box over this cell. The alarm system will soon release mutant poisonous biting bats, so Θ-7 has to leave the warehouse as quickly as possible. You have to help him by writing a program that will determine the minimum number of steps required to reach the exit. Moving to an adjacent box or toppling a box is counted as one step.

Input

The input contains several blocks of test cases. The first line of each block contains three integers: the size $1 \le n \le 8$ of the warehouse and two integers, i and j, that describe the starting position of the secret agent. These numbers are between 1 and n; the row number is given by i and the column number by j. The next n lines describe the warehouse. Each line contains a string of n characters. Each character corresponds to a cell of the warehouse. If the character is ".", then the cell is unoccupied. The characters 2, 3, and 4 correspond to boxes of height 2, 3, and 4, respectively. Finally, the character E shows the location of the exit.

The input is terminated by a block with $n = i = j = 0$.

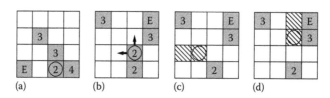

(a)   (b)   (c)   (d)

Figure 11.17   The warehouse.

Output

For each test case, you have to output a single line containing an integer, the minimum number of steps required to reach the exit. If it is not possible to reach the exit, then output the text "Impossible" (without quotation marks).

Sample Input	Sample Output
5 5 3	18
.2..E	
...2.	
4....	
....4	
..2..	
0 0 0	

Source: ACM Central Europe 2005.

IDs for online judges: POJ 2946, UVA 3528.

Hint

The problem requires calculating the shortest path from the starting position of the secret agent to the exit. Θ-7 toppling the box is regarded as a step, so the warehouse can be represented as an unweighted graph. It is suitable that BFS is used to find the shortest path.

Hash technology is used to determine it. A hash table is constructed to store queue pointers pointing to the rear of the queue, that is, pointers pointing to the extended vertices. A hash address is obtained based on the current state a.

A coordinate (x, y) is the position of Θ-7 in the warehouse, $1 \leq x, y \leq 8$. Numbers in cells in the warehouse are in $[0, 4]$, where 0 means the cell is unoccupied, 1 means the cell can't be occupied by other boxes, and 2–4 mean the height of a box occupying the cell. Calculate the hash address for state a as follows. First, from top to bottom, and from left to right, numbers in cells in state a are regarded quinary numbers $(a_{11}, \ldots, a_{1n}, \ldots, a_{nn})_5$. Second, an integer t is obtained as follows. The quinary number is transferred into its corresponding decimal number, and then two digits are added at the end of the corresponding decimal number, and the two digits x and y are the coordinate of Θ-7. Finally, the hash address k for state a is t& (the size of the hash table).

Direct addressing is used to determine whether a state is repeated. For the new extended state b, the corresponding hash address k is calculated. The sequential search is done from $hash[k]$: if $queue[hash[k]]$ and b aren't the same, then $k = (k + 1)$&(the size of t hash table). The process repeats until $queue[hash[k]] == b$ or $hash[k] == 0$.

If there exists a repeated state (there exists a k such that $queue[hash[k]] == b$), a new direction should be selected.

If $hash[k] == 0$, that is, there is no repeated state for b, the state b and the corresponding number of steps for b are added into the queue, and the pointer pointing to the rear is stored into $hash[k]$.

400 ■ *Data Structure Practice: For Collegiate Programming Contests and Education*

1. Hash Technology
   There are two problems for BFS:
   - Storage: Because each step is based on the previous states of the warehouse, previous states of the warehouse should be stored. It is also the reason why the upper limit of the warehouse's size is 8*8.
   - Repeated states: If the current state of the warehouse is the same as a previous state of the warehouse, the search will enter an endless loop.

2. The method extending the vertex at the front of the queue
   A queue is used to store states and the corresponding number of steps, where the state consists of the map of the warehouse and the position of Θ-7 (x, y). State a is obtained from the queue, that is, the front of the queue, and is extended in four directions (up, down, left, right): For cell (x, y), adjacent cells (x', y') in direction d ($0 \leq d \leq 3$) are analyzed as follows.
   - If (x', y') in state a is out of the warehouse, then the direction needn't be considered;
   - If (x', y') in state a is occupied by a box, then a new state b is extended, and (x', y') is the position of Θ-7 in state b;
   - If in state a the box occupying cell (x, y) can't be toppled, that is, the number in the cell $(x, y) < 2$, then the direction needn't be considered;
   - If in state a the height of the box occupying cell (x, y) is k ($k \geq 2$), and the box can be toppled in direction d, that is, there are k consecutive unoccupied cells in direction d of (x, y), then a new state b is extended: the number in (x, y) is set 0, numbers in the k consecutive cells are set 1, and (x', y') is the new position of Θ-7.
   - Then the hash method is used to determine whether state b appeared before. If state b didn't appear before, state b and its corresponding number of steps (the number of steps for state a+1) is added into the queue.
3. Main program

```
Input test case and the initial state a is set up;
State a and the number of steps 0 are added into the queue;
while (the queue isn't empty)   {
     the front of the queue is taken out and as state a;
     for (d=0; d<4; ++d)
          if ( state b can be gotten from state a in direction d)
              if (state b didn't appear in the queue)
              {
                   state b and its corresponding number of steps is
added into the queue;
                   if (state b is the exit)
                   {
                      Output the number of steps;
                      return;
                   }
                   The pointer pointing to the rear of the queue is
added into Hash table;
              }
     }
Output "Impossible."
```

Program

```
#include<iostream>
```

```cpp
using std::cin;
using std::cout;
using std::endl;
#include<cstdio>
#include<cstring>
const long MAXN = 10;           //The upper limit of the size of the
warehouse
const long MAXH = 1048575;    // The upper limit of the length of Hash
table and queue
const long fx[4][2] = {{0, 1}, {1, 0}, {0, -1}, {-1, 0}};  //Four
directions
struct Tmap     //the structure of the state
{
    long data[MAXN][MAXN]; //Map of the warehouse,
```

$$// \text{ where } data[i][j] = \begin{cases} 0 & (i, j) \text{ is empty} \\ 1 & (i, j) \text{ can't be covered by other box} \\ \text{other integer} & \text{the height of box at } (i, j) \end{cases}$$

```cpp
    long x, y;    //the position of Θ- 7
};
long n, x, y, tx, ty; //the size of the warehouse is n*n, the exit is
(tx, ty)
Tmap a;   // the current state
Tmap queue[1048580];    // the queue storing states
long hash[1048580];     // Hash table storing pointers for the queue
long dep[1048580];      // queue storing the number of steps
void init( )            // Input data and set up initial state
{
    char st[MAXN];   //string for the current row
    long i, j;
    memset(a.data, 0xff, sizeof(a.data));   // Initialize cells in the
warehouse -1
    for (i=1; i<=n; ++i)       // input the warehouse row by row
    {
        scanf("%s", st);      // the i-th row
        for (j=0; j<n; ++j)
        {
            if (st[j] == '.') a.data[i][j+1] = 0;
            else if (st[j] == 'E')        // (i, j+1) is exit
            {
                a.data[i][j+1] = 1;  // the position can't be occupied by
other boxes
                tx = i;
                ty = j + 1; // the position of exit
            }
            else a.data[i][j+1] = st[j] - '0'; // the height of the box at (i,
j+1)
        }
    }
    a.x = x;
    a.y = y;    // the start position of Θ- 7
}
```

```
long gethash(const Tmap &a)
// Calculate the hash address for state a: Firstly, from top to down, and
   from left to right, numbers in cells in state a are regards as a
   quinary number (a₁₁,...,a₁ₙ, qaₙₙ)₅. Secondly, an integer t is gotten as
   follows. The quinary number is transferred into its corresponding
   decimal number, then two digits are added at the end of the
   corresponding decimal number, and the two digits x and y are the
   coordinate of Θ- 7. Finally the Hash address for state a is t&(the size
   of Hash table).
{
   long i, j, t = 0;
   for (i=1; i<=n; ++i)
   for (j=1; j<=n; ++j)
     t = (t * 5 + a.data[i][j]) & MAXH;
   t = (t * 10 + a.x) & MAXH;
   t = (t * 10 + a.y) & MAXH;
   return t;
}
bool thesame(const Tmap &a, const Tmap &b)
// Determine whether state a and state b are same or not
{
   long i, j;
// If positions of Θ- 7 in state a and state b aren't same, or two
corresponding cells in state a and state b aren't same; then return
false; else return true
   if ((a.x != b.x) || (a.y != b.y)) return false;
   for (i=1; i<=n; ++i)
   for (j=1; j<=n; ++j)
   if (a.data[i][j] != b.data[i][j]) return false;
   return true;
}
bool get(const Tmap &a)
// Determine whether state a has been appeared in the queue
{
   long k = gethash(a);   // Calculate Hash value k for state a
   while (hash[k] != 0)
   {
      if (thesame(queue[hash[k]], a)) return true;
      k = (k + 1) & MAXH;
   }
   return false;   // state a didn't appear
}
bool run(const Tmap &a, Tmap &b, long d)
//If new state b can be gotten from state a expanded in direction d, then
the new state b and successful mark true are returned; else return false
{
   long x = a.x + fx[d][0];
   long y = a.y + fx[d][1];   //(x, y) is adjacent cell for position of
Θ- 7 in state a in direction d
   b = a;
// If (x, y) is out of the boundary, return false;
   if (a.data[x][y] == -1) return false;
// If (x, y) isn't empty, record the position of Θ- 7 in state b and
return true
```

```
      if (a.data[x][y] > 0)
      {
         b.x = x;
         b.y = y;
         return true;
      }
//if the position of Θ- 7 in state a can't be occupied by other boxes,
then return false
      if (a.data[a.x][a.y] < 2) return false;
//enumerate every height i for box: in direction the i-th cell isn't
  empty, then return false; else the height of the cell in state b is 1
      for (long i=1; i<=a.data[a.x][a.y]; ++i) //enumerate
      {
         if (a.data[a.x+fx[d][0]*i][a.y+fx[d][1]*i] != 0) return false;
         b.data[a.x+fx[d][0]*i][a.y+fx[d][1]*i] = 1;
      }
// the box at the position of Θ- 7 in state a is toppled in direction d,
   the position is empty in state b, (x, y) is the position of Θ- 7, and
   return true
      b.data[a.x][a.y] = 0;
      b.x = x;
      b.y = y;
      return true;
}
void save(long a)          //direct addressing, pointer for queue rear a is
stored in Hash table
{
   long k = gethash(queue[a]);
   while (hash[k] != 0) k = (k + 1) & MAXH;
   hash[k] = a;
}
void solve(void)  //calculate and output the solution
{
   long open = 0;  //Initialize pointers for the front and rear for the
queue
   long closed = 1;
   long i;
   Tmap k, k2;
   memset(hash, 0x00, sizeof(hash)); //Initialize Hash Table
   queue[1] = a; dep[1] = 0; //map of warehouse and the number of steps 0
are added into queues
   while (open < closed ) // while the queue isn't empty
   {
      k = queue[++open]; // the front of the queue
      for (i=0; i<4; ++i)  // 4 directions are enumerated
      if (run(k, k2, i)) //new state k2 can be gotten from state k in
direction i
      if (!get(k2)) //state k2 is different from all states in the queue
      {
          queue[++closed] = k2; //state k2 and number of steps are added
into the queue
          dep[closed] = dep[open] + 1;
          if ((k2.x == tx) && (k2.y == ty)) // state k2 is exit
          {
```

```
                printf("%ld\n", dep[closed]);
                return;
        }
        save(closed); //pointer closed pointing to the rear of the queue
is stored in Hash Table
      }
   }
   printf("Impossible.\n"); //output impossible
}
int main( )
{
   while ((scanf("%ld %ld %ld", &n, &x, &y) != EOF) && n && x && y)
   {
      init( ); // Input data and set up initial state
      solve( ); //calculate and output the solution
   }
   return 0;
}
```

Chapter 12

Algorithms of Minimum Spanning Trees

A tree is a connected undirected simple graph with no circuit. A tree with n vertices has $n-1$ edges. If an edge of a tree is deleted, the tree becomes an unconnected graph. If an edge is added between any two vertices in a tree, a circuit is produced. As mentioned, breadth-first search (BFS) or depth-first search (DFS) for a connected graph will produce a BFS tree or a DFS tree. In this chapter, we focus on finding a spanning tree so that the sum of weights of edges of the tree is minimized in a weighted undirected connected graph. Such a spanning tree is called a minimum spanning tree (MST).

The greedy strategy is used to find an MST. That is, at each step, the added edge can't form a circuit, and based on it, the weight of the added edge should be minimal. Such an added edge is called a safe edge. There are two algorithms of MSTs:

1. Kruskal algorithm
2. Prim algorithm

The two algorithms get an MST by different methods.

12.1 Kruskal Algorithm

Initially, n vertices constitute a forest. Then, edge (u, v) connecting two distinct trees in the forest with the least weight is regarded as the safe edge and is added to the forest. Repeat the process until the MST is obtained.

Based on it, given a weighted connected graph $G(V, E)$, where $|V| = n$ and $|E| = m$, Kruskal algorithm is as follows:

```
Sort edges in ascending order of weight values;
    Suppose initially F is a forest consisting of n trees, and each tree
corresponds to a vertex in graph G;
    Initially the sum of weights of edges of the minimum spanning tree
ans=0;
```

```
    for (int k=1; k<=m; k++)  // Enumerate m edges in Graph G in the order
    { if ( the current edge (i, j) connects two distinct components)
        {  add edge (i, j) into the forest and combine the two
components;
            ans+=the weight of edge (i, j);  }
    }
Output ans;
```

Obviously, if a subtree is regarded as a set, and the root of the subtree is regarded as the representative of the set, then a union–find set can be used to determine whether two vertices belong to one tree and combine two different trees.

The time complexity for Kruskal algorithm is $O(E*\ln E)$. That is, its efficiency depends on the number of edges $|E|$. Therefore, Kruskal algorithm is suitable for sparse graphs.

12.1.1 Constructing Roads

There are N villages, which are numbered from 1 to N, and you should build some roads such that every two villages can connect to each other. We say two villages A and B are connected if and only if there is a road between A and B, or there exists a village C such that there is a road between A and C, and C and B are connected.

We know that there are already some roads between some villages, and your job is the build some roads such that all the villages are connected and the length of all the roads built is minimum.

Input

The first line is an integer N ($3 \leq N \leq 100$), which is the number of villages. Then come N lines, the ith of which contains N integers, and the jth of these N integers is the distance (the distance should be an integer within [1, 1000]) between village i and village j.

Then there is an integer Q ($0 \leq Q \leq N * [N + 1]/2]$). Then come Q lines; each line contains two integers a and b ($1 \leq a < b \leq N$), which means the road between village a and village b has been built.

Output

You should output a line containing an integer, which is the length of all the roads to be built such that all the villages are connected, and this value is minimum.

Sample Input	Sample Output
3	179
0 990 692	
990 0 179	
692 179 0	
1	
1 2	

Source: PKU Monthly, kicc.

ID for online judge: POJ 2421.

Analysis

Villages and roads connecting villages are represented as a weighted graph, where villages are represented as vertices, roads are represented as edges, and the length of a road is the weight of an edge. Obviously, the problem requires adding edges to spanning trees (already built roads between some villages) to construct an MST. Because the number of added edges is less than $N - 1$, Kruskal algorithm is suitable for the problem. The problem requires calculating the length of all the roads to be built.

An adjacency matrix P is used to represent the weighted graph, and an array Fa is used to store every vertex's parent pointer pointing to its parent. By using parent pointers, the root of the tree containing the vertex can be found: from vertex i, repeat using array Fa ($Fa[Fa[\dots Fa[i] \dots]]$) until $x == Fa[x]$. x is the root of the tree containing vertex i, that is, $Fa[i] = x$ $(0 <= i <= n - 1)$.

The algorithm is as follows:

1. Initialization;
       Input a test case and construct an adjacency matrix P to represent the weighted graph;
       N vertices are represented as N distinct spanning trees ($Fa[i]=i$, $0<=i<=n-1$);
       Input Q built roads (a, b), Suppose $Fa[b]=a$ ($1<=a<b<=N$);
       Initialize the minimal length of all the roads to be built ans 0;
2. Compute the minimal length of all the roads to be built ans;
   Sort weights of edges k in ascending order ($1<=k<=1000$) and enumerate these edges:
       If the current enumerated edge (i, j) whose weight is k connects two distinct subtrees ($P[i][j]==k$ && the root of the subtree containing vertex i != the root of the subtree containing vertex j, $0<=i<j<=n-1$), then combine the tree containing vertex i into the tree containing vertex j ($Fa[Fa[i]]=Fa[j]$); and k is added into the length ans (ans += k);
3. Output the length of all the roads to be built ans;

Program

```java
import java.util.*;
import java.io.Reader;
import java.io.Writer;
import java.math.*;
public class Main{
public static void print(String x){   // Output the result
    System.out.print(x);
}
   static int[ ] Fa;        // array Fa is used to store every vertex's
father pointer
   public static int Get_father(int x){    //Get the root of the tree
containing vertex x
    return Fa[x]=Fa[x]==x?x:Get_father(Fa[x]);
}
   public static void main(String[ ] args){
    Scanner input = new Scanner(System.in); //java standard input
    while (input.hasNextInt()){              //Input test cases
repeatedly
```

```
        int N = input.nextInt();              //Input the number of
vertices
        int[ ][ ] P = new int [N+1][N+1];
        for (int i=0;i<N;i++)                 //Input the adjacency matrix
           for (int j=0;j<N;j++)
              P[i][j] = input.nextInt();
        Fa = new int[N+1];                    //Memory for array Fa
        for (int i=0;i<N;i++) Fa[i]=i;
//Input initial m edges for spanning trees, set up parent-child
relationships
        for (int M=input.nextInt();M>0;M--)
         Fa[Get_father(input.nextInt()-1)]=Get_father(input.
nextInt()-1);
        int ans = 0;                 //Initialize the length of roads to be
built
        for (int k=1;k<=1000;k++)    //Enumerate lengths of edges in
ascending order k
        for (int i=0;i<N;i++)      // Enumerate any pair of vertices i and
j
            for (int j=0;j<N;j++)
             if (P[i][j]==k&&Get_father(i)!=Get_father(j)){ //if the
length of edge (i, j) is k, combine the ree containing vertex i into the
tree containing vertex j, and k is added into the length of roads to be
built
               Fa[Fa[i]]=Fa[j];
               ans += k;
             }
        print(ans+"\n");              //Output the length of roads to be
built
      }
   }
}
```

12.2 Prim Algorithm

In Prim algorithm, edges in set A make up a single MST. Initially, A is empty. Repeatedly add edges to A so that at each step an added edge has only one vertex in the tree and contributes the minimum amount possible to the tree's weight (Figure 12.1).

The idea for Prim algorithm is as follows:

```
Suppose r is the starting vertex; and d[i]=min{weight(j, i) | Vertex j is
a vertex in the spanning tree, and vertex i isn't in the spanning tree.
};
In the process of the algorithm, all vertices which aren't in the tree
are sorted in the ascending order of their values in array d and form a
priority queue Q;
Suppose f[u] is the parent of vertex u in the tree. In the process of the
algorithm, the set of edges of the minimum spanning tree A satisfies
A={(u, f[u]) | u∈V-{r}-Q}.
```

When the process of the algorithm ends, priority queue Q is empty, and the set of edges of the MST A satisfies $A = \{(u, f[u]) | u \in V - \{r\}\}$ and

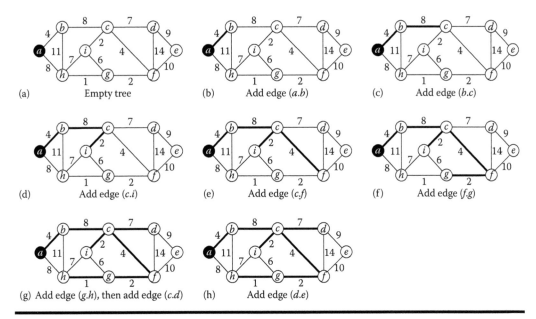

(a) Empty tree
(b) Add edge (*a.b*)
(c) Add edge (*b.c*)

(d) Add edge (*c.i*)
(e) Add edge (*c.f*)
(f) Add edge (*f.g*)

(g) Add edge (*g.h*), then add edge (*c.d*)
(h) Add edge (*d.e*)

Figure 12.1   The Prim algorithm process used to calculate an MST.

$$ans = \sum_{u \in V - \{r\}} weight(u, f[u])$$

The process for Prim algorithm is as follows:

```
for (each v∈G(V))
{ d[v]=∞; f[u]=nil };
d[r]=0; Q=G(V);
    while (Q!=∅)
    {   Get vertex u whose d[u] is the least in Q;   // Add vertex u into
minimum spanning tree
        Q=Q-{u};
        if (u!=r) ans=ans+w[u,f[u]];   //If vertex u isn't the root, then
add the weight into ans
    for (each v∈ the set of vertices adjacent to vertex u)        // renew
the value of d and parent pointer of vertex v which is adjacent to vertex
u
if (v∈Q)&&(w[u,v]<d[v])
   { f[v]=u; d[v]=w[u,v]; }
};
Output the weight of the minimum spanning tree ans;
```

The time complexity for Prim algorithm is $O(V^*\ln V + E^*\ln V)$. Because $|E| < |V|^2$, the upper limit of the time complexity is $O(V^*\ln V + V^2*\ln V)$. That is, the efficiency of Prim algorithm depends on $|V|$. Therefore, Prim algorithm is suitable for dense graphs.

12.2.1 Agri-Net

Farmer John has been elected mayor of his town. One of his campaign promises was to bring internet connectivity to all farms in the area. He needs your help, of course.

Farmer John ordered a high-speed connection for his farm and is going to share his connectivity with the other farmers. To minimize cost, he wants to lay the minimum amount of optical fiber to connect his farm to all the other farms.

Given a list of how much fiber it takes to connect each pair of farms, you must find the minimum amount of fiber needed to connect them all together. Each farm must connect to some other farm such that a packet can flow from any one farm to any other farm.

The distance between any two farms will not exceed 100,000.

Input

The input includes several cases. For each case, the first line contains the number of farms, N $(3 \le N \le 100)$. The following lines contain the $N \times N$ connectivity matrix, where each element shows the distance from one farm to another. Logically, there are N lines of N space-separated integers. Physically, they are limited in length to 80 characters, so some lines continue onto others. Of course, the diagonal will be 0, since the distance from farm i to itself is not interesting for this problem.

Output

For each case, output a single integer length that is the sum of the minimum length of fiber required to connect the entire set of farms.

Sample Input	Sample Output
4	28
0 4 9 21	
4 0 8 17	
9 8 0 16	
21 17 16 0	

Source: USACO.

ID for online judge: POJ 1258.

Analysis

Farms and fibers connecting farms are represented as a weighted graph, where farms are represented as vertices, and John's farm is represented as vertex 0; straight lines connecting farms are represented as edges; and distances between two farms are weights of corresponding edges. Finding the minimum amount of fiber needed to connect them all together is to calculate the MST of the graph. Because the upper limit of the number of vertices is 100, Prim algorithm is suitable for the problem.

Suppose v is the adjacency matrix for the graph. Array *dist* is used to store a priority queue Q, where $dist[i]$ is the distance between vertex i and the spanning tree. Initially, $dist[0] = \infty$, $dist[i] = v[0][i]$ $(1 \le i \le n - 1)$. Array *use* is the flag that a vertex is in the spanning tree or not. Initially, only John's farm is in the spanning tree ($use[0] =$ true), and other vertices aren't in the spanning tree ($use[i] =$ false, $1 \le i \le n - 1$).

$n - 1$ edges are added into the spanning tree as follows:

{ find such a vertex *tmp* connecting the spanning tree that *dist*[*tmp*]=

$\min_{1 \le i \le n-1, usd[i]=false} \{dist[i]\}$);

      The weight of the edge connecting vertex *tmp* and the spanning tree is accumulated (*tot* +=*dist*[*tmp*]);

      Vertex *tmp* is added into the spanning tree (*use*[*tmp*]=true);

      Adjust array *dist* (*dist*[*k*]=min{*dist*[*k*], *v*[*k*][*tmp*] | *use*[*k*]=false },1≤*k*≤n-1);

   };

  Output the weight of the MST *tot*;

Program

```java
import java.util.*;
public class Main {
public static void main(String[ ] args){
      Scanner input = new Scanner(System.in);
      while (input.hasNextInt()){     //Input test cases
          int n=input.nextInt(),tot=0;//Input the number of vertices,
initialize the weight of the spanning tree
          int[ ][ ] v = new int[n][n];  // v is the adjacency matrix for
the graph
          int[ ] dist = new int[n];
          boolean[ ] use = new boolean[n];
          use[0] = true;        //Initially the spanning tree only contains
vertex 0
          for (int i=1;i<n;i++)
             use[i] = false;
          for (int i=0;i<n;i++)       // Input the adjacency matrix for the
graph
             for (int j=0;j<n;j++)
                v[i][j] = input.nextInt();
          dist[0] = 0x7FFFFFFF;      //Initialize array dist
          for (int i=1;i<n;i++)
             dist[i] = v[0][i];
          for (int i=1;i<n;i++){   // n-1 edges for the spanning tree
             int tmp = 0;          //  Find vertex tmp
             for (int k=1;k<n;k++)
                if (dist[k]<dist[tmp]&&!use[k]) tmp = k;
             tot += dist[tmp]; // Add the weight, and vertex tmp into the
spanning tree
             use[tmp] = true;
             for(int k=1;k<n;k++)    //Adjust array dist
                if (!use[k])
                dist[k] = min(dist[k],v[k][tmp]);
          }
          System.out.println(tot);//Output the weight of the minimum
spanning tree
      }
   }
   private static int min(int i, int j) {
      if (i<j) return i;
      else  return j;
   }
}
```

12.3 Problems

12.3.1 Network

Andrew is working as system administrator and is planning to establish a new network in his company. There will be N hubs in the company, and they can be connected to each other using cables. Since each worker of the company must have access to the whole network, each hub must be accessible by cables from any other hub (with possibly some intermediate hubs).

Since cables of different types are available and shorter ones are cheaper, it is necessary to make such a plan of hub connection that the maximum length of a single cable is minimal. There is another problem—not each hub can be connected to any other one because of compatibility problems and building geometry limitations. Of course, Andrew will provide you with all the necessary information about possible hub connections.

You are to help Andrew find the way to connect hubs so that all of the above conditions are satisfied.

Input

The first line of the input contains two integer numbers: N, the number of hubs in the network ($2 \leq N \leq 1000$), and M, the number of possible hub connections ($1 \leq M \leq 15{,}000$). All hubs are numbered from 1 to N. The following M lines contain information about possible connections—the numbers of two hubs, which can be connected, and the cable length required to connect them. Length is a positive integer number that does not exceed 10^6. There will be no more than one way to connect two hubs. A hub cannot be connected to itself. There will always be at least one way to connect all hubs.

Output

Output first the maximum length of a single cable in your hub connection plan (the value you should minimize). Then output your plan: first output P, the number of cables used, and then output P pairs of integer numbers, the numbers of hubs connected by the corresponding cable. Separate numbers by spaces or line breaks.

Sample Input	Sample Output
4 6	1
1 2 1	4
1 3 1	1 2
1 4 2	1 3
2 3 1	2 3
3 4 1	3 4
2 4 1	

Source: ACM Northeastern Europe 2001, Northern Subregion.

IDs for online judges: POJ 1861, ZOJ 1542.

Hint

N hubs are represented as vertices, and cables between hubs are represented as edges. There will be no more than one way to connect two hubs. A hub can't be connected to itself. There will always be at least one way to connect all other hubs. Therefore, connecting hubs can be represented as a weighted connected graph. Because hub connection must be the cheapest, it is a problem for the MST. And because the maximum length of a single cable is minimal (it can be proved), the greatest weight of the edge in the MST is the maximum length of a single cable.

The reason for using Kruskal algorithm is as follows:

1. The upper limit of the number of vertices is 1,000, and the upper limit of the number of edges is 15,000. Therefore, the graph is regarded as a sparse graph.
2. The problem requires calculating the maximum length of a single cable in the hub connection plan. In Kruskal algorithm, the last edge added into the MST is the maximum length of a single cable.

12.3.2 Truck History

Advanced Cargo Movement, Ltd. uses trucks of different types. Some trucks are used for vegetable delivery, others for furniture, and still others for bricks. The company has its own code describing each type of truck. The code is simply a string of exactly seven lowercase letters (each letter on each position has a very special meaning, but that is unimportant for this task). At the beginning of the company's history, just a single truck type was used, but later other types were derived from it, and then from the new types other types were derived, and so on.

Today, ACM is rich enough to pay historians to study its history. One thing historians tried to find out is the so-called derivation plan, that is, how the truck types were derived. They defined the distance of truck types as the number of positions with different letters in truck type codes. They also assumed that each truck type was derived from exactly one other truck type (except for the first truck type, which was not derived from any other type). The quality of a derivation plan was then defined as

$$1/\sum_{(t_0,t_d)} d(t_0,t_d)$$

where the sum goes over all pairs of types in the derivation plan such that t_0 is the original type, t_d is the type derived from it, and $d(t_0, t_d)$ is the distance of the types.

Since historians failed, you are to write a program to help them. Given the codes of truck types, your program should find the highest possible quality of a derivation plan.

Input

The input consists of several test cases. Each test case begins with a line containing the number of truck types, N, $2 \leq N \leq 2000$. Each of the following N lines of input contains one truck type code (a string of seven lowercase letters). You may assume that the codes uniquely describe the trucks; that is, no two of these N lines are the same. The input is terminated with zero at the place of number of truck types.

Output

For each test case, your program should output the text "The highest possible quality is 1/Q," where 1/Q is the quality of the best derivation plan.

Sample Input	Sample Output
4	The highest possible quality is 1/3.
aaaaaaa	
baaaaaa	
abaaaaa	
aabaaaa	
0	

Source: CTU Open 2003.

IDs for online judges: POJ 1789, ZOJ 2158.

Hint

Suppose each truck is represented as a vertex, the type of truck i is *code[i]*, and the weight of the edge connecting vertex i and vertex j is

$$\sum_{k=1}^{7} code[i][k]! = code[j][k], \ 0 \le i, j \le n-1$$

Because each truck type was derived from exactly one other truck type (except for the first truck type, which was not derived from any other type), the graph is a weighted connected undirected graph.

Based on the quality of a derivation plan

$$\left(\frac{1}{\displaystyle\sum_{(t_0,t_d)} d(t_0,t_d)} \right)$$

the program should find the highest possible quality of a derivation plan. Therefore,

$$\sum_{(t_0,t_d)} d(t_0,t_d)$$

must be minimal. All trucks should be included, and thus it is a problem for the MST.
The reason for using Prim algorithm is as follows:

1. Based on the quality of a derivation plan, the weighted connected graph is a dense graph.
2. The upper limit of the edge's weight is 7. In priority queue Q, vertices that aren't in the spanning tree are sorted based on distances. All vertices with distance i are stored in linear list $Q[i]$. It can improve the time complexity.

12.3.3 Slim Span

Given an undirected weighted graph G, you should find one of the spanning trees specified as follows.

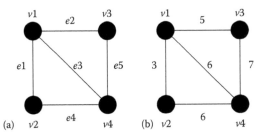

Figure 12.2 A graph G and the weights of the edges.

The graph G is an ordered pair (V, E), where V is a set of vertices $\{v_1, v_2, \ldots, v_n\}$ and E is a set of undirected edges $\{e_1, e_2, \ldots, e_m\}$. Each edge $e \in E$ has its weight $w(e)$.

A spanning tree T is a tree (a connected subgraph without cycles) that connects all the n vertices with $n - 1$ edges. The slimness of a spanning tree T is defined as the difference between the largest weight and the smallest weight among the $n - 1$ edges of T.

For example, a graph G in Figure 12.2a has four vertices $\{v_1, v_2, v_3, v_4\}$ and five undirected edges $\{e_1, e_2, e_3, e_4, e_5\}$. The weights of the edges are $w(e_1) = 3$, $w(e_2) = 5$, $w(e_3) = 6$, $w(e_4) = 6$, and $w(e_5) = 7$, as shown in Figure 12.2b.

There are several spanning trees for G. Four of them are depicted in Figure 12.3a–d. The spanning tree T_a in Figure 12.3a has three edges whose weights are 3, 6, and 7. The largest weight is 7 and the smallest weight is 3, so that the slimness of the tree T_a is 4. The slimnesses of spanning trees T_b, T_c, and T_d shown in Figure 12.3b–d are 3, 2, and 1, respectively. You can easily see the slimness of any other spanning tree is greater than or equal to 1; thus, the spanning tree T_d in Figure 12.3d is one of the slimmest spanning trees whose slimness is 1.

Your job is to write a program that computes the smallest slimness.

Input

The input consists of multiple data sets, followed by a line containing two zeros separated by a space. Each data set has the following format:

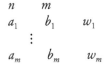

$$
\begin{array}{ccc}
n & m & \\
a_1 & b_1 & w_1 \\
& \vdots & \\
a_m & b_m & w_m
\end{array}
$$

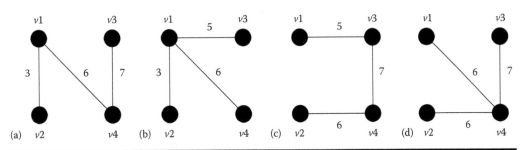

Figure 12.3 Examples of the spanning trees of G.

Every input item in a data set is a nonnegative integer. Items in a line are separated by a space. n is the number of the vertices and m the number of the edges. You can assume $2 \leq n \leq 100$ and $0 \leq m \leq n(n-1)/2$. a_k and b_k $(k = 1, \ldots, m)$ are positive integers less than or equal to n, which represent the two vertices v_{ak} and v_{bk} connected by the kth edge e_k. w_k is a positive integer less than or equal to 10,000, which indicates the weight of e_k. You can assume that the graph $G = (V, E)$ is simple; that is, there are no self-loops (that connect the same vertex) or parallel edges (two or more edges whose ends are the same two vertices).

Output

For each data set, if the graph has spanning trees, the smallest slimness among them should be printed. Otherwise, −1 should be printed. An output should not contain extra characters.

Sample Input	Sample Output
4 5	1
1 2 3	20
1 3 5	0
1 4 6	−1
2 4 6	−1
3 4 7	1
4 6	0
1 2 10	1686
1 3 100	50
1 4 90	
2 3 20	
2 4 80	
3 4 40	
2 1	
1 2 1	
3 0	
3 1	
1 2 1	
3 3	
1 2 2	
2 3 5	
1 3 6	

Sample Input	Sample Output
5 10	
1 2 110	
1 3 120	
1 4 130	
1 5 120	
2 3 110	
2 4 120	
2 5 130	
3 4 120	
3 5 110	
4 5 120	
5 10	
1 2 9384	
1 3 887	
1 4 2778	
1 5 6916	
2 3 7794	
2 4 8336	
2 5 5387	
3 4 493	
3 5 6650	
4 5 1422	
5 8	
1 2 1	
2 3 100	
3 4 100	
4 5 100	
1 5 50	
2 5 50	

(Continued)

Sample Input	Sample Output
3 5 50	
4 1 150	
0 0	

Source: ACM Japan 2007.

IDs for online judges: POJ 3522, UVA 3887.

Hint

Suppose a one-dimensional sequence x, y, and w is used to store edges, where the ith edge is (x_i, y_i), and its weight is w_i $(1 <= i <= m)$. And a one-dimensional sequence *fa* is used to store all vertices' parent pointers. Based on it, we can get the root of the subtree containing vertex i: trace through the parent pointer *fa* ($fa[fa[... fa[i] ...]]$) until $x == fa[x]$, and then vertex x is the root of the subtree containing vertex i, and $fa[i] = x$ $(0 <= i <= n-1)$. Initially, $fa[i] = i$ $(0 <= i <= n -1)$.

The algorithm is as follows:

1. Determine whether the graph has a spanning tree with $n-1$ edges or not.
   Initially, for n vertices, each vertex is a tree (fa[i]=i, 0<=i<=n-1) and the number of edges of the spanning tree tot=0;
   For each edge:
   { Calculate fa[x_i] and fa[y_i];
       If vertex x_i and vertex y_i belong to different trees
   (fa[x_i]!=fa[y_i]),then add edge (x_i, y_i) (++tot); and the tree containing vertex x_i is added into the tree containing vertex y_i
   (fa[fa[x_i]]=fa[y_i]);
       }
   If tot!=n-1, then the graph has no spanning tree, output -1 and exceed; else, calculate the smallest slimness among spanning trees.
2. Kruskal algorithm is used to get the spanning tree with the smallest slimness.
   Sort M edges in ascending order of weights of edges (the first key) and the number of edges (the second key);
   Initialize the smallest slimness among spanning trees *ans* as ∞;
   Enumerate edge i with the smallest weight in the spanning tree
   (0<=i<=m-1):
       { if (i == 0 || w_i != w_{i-1})
   { n vertices are n distinct trees (fa[j]=j, 0<=j<=n-1), the number of edges of the spanning tree tot=0;
           Enumerate edge k with the largest weight in the spanning tree (i<=k<=m-1):
               { Calculate fa[x_k] and fa[y_k];
               if (x_k and y_k belong to different subtrees (fa[x_k]!=fa[y_k]))
   { add edge k(++tot);
   add the tree containing x_k into the tree containing y_k
   (fa[fa[x_k]]=fa[y_k]);
   If a spanning tree is gotten (tot==n-1), then adjust the smallest slimness ans=min(ans, w_k-w_i), and break;
               }
       }

```
        If a spanning tree isn't gotten (tot!=n-1), then break;
    }
    Output the smallest slimness ans;
```

12.3.4 The Unique MST

Given a connected undirected graph, tell if its MST is unique.

Definition 1 (spanning tree): Consider a connected, undirected graph $G = (V, E)$. A spanning tree of G is a subgraph of G, say, $T = (V', E')$, with the following properties:

1. $V' = V$.
2. T is connected and acyclic.

Definition 2 (MST): Consider an edge-weighted, connected, undirected graph $G = (V, E)$. The MST $T = (V, E')$ of G is the spanning tree that has the smallest total cost. The total cost of T means the sum of the weights on all the edges in E'.

Input

The first line contains a single integer t ($1 \le t \le 20$), the number of test cases. Each case represents a graph. It begins with a line containing two integers n and m ($1 \le n \le 100$), the number of nodes and edges. Each of the following m lines contains a triple (x_i, y_i, w_i), indicating that x_i and y_i are connected by an edge with weight $= w_i$. For any two nodes, there is at most one edge connecting them.

Output

For each input, if the MST is unique, print the total cost of it; otherwise, print the string "Not Unique!"

Sample Input	Sample Output
2	3
3 3	Not Unique!
1 2 1	
2 3 2	
3 1 3	
4 4	
1 2 2	
2 3 2	
3 4 2	
4 1 2	

Source: POJ Monthly, June 27, 2004. srbga@POJ.

ID for online judge: POJ 1679.

Hint

If the MST is unique, then increasing the weight of any edge in the MST will lead to increasing the weight of MST. Based on Kruskal algorithm, if MST isn't unique, there exists an edge in the MST so that increasing the weight of the edge in the MST can't change the weight of MST.

The algorithm is as follows:

1. Kruskal algorithm is used to get the number of edges *tot* and the sum of edges' weights *ans* for the MST. Numbers of *tot* edges are stored into array *res* in ascending order of weights.
   Set the unique flag of the MST *unique*=(*tot* == *N*-1);
2. For each edge *c* in array *res* (1<=*c*<=*tot*):
   { For edge *res*[*c*], its weight increases 1 and form a new graph *G'*;

   Kruskal algorithm is used to get the number of edges *ttot* and the sum of edges' weights *tans* for the MST for *G'*;
   If (*tans*==ans && *ttot* == *tot*) then MST is not unique (*unique*=false), and goto 3;
   Restore *G* (For edge *res*[*c*], its weight in *G'* decreases 1);
   }
3. If MST is unique (*unique*==true), then output *ans*; else output "Not Unique!".

12.3.5 *Highways*

The island nation of Flatopia is perfectly flat. Unfortunately, Flatopia has no public highways. So the traffic is difficult in Flatopia. The Flatopian government is aware of this problem. They're planning to build some highways so that it will be possible to drive between any pair of towns without leaving the highway system.

Flatopian towns are numbered from 1 to *N*. Each highway connects exactly two towns. All highways follow straight lines. All highways can be used in both directions. Highways can freely cross each other, but a driver can only switch between highways at a town that is located at the end of both highways.

The Flatopian government wants to minimize the length of the longest highway to be built. However, they want to guarantee that every town is reachable by highway from every other town.

Input

The first line of input is an integer *T*, which tells how many test cases followed.

The first line of each case is an integer *N* (3 ≤ *N* ≤ 500), which is the number of villages. Then come *N* lines, the *i*th of which contains *N* integers, and the *j*th of these *N* integers is the distance (the distance should be an integer within [1, 65,536]) between village *i* and village *j*. There is an empty line after each test case.

Output

For each test case, you should output a line containing an integer, which is the length of the longest road to be built such that all the villages are connected, and this value is mimimum.

Sample Input	Sample Output
1	692
3	
0 990 692	
990 0 179	
692 179 0	

Source: POJ Contest, Author: Mathematica@ZSU.

ID for online judge: POJ 2485.

Hint

Flatopia is represented as a weighted complete graph, where towns are represented as vertices, straight lines between towns are represented as edges, and distances between towns are represented as weights of corresponding edges. Distances are integers within [1, 65,536].

The Flatopian government wants to minimize the length of the longest highway to be built. And they want to guarantee that every town is reachable by highway from every other town. Therefore, it is a problem for an MST. There are two methods to calculate the length of the longest road to be built.

1. Prim algorithm is used to calculate the MST and the length of the longest road to be built.
   Because Flatopia is represented as a weighted complete graph, Kruskal algorithm isn't suitable for calculating the MST. In Prim algorithm, if an array is used as a priority queue Q, the time complexity is $O(V^3)$, and if a heap is used as a priority queue Q, the time complexity is $O(V*\log_2 V + V^2*\log_2 V)$.
2. DFS and binary search are used to calculate the length of the longest road to be built.
   Suppose g is the flag whether a vertex has been visited or not, where

$$g[i] = \begin{cases} true & \text{Vertex } i \text{ has been visited.} \\ false & \text{Vertex } i \text{ hasn't been visited.} \end{cases}, 0 <= i <= n-1$$

and v is an adjacency matrix, where $v[i][j]$ is the length of edge (i, j), $0 <= i, j <= n-1$.
3. Calculate the number of reachable vertices from vertex c where the weight of the edge is less than or equal to up.
   In graph G, edges whose weights exceed up are deleted and a new graph G' is formed. DFS can be used to calculate how many vertices can be reachable from vertex c. The algorithm is as follows:

```
int dfs(c, up, tot){ // Calculate how many vertices can be reachable
from vertex c through edges whose weight isn't greater than up. tot
is the size of the adjacency matrix.
    int ans = 1;  // visit vertex c
    g[c] = true;
```

```
        for (int i=0; i<tot; i++) // Enumerate all vertices which are
    adjacent to vertex i, and weights of edges aren't greater than up.
    The number of visited vertices is added into ans.
            if (v[i][c]<=up &&!g[i])
                ans += dfs(i, up, tot);
            return ans;
    }
```

4. Binary search is used to calculate the length of the longest road to be built.

Suppose the interval for the length of the longest road to be built is [*l*, *r*]. Initially, the interval is [1, 65,536].

Binary search is used:

Calculate

$$min = \left\lfloor \frac{l+r}{2} \right\rfloor$$

If all *n* vertices can be visited through roads whose lengths are less than *min* (*dfs*(0, *mid*, *N*) == *N*), then the value of the longest length is in the left subinterval, and *r* = *mid*; else, the value of the longest length is in the right subinterval, and *l* = *mid*.

Repeat the process until *l* = *r*.

Output the length of the longest road to be built *r*.

The time complexity for the binary search is $O(\log_2 65536) \approx O(16)$, and the time complexity for each DFS is $O(E)$. Therefore, the total time complexity is $O(16*E)$. Because edges whose weights are less than *min* aren't visited, the actual running time is far lower than this number.

Chapter 13

Algorithms of Best Paths

Given a weighted, directed graph $G = (V, E)$, each edge is with a real-valued weight. The weight of path $P = (v_0, v_1, \ldots, v_k)$ is the sum of weights of its constituent edges:

$$w(p) = \sum_{i=1}^{k} w(v_{i-1}, v_i)$$

The weight of the shortest path (longest path) from vertex u to vertex v is defined as follows:

$$\delta(u,v) = \begin{cases} \min(\max)\{w(p) \mid u \xrightarrow{p} v\} & \text{if there is a path } p \text{ from } u \text{ to } v \\ \infty & \text{otherwise} \end{cases}$$

The best path from vertex u to vertex v is defined as any path with weight $w(p) = \delta(u, v)$. In this chapter, three kinds of algorithms are introduced:

1. The Warshall algorithm is used to get the transitive closure for a graph.
2. The Floyd–Warshall algorithm is used to get all-pairs best paths in a graph.
3. The Dijkstra algorithm, Bellman–Ford algorithm, and shortest path faster algorithm (SPFA) are used to get single-source shortest paths in a graph.

13.1 Warshall Algorithm and Floyd–Warshall Algorithm

The Warshall algorithm is used to compute the transitive closure of a relation for a graph.

Suppose relation R is represented by digraph G. All vertices in G are $v_1, v_2, \ldots, v_n$. The graph G' for the transitive closure $t(R)$ can be obtained from G as follows. If there exists a path from vertex v_i to vertex v_j in G, then an arc from v_i to v_j is added in G'. The adjacency matrix A for G' is defined as follows. If there exists a path from vertex v_i to vertex v_j, then $A[i][j] = 1$, and vertex v_j is reachable from vertex v_i; otherwise, $A[i][j] = 0$, and vertex v_j isn't reachable from vertex v_i. That

423

is, computing the transitive closure of a relation is to determine whether every pair of vertices are reachable or not. It is a problem of transitive closure for a graph.

Suppose there is a sequence of square matrices order n $A^{(0)}$, ..., $A^{(n)}$, where each element in square matrices is 0 or 1. $A^{(0)}$ is the adjacency matrix for digraph G. For $1 \leq k \leq n$, $A^{(k)}[i][j] = 1$ represents that there exist paths from v_i to v_j passing just v_1, ..., v_k, and $A^{(k)}[i][j] = 0$ represents there is no such a path.

The Warshall algorithm is as follows:

```
A⁽⁰⁾ is the adjacency matrix for digraph G.
for (k=1; k<=n; k++)
  for (i=1; i<=n; i++)
    for (j=1; j<=n; j++)
      A⁽ᵏ⁾[i][j]= =(A⁽ᵏ⁻¹⁾[i][k] & A⁽ᵏ⁻¹⁾[k][j]) | A⁽ᵏ⁻¹⁾[i][j];
```

The Warshall algorithm can be used not only to compute the transitive closure of a graph, but also to solve the problem that there exists a length limit for arcs in a graph. The frogger problem is an example.

13.1.1 *Frogger*

Freddy Frog is sitting on a stone in the middle of a lake. Suddenly, he notices Fiona Frog, who is sitting on another stone. He plans to visit her, but since the water is dirty and full of tourists' sunscreen, he wants to avoid swimming and instead reach her by jumping. Unfortunately, Fiona's stone is out of his jump range. Therefore, Freddy considers using other stones as intermediate stops and reaches her by a sequence of several small jumps. To execute a given sequence of jumps, a frog's jump range obviously must be at least as long as the longest jump occurring in the sequence. The frog distance (humans also call it minimax distance) between two stones is therefore defined as the minimum necessary jump range over all possible paths between the two stones.

You are given the coordinates of Freddy's stone, Fiona's stone, and all other stones in the lake. Your job is to compute the frog distance between Freddy's and Fiona's stone.

Input

The input will contain one or more test cases. The first line of each test case will contain the number of stones n ($2 \leq n \leq 200$). The next n lines each contain two integers, x_i,= and y_i ($0 \leq x_i$, $y_i \leq 1000$), representing the coordinates of stone i. Stone 1 is Freddy's stone, stone 2 is Fiona's stone, and the other $n - 2$ stones are unoccupied. There's a blank line following each test case. Input is terminated by a value of zero (0) for n.

Output

For each test case, print a line saying "Scenario #x" and a line saying "Frog Distance = y," where x is replaced by the test case number (they are numbered from 1) and y is replaced by the appropriate real number, printed to three decimals. Put a blank line after each test case, even after the last one.

Sample Input	Sample Output
2	Scenario #1
0 0	Frog Distance = 5.000

Sample Input	Sample Output
3 4	
	Scenario #2
3	Frog Distance = 1.414
17 4	
19 4	
18 5	
0	

Source: Ulm Local Contest 1997.

IDs for online judges: POJ 2253, ZOJ 1942, UVA 534.

Analysis

Stones in the lake are represented as vertices, and relations between stones are represented as edges; weights of edges are Euclidean distances between vertices. Suppose vertex 0 is the stone on which Freddy initially sits, and vertex 1 is the stone on which Fiona initially sits. The key to the problem is to determine whether vertex 1 is reachable from vertex 0 through a path in which the length of edges isn't greater than K.

Suppose the weighted adjacency matrix is L, where the length of the edge connecting (x_i, y_i) and (x_j, y_j) is

$$L[i][j] = \sqrt{(x_i - x_j)^2 + (y_i - y_j)^2}, \ 0 \leq i, j \leq n-1$$

And *con* is the reachability matrix, where *con*[i][j] shows whether there is a path from vertex i to vertex j and the length of edges in the path isn't greater than K. The Warshall algorithm can be used to compute the reachability matrix *con*; that is, after edges whose lengths are greater than K are deleted, the connectivity of a graph is computed.

Initially, the reachability matrix *con* is as follows:

$$con[i][j] = \begin{cases} false & L[i, j] > K \\ true & L[i, j) \leq K \end{cases} \quad (0 \leq i, j \leq n-1)$$

Then the Warshall algorithm is used to compute the transitive closure of the graph:

```
for (int k=0; k< n; k++)        // Enumeration of intermediate vertices
    for (int i=0; i< n; i++)    // Enumeration starting vertices and
terminal vertices in paths
        for (int j=0; j<n; j++)
            con[i][j] |= con[i][k]&con[k][j];   // Determine whether there is
such a path or not from vertex i to vertex j.
```

Binary search is used to compute frog distance. Suppose the interval for frog distance is [1, r]. Initially l=0, r=10^5.

```
while (r-l>=10⁻⁵) {
```
$$K=\left\lfloor\frac{1+r}{2}\right\rfloor;$$ // computing the intermediate value for the interval
```
    the reachability matrix con under the upper limit K;
    if (con[0][1]) r =K; // Binary search
        else l =K;
    }
Output Frog Distance r;
```

Program

```java
import java.util.*;
import java.math.*;
public class Main {
public static void main(String[ ] args){
   Scanner input = new Scanner(System.in); // java standard input
   int N,testcase = 0;        // Initially the numbers of stones and test
cases are 0
   boolean[ ][ ] con=new boolean[1<<9][1<<9;  // Under the upper limit
of the edges' length, the connectivity from vertex i to vertex j is
con[i][j]
   double[ ][ ] L = new double[1<<9][1<<9;    // The length of edge (i,
j) is dis[i][j]
   while ((N=input.nextInt())!=0){              // Input the number of
stones
      double[ ] x = new double [N];       // Coordinates of stones
      double[ ] y = new double [N];
      for (int i=0;i<N;i++){                 //Input coordinates of stones
         x[i] = input.nextDouble();
         y[i] = input.nextDouble();
      }
      double l = 0, r = 1e5;                // Initialize the interval
      for (int i=0; i<N; i++)               // weighted adjacency matrix L
          for (int j=0; j<N; j++)
             L[i][j] = Math.sqrt((x[i]-x[j])*(x[i]-x[j])+(y[i]-y[j])*
(y[i]-y[j]));
      while (r-l>=1e-5){
      double mid = (l+r)/2;                 // Intermedia value of the
interval
      for (int i=0;i<N;i++)
//con[i][j]=
```
$$\begin{cases} false & \text{The length of edge (i,j) is greater than the intermedia value} \\ true & \text{The length of edge (i,j) is less than the intermedia value} \end{cases}$$
```java
      for (int j=0; j<N; j++)
        if (L[i][j]>mid) con[i][j] = false;
          else con[i][j] = true;
      for (int k=0; k<N; k++)//Compute the reachability matrix con when
the length of edges isn't greater than mid
        for (int i=0; i<N; i++)
          for (int j=0; j<N; j++)
            con[i][j] |= con[i][k]&con[k][j];
      if (con[0][1]) r = mid; //If the length of edges is less than mid,
frog distance is in the left interval; else in the right interval
          else l = mid;
```

```
        }
        System.out.println("Scenario #"+(++testcase)); // Output the result
        System.out.println("Frog Distance = "+ BigDecimal.valueOf(1).
setScale(3,RoundingMode.HALF_UP));
        System.out.println("");
        }
    }
}
```

Based on the Warshall algorithm, the Floyd–Warshall algorithm is used to find the best paths between each pair of vertices in a weighted graph. In the Floyd–Warshall algorithm, the boolean operator '&' is changed into the arithmetic operator '+', and boolean calculation '|' is changed into comparing $A^{(k-1)}[i][k] + A^{(k-1)}[k][j]$ with $A^{(k-1)}[i][j]$. That is, the Floyd–Warshall formula is as follows:

$$A^{(0)}[i][j] = \text{adjacency matrix } M$$
$$A^{(k)}[i][j] = \min(\max)\{A^{(k-1)}[i][k] + A^{(k-1)}[k][j], A^{(k-1)}[i][j]\}$$

where $i, j, k = 1 .. n$.

That is, $A^{(k)}[i][j]$ is the length of the best path from vertex v_i to v_j passing just $v_1, \ldots, v_k$. $A^{(n)}[i][j]$ is the length of the best paths from vertex v_i to v_j.

The Floyd–Warshall algorithm can be used to compute best paths for all pairs in a graph. Its time complexity is $O(n^3)$. If the shortest path needs to be calculated, there must be no negative weighted circuit. And if the longest path needs to be calculated, there must be no positive weighted circuit. Otherwise, it will lead to an endless loop.

13.1.2 Arbitrage

Arbitrage is the use of discrepancies in currency exchange rates to transform one unit of a currency into more than one unit of the same currency. For example, suppose that US$1 buys £0.5, £1 buys F10.0, and F1 buys US$0.21. Then, by converting currencies, a clever trader can start with US$1 and buy 0.5*10.0*0.21 = US$1.05, making a profit of 5%.

Your job is to write a program that takes a list of currency exchange rates as input and then determines whether arbitrage is possible or not.

Input

The input will contain one or more test cases. On the first line of each test case there is an integer n ($1 \leq n \leq 30$), representing the number of different currencies. The next n lines each contain the name of one currency. Within a name no spaces will appear. The next line contains one integer m, representing the length of the table to follow. The last m lines each contain the name c_i of a source currency, a real number r_{ij} that represents the exchange rate from c_i to c_j, and a name c_j of the destination currency. Exchanges that do not appear in the table are impossible.

Test cases are separated from each other by a blank line. Input is terminated by a value of zero (0) for n.

Output

For each test case, print one line telling whether arbitrage is possible or not in the format "Case case: Yes" and "Case case: No," respectively.

Sample Input	Sample Output
3	Case 1: Yes
	Case 2: No
USDollar	
BritishPound	
FrenchFranc	
3	
USDollar 0.5 BritishPound	
BritishPound 10.0 FrenchFranc	
FrenchFranc 0.21 USDollar	
3	
USDollar	
BritishPound	
FrenchFranc	
6	
USDollar 0.5 BritishPound	
USDollar 4.9 FrenchFranc	
BritishPound 10.0 FrenchFranc	
BritishPound 1.99 USDollar	
FrenchFranc 0.09 BritishPound	
FrenchFranc 0.19 USDollar	
0	

Source: Ulm Local Contest 1996.

IDs for online judges: POJ 2240, ZOJ 1092, UVA 436.

Analysis

The currency exchange is represented as a weighted, directed graph. A currency is represented as a vertex. A currency exchange is represented as an arc. Vertex i represents the ith currency c_i, arc (i, j) represents currency exchange from c_i to c_j, and its weight r_{ij} represents the exchange rate from c_i to c_j, $1 <= i, j <= n$. Suppose $dist[i][j]$ is the exchange rate from c_i to c_j. If currency c_i can be changed into currency c_k, and currency c_k can be changed into currency c_j, then the exchange rate from c_i to c_j through c_k is $dist[i][k]*dist[k][j]$. It can be regarded as the length of the path from vertex *i* to vertex *j* through vertex *k*. All possibilities of currency exchanges should be enumerated. Therefore, the problem is to compute the all pairs' longest paths in a graph. In the graph there are circuits. In order to avoid an endless loop, there must be a restriction $(i\ != j\ \&\&\ j\ != k\ \&\&\ k\ != i)$. The process is as follows:

```
for (int k=1; k<= n; k++)          // Enumerate intermedia vertex k
for (int i=1; i<= n; i++)          //Enumerate all pairs (i, j)
 for (int j=1; j<= n; j++)
  if (i!=j&&j!=k&&k!=i)
      if (dist[i][k]*dist[k][j]>dist[i][j])
          dist[i][j]= dist[i][k]*dist[k][j];
```

Then all currencies pairs are enumerated. If there exists such a currency *i*, by converting currencies currency $i \rightarrow$ currency $j_1 \rightarrow$ ...... $\rightarrow$ currency $j_m \rightarrow$ currency *i* $(m \geq 1)$, and a profit can be made, then arbitrage is possible. If there is no such a currency, arbitrage is impossible.

```
  flag = 0;              //The flag whether arbitrage is possible or not
  for (int i=1; i<= n; i++) //Enumerate all currencies
      for (int j=1; j<= n; j++) //Enumerate intermedia currencies
          if (dist[i][j]*dist[j][i]>1) flag = 1; //If making a profit,
arbitrage is possible, and solution to the current test
case(flag?"Yes":"No").
```

Program

```cpp
#include<iostream>
#include<cstdio>
#include<cstring>
using namespace std;         // C++ Standard Library
const int MaxN = 50;         //The upper limit of the number of currencies
const int MaxL = 1005;       //The upper limit of the length of currency
names
char str[MaxN][MaxL],strA[MaxL],strB[MaxL]; //str is used to store
currency names and their length, source, strA is the source currency
string, and strB is the target currency string
long double dist[MaxN][MaxN];    //Martix dist is used to deal with
Currency exchange
int N,M;                     //Numbers of currencies and Currency exchanges
int find(char *_str){
    for (int i=1; i<=N; i++)
       if (strlen(_str)==strlen(str[i])&&strcmp(_str,str[i])==0) return
i;
    return 0;
}
int main(){
    while(scanf("%d",&N)&&N){    // the number of different
currencies(vertices)
            static int cnt = 0;
        for (int i=1;i<=N;i++)
           for (int j=1;j<=N;j++)
              dist[i][j] = 0;
        for (int i=1;i<=N;i++)       //Input names of currencies
           scanf("%s",str[i]);
```

```
        scanf("%d",&M);                //Input the number of arcs
        for (int i=1;i<=M;i++){    //Input arcs' information (vertices
are source currency and target currency, weight is currency exchange
rate)
            double w;
            scanf("%s %lf %s",strA,&w,strB);
            dist[find(strA)][find(strB)] = w;
    }
    for (int k=1;k<=N;k++)       //Compute all pairs longest paths
        for (int i=1;i<=N;i++)
            for (int j=1;j<=N;j++)
              if (i!=j&&j!=k&&k!=i)
                  if (dist[i][k]*dist[k][j]>dist[i][j])
                      dist[i][j] = dist[i][k] * dist[k][j];
        bool flag = 0;              // whether arbitrage is possible or not
        for (int i=1;i<=N;i++)
            for (int j=1;j<=N;j++)
            if (dist[i][j]*dist[j][i]>1) flag = 1;
        printf("Case %d: %s\n",++cnt,flag?"Yes":"No"); // Output the result
    }
    return 0;
}
```

13.2 Dijkstra's Algorithm

Dijkstra's algorithm is used to solve the single-source shortest-path problem in a weighted, directed graph $G(V, E)$ for the case in which all arcs' weights are nonnegative. That is, for each arc $(u, v) \in E$, $w(u, v) \geq 0$.

Dijkstra's algorithm is as follows:

```
void Dijkstra(int r);        //Dijkstra's algorithm: shortest-paths from
vertex r to other vertices
{ for (i=0; i<n; i++)
      dist[i]=∞;
  dist[r]=0;                              // the length for the shortest-paths
for r is 0
  S=∅;
  Q is a min-priority queue used to store n vertices;
  while (Q≠∅)                           //if Q isn't empty
  { vertex u isn't in S and dist[u] is minimal;
    S=S∪{u};               //u is added into the set of vertices S known
the shortest paths
    for ( each vertex v not in S)
      if (dist[u]+w_uv<dist[v])
        dist[v]=dist[u]+ w_uv;
  }
}
```

If the min-priority queue is implemented by an array, the time complexity of Dijkstra's algorithm is $O(V^2 + E) \approx O(V^2)$. If the min-priority queue is implemented by a binary min-heap, the time complexity of Dijkstra's algorithm is $O((V + E)*\ln V) \approx O(E*\ln V)$. If the graph is a sparse graph, the min-priority queue implemented by a binary min-heap is suitable.

13.2.1 Toll

Sindbad the Sailor sold 66 silver spoons to the sultan of Samarkand. The selling was quite easy, but delivering was complicated. The items were transported over land, passing through several towns and villages. Each town and village demanded an entry toll. There were no tolls for leaving. The toll for entering a village was simply one item. The toll for entering a town was 1 piece per 20 items carried. For example, to enter a town carrying 70 items, you had to pay 4 items as toll. The towns and villages were situated strategically between rocks, swamps, and rivers, so you could not avoid them (Figure 13.1).

Predicting the tolls charged in each village or town is quite simple, but finding the best route (the cheapest route) is a real challenge. The best route depends upon the number of items carried. For numbers up to 20, villages and towns charge the same. For large numbers of items, it makes sense to avoid towns and travel through more villages, as illustrated in Figure 13.2.

You must write a program to solve Sindbad's problem. Given the number of items to be delivered to a certain town or village and a road map, your program must determine the total number of items required at the beginning of the journey that uses the cheapest route.

Input

The input consists of several test cases. Each test case consists of two parts: the road map followed by the delivery details.

The first line of the road map contains an integer n, which is the number of roads in the map ($0 \leq n$). Each of the next n lines contains exactly two letters representing the two end points of a road. A capital letter represents a town; a lowercase letter represents a village. Roads can be traveled in either direction.

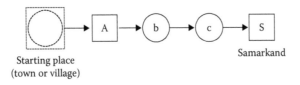

Starting place
(town or village)

Samarkand

Figure 13.1 To reach Samarkand with 66 spoons, traveling through a town followed by two villages, you must start with 76 spoons.

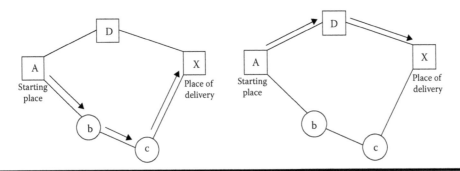

Figure 13.2 The best route to reach X with 39 spoons, starting from A, is A→b→c→X, shown with arrows in the figure on the left. The best route to reach X with 10 spoons is A→D→X, shown in the figure on the right. The figures display towns as squares and villages as circles.

Following the road map is a single line for the delivery details. This line consists of three things: an integer p $(0 < p \le 1000)$ for the number of items that must be delivered, a letter for the starting place, and a letter for the place of delivery. The road map is always such that the items can be delivered.

The last test case is followed by a line containing the number -1.

Output

The output consists of a single line for each test case. Each line displays the case number and the number of items required at the beginning of the journey. Follow the output format in the example given below.

Sample Input	Sample Output
1	Case 1: 20
a Z	Case 2: 44
19 a Z	
5	
A D	
D X	
A b	
b c	
c X	
39 A X	
−1	

Source: ACM World Finals Beverly Hills 2002–2003.

ID for online judge: UVA 2730.

Hint

The road map is represented as an undirected graph. A village or a town is represented as a vertex, where town A is represented as vertex 1, ..., town Z is represented as vertex 26; village a is represented as vertex 27, ..., village z is represented as vertex 52. Roads between towns or villages are represented as undirected edges.

If integer p (the number of items that must be delivered) is regarded as the length of the longest path from the starting place *from* to the place of delivery *to*, then the problem requires computing the number of items required at the starting place *from* such that p items are delivered to the place of delivery *to*. Binary search is used to compute the number of items required at the starting place *from*. Suppose the interval is $[l, r]$:

```
Suppose the initial interval is [p, 2²⁰];
while (l!=r) {
```

$$mid=\left\lfloor \frac{1+r-1}{2} \right\rfloor ;$$

```
    if (the number of items mid at the starting place from makes the
number of items delivered to the place of delivery to >=p)
        r=mid;
    else
        l = mid+1;
}
```

Therefore, the key to the problem is to compute the number of items at the place of delivery when the number of items k is at the beginning place. Because edges' weights are nonnegative, and the number of vertices isn't greater than 52, Dijkstra's algorithm can be used to solve the problem, and the priority queue is implemented by an array.

Suppose $g[\ ]$ is the array for the longest paths. Obviously, $g[from] = k$. $flag[\ ]$ is used to store flags of vertices in the priority queue. Initially, all vertices are in the priority queue. The process is as follows:

```
while (true) {
    In the priority queue, g[next]=w=  max      {g[i]} ;
                                     1≤i≤52,flag[i]=true
    if (the priority queue is empty)
        break;
    flag[next]=false; // Vertex next is taken out from the priority queue
    for (int i = 1; i <= 52; i++)
    { if ( Vertex i is adjacent to vertex next)
        { if i<=26) //Vertex i is a town
            l=w-(w-1)/20-1;
          else l=w-1;
          g[i]=max{1, g[i]};
        };
    };
Return the number of items at the place of delivery g[to];
```

Program

```java
import java.util.*;
import java.math.*;
public class Uva2730 {
    static int Tot;                          //The number of the test case
    static boolean go[ ][ ];                      // adjacency matrix
    static int turn(char x) {                      // numbers of vertices
        return x < 'a' ? x - 64 : x - 70;
    }
    static int check(int from, int to, int o) {    //at the starting place
from, o items. Computing how many items at the place of delivery to
        int g[ ] = new int[55];
        boolean flag[ ] = new boolean[55];
        g[from] = o;
        while (true) {
            int w = 0, next = -1;   // Find the next vertex in the priority
queue
            for (int i = 1; i <= 52; i++)
                if (!flag[i] && (next == -1 || w < g[i])) {
```

```
                next = i;
                w = g[i];
            }
        if (next == -1)                    // the priority queue is empty
            break;
        flag[next] = true;                 //Vertex next is taken out
        for (int i = 1; i <= 52; i++)      // Relaxation
            if (go[next][i])
                g[i] = Math.max(w - (i < 27 ? (w - 1) / 20 + 1 : 1),
g[i]);
        }
        return g[to];
    }
    public static void main(String args[ ]) {
        Scanner input = new Scanner(System.in);
        int tot = 0;                       //initialize the number of test case
        while (input.hasNextInt()) {
            int T = input.nextInt();       // number of edges
            if (T == -1)                   //test case ends
                break;
            go = new boolean[55][55];
        for (int i = 0; i < T; i++) { // construct an adjacency matrix for
the undirected graph
            int x=turn(input.next().charAt(0)),y=turn(input.next().
charAt(0));
            go[x][y] = go[y][x] = true;
            }
            Tot=input.nextInt();           // the delivery details
            int from=turn(input.next().charAt(0)),to=turn(input.next().
charAt(0));
            int l = Tot, r = 1 << 20;      //initial interval is [Tot,2²⁰]
            while (l != r) {
                int mid = (l + r - 1) >> 1;
                if (check(from, to, mid) >= Tot)
                    r = mid;
                else
                    l = mid + 1;
            }
        System.out.println("Case " + ++tot + ": " + l);//Output the number
of items at the starting place
        }
    }
}
```

13.3 Bellman–Ford Algorithm

The Bellman–Ford algorithm is used to calculate the single-source shortest paths in a weighted, directed graph in which edges' weights may be negative. Like Dijkstra's algorithm, the Bellman–Ford algorithm also uses relaxation. For each vertex $v \in V$, the estimate $dist[v]$ on the weight of a shortest path from the source s to v is progressively decreased until it achieves the actual shortest-path weight. If there exist negative-weight cycles in the graph, the Bellman–Ford algorithm should report that there aren't shortest paths.

```
Bool Bellman_Ford(int s)          //Bellman-Ford algorithm is used to
compute the shortest paths from source s to other vertices
{
    for (i=0; i<n; i++)                    //Initialization
      { dist[i]=∞; π[i]=nil ;}
    dist[s]=0;
    for (i=1; i<n; i++)               // n-1 iterations
      for (each(u, v)∈E )            // For each edge, relaxation is used
      if ( dist[v] -w_uv >dist[u] )
      { dist[v]=dist[u]+ w_uv; π[v]=u; }
    for ( each (u, v)∈E)             //If there exists negative-weight
cycle, return false
        if ( dist[v] -w_uv >dist[u] ) return false;
    return true;
}
```

The reason why there are $n - 1$ iterations in the Bellman–Ford algorithm is that if there is a shortest past between two vertices, each vertex will appear at most one time in the path; that is, there are at most $n - 1$ edges in the path.

The time complexity for the Bellman–Ford algorithm is $O(VE)$.

Sometimes the algorithm can get the solution before $n - 1$ iterations. If a vertex's estimate isn't renewed, then it can't be renewed in the next iteration. Therefore, the program can be optimized.

13.3.1 Minimum Transport Cost

There are N cities in Spring country. Between each pair of cities there may be one transportation track or none. Now there is some cargo that should be delivered from one city to another. The transportation fee consists of two parts:

1. The cost of the transportation on the path between these cities
2. A certain tax that will be charged whenever any cargo passes through one city, except for the source and destination cities

You must write a program to find the route that has the minimum cost.

Input

The data of path cost, city tax, and source and destination cities is given in the input file, which is of the form:

$$
\begin{matrix}
a_{11} & a_{12} & \cdots & a_{1n} \\
a_{21} & a_{22} & \cdots & a_{2n} \\
\cdots & \cdots & \ddots & \cdots \\
a_{n1} & a_{n2} & \cdots & a_{nn} \\
b_1 & b_2 & \cdots & b_n
\end{matrix}
$$

$$
\begin{array}{cc}
c & d \\
e & F \\
\cdots & \cdots \\
g & h
\end{array}
$$

where a_{ij} is the transport cost from city i to city j, $a_{ij}=-1$ indicates there is no direct path between city i and city j, b_i represents the tax of passing through city i, and the cargo is to be delivered from city c to city d, city e to city f, ..., city g to city h.

Output

You must output the sequence of cities passed by and the total cost, which is of the form:

From c to d :
   Path: $c \rightarrow c_1 \rightarrow \ \rightarrow c_k \rightarrow d$
Total cost : ...

   ...

From e to f:
   Path: $e \rightarrow e_1 \rightarrow \ \rightarrow e_k \rightarrow f$
Total cost : ...

Sample Input	Sample Output
0 3 22 –1 4	From 1 to 3 :
3 0 5 –1 –1	Path: 1→5→4→3
22 5 0 9 20	Total cost : 21
–1 –1 9 0 4	
4 –1 20 4 0	From 3 to 5 :
5 17 8 3 1	Path: 3→4→5
1 3	Total cost : 16
3 5	
2 4	From 2 to 4 :
	Path: 2→1→5→4
	Total cost : 17

Source: ACM 1996 Asia Regional Shanghai.

ID for online judge: UVA 523.

Analysis

The problem requires outputting the route with the minimum cost. For each vertex in the route, the number of the next city that needs to charge tax should be stored. In the route, the last city is the destination city and needn't charge tax. Therefore, there must be pointers pointing to next cities in the route.

Suppose the route with the minimum cost is *router*, where *route[i][j].cost* is the minimum cost of the path from *i* to *j*. In the route, the first city that charges tax is *route[i][j].nexthop*; that is, the first edge is (*i*, *route[i][j].nexthop*). Initially, *route[i][i].cost* = 0, *route[i][i].nexthop* = *i*.

If a_{ij} >= 0, then *route[i][j].cost* = a_{ij}, *route[i][j].nexthop* = *j*; else, *route[i][j].cost* = ∞, *route[i][j].nexthop* = −1 (0 <= *i*, *j* <= n − 1).

1. The Bellman–Ford algorithm is used to compute the route that has the minimum cost *router*. The relaxation is to reduce the cost by changing the first city that needs to charge tax.

```
for(l=N-1; l--;)              // n-1 iterations
{
    for(i=N-1; i--;)    // Enumerate the starting vertex i in the route
    {
        for(j=N-1; j--;)    // Enumerate cities charging tax j
        { if( i == j || cost[i][j]<=0 ) continue;
            for(k =N-1; k--;) // Enumerate destination city k in the
route
            { if ( ( route[j][k].cost+b_j+a_ij<route[i][k].cost ) || (
route[j][k].cost+b_j+a_ij == route[i][k].cost && route[i][k].nexthop>j
) ) //Relaxation
                { route[i][k].cost = route[j][k].cost +b_j+a_ij;
                  route[i][k].nexthop = j; }
            }
        }
    }
}
```

2. Find the route with the minimum cost between cities.
   Suppose the starting city is *from*, the destination city is *to*, and the minimum cost from *from* to *to* is *route[from][to].cost*; the sequence of cities passed is *from* $f_1 f_2 \ldots f_k$ *to*, where cities $f_1 f_2 \ldots f_k$ need to charge tax. First, the starting city *from* is output, and then along each vertex's *nexthop* cities that need to charge tax are output:

```
f_1=route[from][to].nexthop; f_2=route[f_1][to].nexthop;
.........;
f_k=route[f_{k-1}][to].nexthop;
```

Finally, the destination city is output: *tot* (= *route[f_k][to].nexthop*).

Program by the Problemsetter

```
#include <stdio.h>
#include <values.h>
#define TWN_MX 270                      //the upper limit of vertices
#define INFTY (MAXINT/8)
#define MAX(a,b) ((a)>(b)?(a):(b))
#define MIN(a,b) ((a)<(b)?(a):(b))
typedef struct t_route              // srtuct for the route
{
  int nexthop;          //the first city in the route which need charge tax
  int cost;                     //the minimum cost of the path from i to j
} t_route;
```

```
t_route route[TWN_MX][TWN_MX];
int cost[TWN_MX][TWN_MX];            //Cost matrix, diagonal is tax
int N;
void init_route()          // the minimum cost between each pair of cities
{
  int i,j,k, l;
  int tmp_do_k, cur_do_k;
  for(i=N;i--;)
  for(j=N;j--;)
  {
    if(j == i)            // diagonal element: cost is 0, should charge tax
    {
      route[i][i].cost = 0;
      route[i][i].nexthop = i;
      continue;
    }
    if( cost[i][j] >= 0 )  //there exists an edge (i, j)
    {
      route[i][j].cost = cost[i][j];
      route[i][j].nexthop = j;
    }
    else            //there is no edge between i and j, cost is ∞, no city
charge tax
    {
      route[i][j].cost = INFTY;
      route[i][j].nexthop = -1;
    }
  }
  for(l=N-1;l--;)
  {
      for(i=N;i--;)           //starting city i is enumerated
    {
      for(j=N;j--;)    //the first city j which charge tax is enumerated
      {
        if( i==j||cost[i][j]<=0) continue; //if there is no edge between
i and j, the next edge is enumerayed
        for(k = N;k--;)                //destination cities are enumerated
        {   //from i to k, current cost tmp_do_k, known cost cur_do_k
          tmp_do_k = route[j][k].cost + cost[j][j] + cost[i][j];
          cur_do_k = route[i][k].cost;
          if( tmp_do_k < cur_do_k ||
             ( tmp_do_k == cur_do_k && route[i][k].nexthop > j))
          { // record the current route
            route[i][k].cost = tmp_do_k;
            route[i][k].nexthop = j;
          }
        }
      }
    }
  }
}
int read_zad()                          //Input cost matrix and taxes
{
  int i,j;
```

```
  scanf("%i", &N);                              //number of cities
  if( N == 0 ) return 0;
  for(i=0;i<N;i++)                              //cost matrix
   for(j=0;j<N;j++)  scanf("%i", &cost[i][j]);
   for(i=0;i<N;i++) scanf("%i", &cost[i][i]);  //cities' taxes
   return 1;
}
void solve()                                    //answer the minimum cost
{
  int from, to, f, lastf;
  while(1)
  {
    scanf("%i %i", &from, &to);                 // the starting city and the
destination city
    if( from == -1 ) return;
    printf("From %i to %i :\n", from, to);
    from--;              // numbers for the starting city and the destination
city
    to--;
    printf("Path: %i",from + 1);               //output the starting city
    lastf = from;
if(from != to )
    {
        for(  f = route[from][to].nexthop; f != to;f = route[f][to].
nexthop)
        { if(f == lastf)
          { printf("*");
            break;
          }
         printf("-->%i", f + 1); //output city which charge tax
         lastf= f;
      }
      printf("-->%i\n", f + 1); //destination city
  }
    else printf("\n");
  printf("Total cost : %i\n\n", route[from][to].cost);  //minimum cost
  }
}
int main()
{
  while(read_zad())       //repeatedly input cost matrix and taxes until 0
  {
    init_route();                     //minimum cost
    solve();                          //answer the inqury
  }
  return 0;
}
```

13.4 Shortest Path Faster Algorithm (SPFA Algorithm)

The Shortest Path Faster Algorithm (SPFA) is an improvement of the Bellman–Ford algorithm that computes single-source shortest paths in a weighted, directed graph. SPFA is suitable for random sparse graphs and graphs containing negative-weight edges.

```
void spfa(int s)    // SPFA is used to compute shortest paths from single-
source s to other vertices
{
  Queue Q is empty;
for (i=0; i<101; i++)       //Initialization
{ dist[i] =∞; π[i]=nil;  }
  dist[s]=0;
Add s into queue Q;
while ( Q is not empty)
{
  Delete the front element x from Q;
    for(i=1; i<=n; i++)     // Relaxation
    if (dist[i]-w_{xi}>dist[x])
      { dist[i]=dist[x]+w_{xi}; π[i]=x;
        if (Vertex i is not in Q)
          Vertex i is added into Q;
    }
 }
}
```

Q is a queue. The time complexity for deleting element u from Q and visiting its adjacent vertices is $O(d)$, where d is the out-degree of vertex u. The average out-degree for a vertex is E/V. Therefore, the time complexity of dealing with a vertex is $O(E/V)$. Suppose the number of adding vertices to Q is h, and it is related to edges' weights. Suppose $h = kV$. Then the time complexity of SPFA is $T = O(h(E/V)) = O((h/V)E) = O(kE)$. In general, k is a little constant. Therefore, the time complexity of SPFA is $O(E)$.

13.4.1 Longest Paths

It is a well-known fact that some people do not have their social abilities completely enabled. One example is the lack of talent for calculating distances and intervals of time. This causes some people to always choose the longest way to go from one place to another, with the consequence that they are late to whatever appointments they have, including weddings and programming contests. This can be highly annoying for their friends.

César has this kind of problem. When he has to go from one point to another, he realizes that he has to visit many people, and thus always chooses the longest path. One of César's friends, Felipe, has understood the nature of the problem. Felipe thinks that with the help of a computer, he might be able to calculate the time that César is going to need to arrive to his destination. That way, he could spend his time in something more enjoyable than waiting for César.

Your goal is to help Felipe develop a program that computes the length of the longest path that can be constructed in a given graph from a given starting point (César's residence). You can assume that the graph has no cycles (there is no path from any node to itself), so César will reach his destination in a finite time. In the same line of reasoning, nodes are not considered directly connected to themselves.

Input

The input consists of a number of cases. The first line of each case contains a positive number n ($1 < n \leq 100$) that specifies the number of points that César might visit (i.e., the number of nodes in the graph).

A value of $n = 0$ indicates the end of the input.

After this, a second number s is provided, indicating the starting point in César's journey ($1 \leq s \leq n$). Then, you are given a list of pairs of places p and q, one pair per line, with the places on each line separated by white space. The pair $p \; q$ indicates that César can visit q after p.

A pair of zeros (0 0) indicates the end of the case.

As mentioned before, you can assume that the graphs provided will not be cyclic.

Output

For each test case you have to find the length of the longest path that begins at the starting place. You also have to print the number of the final place of such longest path. If there are several paths of maximum length, print the final place with the smallest number.

Print a new line after each test case.

Sample Input	Sample Output
2	Case 1: The longest path from 1 has length 1, finishing at 2.
1	
1 2	Case 2: The longest path from 3 has length 4, finishing at 5.
0 0	
5	Case 3: The longest path from 5 has length 2, finishing at 1.
3	
1 2	
3 5	
3 1	
2 4	
4 5	
0 0	
5	
5	
5 1	
5 2	
5 3	
5 4	
4 1	
4 2	
0 0	
0	

ID for online judge: UVA 10000.

Analysis

The graph is a directed acyclic graph. Each edge's weight is 1. The goal of the problem is to find the length of the longest path that begins at the starting place. Because there is no circuit in the graph, SPFA can be used to solve the problem.

Program

```
#include<iostream>
#include<cstring>
using namespace std;
```

```
const int maxn=110;
int n,x,y,st,en,length,r,l,a[maxn][maxn], dis[maxn],h[100*maxn];    //
number of vertices n, starting vertex st, destination of the longest path
en, the length of the longest path length, queue is h[ ], the front and
the rear are l and r respectively, adjacent matrix a[ ][ ], and estimate
for the longest path is dis[ ].
bool v[maxn];
int main(){
      int ca=0;
      cin>>n;                //the number of vertices for the first test case
      while (n){
            memset(a,0,sizeof(a));      //Initialization
            memset(v,0,sizeof(v));
            memset(dis,0,sizeof(dis));
            memset(h,0,sizeof(h));
            length=-1;
            cin>>st;                    // starting vertex
            cin>>x>>y;                  // the first edge
            while (x||y){               // construct the adjacent matrix
                  a[x][y]=1;
                  cin>>x>>y;
            }
            l=r=1;                      // starting vertex enter the queue
            h[1]=st;
            v[st]=1;
            while(1){
                  int x=h[l];
                  if (dis[x]>length||(dis[x]==length&&x<en)){
                        en=x;
                        length=dis[x];
                  }
                  for (int y=1;y<=n;y++) if (a[x][y]){ //relaxation
                      if (dis[x]+1>dis[y]) {
                        dis[y]=dis[x]+1;
                        if (!v[y]) { // adjacent vertex y isn't visited, y
is added into the queue
                              h[++r]=y;
                              v[y]=1;
                        }
                     }
                  }
                  if (l==r) break;      // the queue is empty
                  v[x]=0;
                  l++;
            }
            //serial number priority: little
            for (int i=1;i<=n;i++) if (dis[i]>length){   // the length of
the longest path is the first key, and the serial number is the second key
                  length=dis[i];
                  en=i;
            }
            cout<<"Case "<<++ca<<": The longest path from "<<st;
            cout<<" has length "<<length<<", finishing at
"<<en<<"."<<endl;
```

```
        cout<<endl;
        cin>>n;
    }
    system("pause");
}
```

13.5 Problems

13.5.1 Knight Moves

A friend of yours is doing research on the traveling knight problem (TKP), where you are to find the shortest closed tour of knight moves that visits each square of a given set of n squares on a chessboard exactly once. He thinks that the most difficult part of the problem is determining the smallest number of knight moves between two given squares and that, once you have accomplished this, finding the tour would be easy.

Of course you know that it is vice versa. So you offer him to write a program that solves the "difficult" part.

Your job is to write a program that takes two squares, a and b, as input and then determines the number of knight moves on a shortest route from a to b.

Input

The input will contain one or more test cases. Each test case consists of one line containing two squares separated by one space. A square is a string consisting of a letter (a–h) representing the column and a digit (1–8) representing the row on the chessboard.

Output

For each test case, print one line saying "To get from xx to yy takes n knight moves."

Sample Input	Sample Output
e2 e4	To get from e2 to e4 takes 2 knight moves.
a1 b2	To get from a1 to b2 takes 4 knight moves.
b2 c3	To get from b2 to c3 takes 2 knight moves.
a1 h8	To get from a1 to h8 takes 6 knight moves.
a1 h7	To get from a1 to h7 takes 5 knight moves.
h8 a1	To get from h8 to a1 takes 6 knight moves.
b1 c3	To get from b1 to c3 takes 1 knight move.
f6 f6	To get from f6 to f6 takes 0 knight moves.

Source: Ulm Local Contest 1996.

IDs for online judges: POJ 2243, ZOJ 1091, UVA 439.

Hint

A chessboard is represented as a graph, where each square on the chessboard is regarded as a vertex, a knight move between two vertices in one step is regarded as an edge, and the weight of an edge is 1. Therefore, the TKP is to find the shortest paths in a graph. Suppose w is the shortest-path matrix, where $w[x_1][y_1][x_2][y_2]$ is the smallest number of knight moves between square (x_1, y_1) and square (x_2, y_2). Initially,

$$w[x_1][y_1][x_2][y_2] = \begin{cases} \infty & \text{otherwise} \\ 1 & \text{From } (x_1,y_1), \text{ a knight can move to } (x_2,y_2) \text{ in one step} \end{cases}$$

By an offline method, the Floyd–Warshall algorithm is used to compute the shortest paths between any pair of vertices.

```
for (int kx=1; kx<=8; kx++)   //Enumeration for Intermedia vertices'
coordinates
  for (int ky =1; ky <=8; ky++)
    for (int ix=1; ix<=8; ix++)      // Enumeration for the starting place
      for (int iy=1; iy<=8; iy++)
        for (int jx=1; jx<=8; jx++)     //Enumeration for terminal
vertices
          for (int jy=1; jy<=8; jy++)
            if (w[ix][ iy][ kx][ ky]+w[kx][ ky][ jx][ jy]<w[ix][ iy][ jx]
[ jy])
  w[ix][ iy][ jx][ jy]=w[ix][ iy][ kx][ ky]+w[kx][ ky][ jx][ jy];
```

For each test case $a_1b_1\ a_2b_2$, we can get its solution $w[a_1 - 96][b_1 - 48][a_2 - 96][b_2 - 48]$ from the shortest-path matrix w directly.

13.5.2 Big Christmas Tree

Christmas is coming to KCM city. Suby, a loyal civilian in KCM city, is preparing a big neat Christmas tree. The simple structure of the tree is shown in Figure 13.3.

The tree can be represented as a collection of numbered nodes and some edges. The nodes are numbered 1 through n. The root is always numbered 1. Every node in the tree has its weight. The weights can be different from each other. Also, the shape of every available edge between two nodes is different, so the unit price of each edge is different. Because of a technical difficulty, the price of an edge will be (sum of weights of all descendant nodes) × (unit price of the edge).

Suby wants to minimize the cost of the whole tree among all possible choices. Also, he wants to use all nodes because he wants a large tree. So he decided to ask you for help in solving this task by finding the minimum cost.

Input

The input consists of T test cases. The number of test cases T is given in the first line of the input file. Each test case consists of several lines. Two numbers, v and e ($0 \le v, e \le 50,000$), are given in the first line of each test case. On the next line, v positive integers w_i indicating the weights of v nodes are given in one line. On the following e lines, each line contains three positive integers, a, b, and c, indicating the edge that is able to connect two nodes a and b, and unit price c.

All numbers in input are less than 2^{16}.

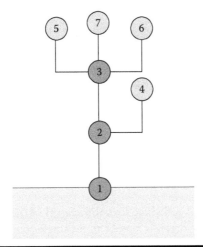

Figure 13.3   Christmas tree.

Output

For each test case, output an integer indicating the minimum possible cost for the tree in one line. If there is no way to build a Christmas tree, print "No Answer" in one line.

Sample Input	Sample Output
2	15
2 1	1210
1 1	
1 2 15	
7 7	
200 10 20 30 40 50 60	
1 2 1	
2 3 3	
2 4 2	
3 5 4	
3 7 2	
3 6 3	
1 5 9	

Source: POJ Monthly, September 29, 2006.

ID for online judge: POJ 3013.

Hint

The Christmas tree is an inverted tree whose root is at the bottom. Because there may exist different edges whose unit prices are different between two vertices, there may exist different paths from the root to a vertex. Therefore, a graph is used to represent the problem.

The unit price of each edge is regarded as the weight of the edge. For the path from vertex i to the root, the sum of weights of edges in the path is called the length of the path. Because the price of an edge is (sum of weights of all descendant nodes) × (unit price of the edge), the cost of the whole tree is

$$\sum_{i=1}^{n} \left(\text{the length of the path from the root to vertex } i\right) * \left(\text{the weight of vertex } i\right)$$

In order to minimize the cost of the whole tree, the lengths of the paths from the root to the vertices must be minimized. Therefore, it is a problem for single-source shortest paths.

Suppose the number of vertices is n and the number of edges is m. Array *weight* is used to store vertices' weights, where the weight of vertex i is *weight*[i]. Linear list E is used to store M edges, where the kth edge is $(E[k].l, E[k].r)$ and the weight of the edge is $E[k].val$. Array *dist* is used to store the sum of edges' weights, where the minimal sum of edges' weights from the root (vertex 0) to vertex i is *dist*[i]. Elements for priority queue Q are used to store the number of vertices and the sum of edges' weights from the root to vertices.

The Dijkstra algorithm is used to calculate *dist*:

```
dist[1]= dist[2]=…dist[n-1]=∞;
    Add vertex 0 and dist[0] to priority queue Q;
    while (Q is not empty) {        // Repeat until Q is empty
        Delete vertex x whose dist[x] is minimal from Q;
        If (x.val<=dist[x.l])
        {
            for (int i = start[x.l]; i < 2 * M && E[i].l == x.l; i++) //
Relaxation
            if (dist[E[i].r] > x.val + E[i].val) {
                dist[E[i].r] = x.val + E[i].val;
                Add vertex E[i].r and dist[E[i].r] to Q;
            }
        }
    }
```

If all vertices are reachable from vertex 0, that is, $dist[i] \neq \infty$, $0 <= i <= n - 1$, then the minimum cost for the tree is

$$\sum_{i=0}^{n-1} dist[i] * weight[i]$$

otherwise, output "No Answer."

13.5.3 Stockbroker Grapevine

Stockbrokers are known to overreact to rumors. You have been contracted to develop a method of spreading disinformation among the stockbrokers to give your employer the tactical edge in the stock market. For maximum effect, you have to spread the rumors in the fastest possible way.

Unfortunately for you, stockbrokers only trust information coming from their trusted sources. This means you have to take into account the structure of their contacts when starting a rumor. It takes a certain amount of time for a specific stockbroker to pass the rumor on to each of his colleagues. Your task will need to write a program that tells you which stockbroker to choose as your starting point for the rumor, as well as the time it will take for the rumor to spread throughout the stockbroker community. This duration is measured as the time needed for the last person to receive the information.

Input

Your program will input data for different sets of stockbrokers. Each set starts with a line with the number of stockbrokers. Following this is a line for each stockbroker that contains the number of people who they have contact with, who these people are, and the time taken for them to pass the message to each person. The format of each stockbroker line is as follows: the line starts with the number of contacts (*n*), followed by *n* pairs of integers, one pair for each contact. Each pair lists first a number referring to the contact (e.g., a 1 means person 1 in the set), followed by the time in minutes taken to pass a message to that person. There are no special punctuation symbols or spacing rules.

Each person is numbered 1 through to the number of stockbrokers. The time taken to pass the message on will be between 1 and 10 minutes (inclusive), and the number of contacts will range between 0 and one less than the number of stockbrokers. The number of stockbrokers will range from 1 to 100. The input is terminated by a set of stockbrokers containing zero people.

Output

For each set of data, your program must output a single line containing the person who results in the fastest message transmission, and how long before the last person will receive any given message after you give it to this person, measured in integer minutes.

It is possible that your program will receive a network of connections that excludes some persons; that is, some people may be unreachable. If your program detects such a broken network, simply output the message "disjoint." Note that the time taken to pass the message from person A to person B is not necessarily the same as the time taken to pass it from B to A, if such transmission is possible at all.

Sample Input	Sample Output
3	3 2
2 2 4 3 5	3 10
2 1 2 3 6	
2 1 2 2 2	
5	
3 4 4 2 8 5 3	
1 5 8	
4 1 6 4 10 2 7 5 2	

(Continued)

Sample Input	Sample Output
0	
2 2 5 1 5	
0	

Source: ACM Southern African 2001.

IDs for online judges: POJ 1125, ZOJ 0182, UVA 2241.

Hint

Stockbrokers are represented as vertices, relationships between stockbrokers are represented as arcs, and an arc's weight is the time taken for one stockbroker to pass the message to the other stockbroker. A weighted, directed graph G is constructed based on it.

The problem requires you to find the stockbroker who results in the fastest message transmission. If stockbroker i is chosen as the starting point for the rumor, the time it will take for the rumor to spread throughout the stockbroker community is the maximum of the shortest paths from vertex i to other vertices

$$d_i = \max_{1 \le j \le n}\{dist[i][j]\}$$

where $dist[i][j]$ is the length of the shortest path from vertex i to vertex j.

Obviously, the earliest time it will take for the rumor to spread throughout the stockbroker community is

$$ans = \min_{1 \le i \le n}\{d_i\}$$

And the chosen stockbroker t satisfies the condition $ans = d_t$. Therefore, the algorithm is as follows:

```
Floyd algorithm is used to compute the matrix for all-pairs shortest
paths dist[ ][ ];
ans=∞;   t=-1;    // Initialization
For each row i in dist[ ][ ] (1≤i≤n)
  { Compute di= max{dist[i][j] }
            1≤j≤n
     if (ans>di) {  // Adjust ans and note down vertex i
         ans=di;
         t = i;
     }
  }
if (ans ==∞)
    Output "disjoint";
else
    Output vertex t and the length of the shortest path ans;
```

13.5.4 Domino Effect

Did you know that you can use domino bones for other things besides playing dominos? Take a number of dominos and build a row by standing them on end with only a small distance in between. If you do it right, you can tip the first domino and cause all others to fall down in succession (this is where the phrase "domino effect"' comes from).

While this is somewhat pointless with only a few dominos, some people went to the opposite extreme in the early 1980s. Using millions of dominos of different colors and materials to fill whole halls with elaborate patterns of falling dominos, they created (short-lived) pieces of art. In these constructions, usually not only one, but several rows of dominos were falling at the same time. As you can imagine, timing is an essential factor here.

It is now your task to write a program that, given such a system of rows formed by dominos, computes when and where the last domino falls. The system consists of several key dominos connected by rows of simple dominos. When a key domino falls, all rows connected to the domino will also start falling (except for the ones that have already fallen). When the falling rows reach other key dominos that have not fallen yet, these other key dominos will fall as well and set off the rows connected to them. Domino rows may start collapsing at either end. It is even possible that a row is collapsing on both ends, in which case the last domino falling in that row is somewhere between its key dominos. You can assume that rows fall at a uniform rate.

Input

The input file contains descriptions of several domino systems. The first line of each description contains two integers: the number n of key dominos ($1 \leq n < 500$) and the number m of rows between them. The key dominos are numbered from 1 to n. There is at most one row between any pair of key dominos and the domino graph is connected; that is, there is at least one way to get from a domino to any other domino by following a series of domino rows.

The following m lines each contain three integers, a, b, and l, stating that there is a row between key dominos a and b that takes l seconds to fall down from end to end.

Each system is started by tipping over key domino number 1.

The file ends with an empty system (with $n = m = 0$), which should not be processed.

Output

For each case, output a line stating the number of the case ("System #1," "System #2," etc.). Then output a line containing the time when the last domino falls, exact to one digit to the right of the decimal point, and the location of the last domino falling, which is either at a key domino or between two key dominos (in this case, output the two numbers in ascending order). Adhere to the format shown in the output sample. The test data will ensure there is only one solution. Output a blank line after each system.

Sample Input	Sample Output
2 1	System #1
1 2 27	The last domino falls after 27.0 seconds, at key domino 2.
3 3	
1 2 5	System #2
1 3 5	The last domino falls after 7.5 seconds, between key dominos 2 and 3.

(Continued)

Sample Input	Sample Output
2 3 5	
0 0	

Source: ACM Southwestern European Regional Contest 1996.

IDs for online judges: POJ 1135, ZOJ 1298, UVA 318.

Hint

A weighted undirected graph $G(V, E)$ is constructed as follows. Key dominos are represented as vertices. If there is a row between key dominos a and b that takes l seconds to fall down from end to end, there is an edge connecting vertex a and vertex b, whose weight is l, that is, $w_{ab} = w_{ba} = l$.

From vertex 1 (the first tipped domino), the algorithm calculating single-source shortest paths is used to compute the lengths of the shortest paths from vertex 1 to other vertices $dist[\ ]$.

There are two cases for domino effect.

Case 1: If it is a key domino that falls in the end, then the time that all dominos are tipped is

$$max1 = \max_{1 \le i \le n}\{dist[i]\}$$

and the last tipped key domino is *index* ($max1 = dist[index]$).

Cases 2: For a row between key dominos i and j, if the last tipped domino falls between key domino i and key domino j, then the time that all dominos in the row between key dominos i and j are tipped is

$$\frac{dist[i] + dist[j] + w_{ij}}{2}$$

Obviously, for all rows, if the last domino falls between its key dominos, then the time that all dominos are tipped is

$$max2 = \max_{1 \le i < j \le n}\{\frac{dist[i] + dist[j] + w_{ij}}{2}\Big|(i, j) \in E\}$$

If the last domino falls between key domino *index1* and key domino *index2*,

$$max2 = \frac{dist[index1] + dist[index2] + w_{indx1,index2}}{2}$$

Obviously, if $max1 \ge max2$, then the last tipped domino is a key domino, and the time that all dominos fall is $max1$; else, the last tipped domino is between two key dominos, and the time that all dominos fall is $max2$.

Program

```cpp
#include <iostream>
#include <algorithm>
#include <cmath>
#include <cstdio>
#include <cstdlib>
#include <cstring>
#include <string>
#define MAX 600                     //The upper limit of the number of vertices
#define INF 0x7FFFFFFF                   //Infinity
using namespace std;
int n,m;                //the number of vertices n, the number of edges m
int edge[MAX][MAX],visit[MAX],dist[MAX];// adjacency matrix edge[ ][ ],
visited mark visit[ ], the length of the shortes paths dist[ ]
void dijkstra(int u0)          // calculating single-source shortes paths
from u0
{
    int i,j,v;
    for(i=1; i<=n; i++) dist[i] = edge[u0][i] ;      //Initialization
    visit[u0] = 1;                           //u0: visited mark
    for(j=0; j<n-1; j++)
    {
        int min = INF;
        for(i=1; i<=n; i++)
        {
            if(!visit[i] && min > dist[i]) { min = dist[i]; v = i; }
        }
        visit[v] = 1;                    // visited mark
        for(i=1; i<=n; i++)              // adjust distances
        {
            if(!visit[i] && edge[v][i]<INF && edge[v][i]+dist[v]<dist[i])
dist[i]=edge[v][i]+ dist[v];
        }
    }
}
int main()
{
    int i,j,a,b,c,tt = 1;
    while(scanf("%d%d",&n,&m))    //numbers of vertices and rows
    {
        if(n==0 && m==0) break;
        for(i=1; i<=n; i++)                     //Initialize edge[ ][ ]
            for(j=1; j<=n; j++)
            {
                if(i == j) edge[i][j] = 0;
                else edge[i][j] = INF;
            }
            for(i=0; i<m; i++)                //Construct edge[ ][ ]
            {
            scanf("%d%d%d",&a,&b,&c);
            edge[a][b] = c; edge[b][a] = c;
        }
        memset(visit,0,sizeof(visit));
```

```
        dijkstra(1);                    // calculating single-source shortes
paths from vertex 1
              double max1 = -10000000;
        int index;
        for(i=1; i<=n; i++)       //Case 1: the last domino which falls is
a key domino index
          {
            if(max1 < dist[i])  { max1 = dist[i]*1.0; index = i; }
          }

        double max2 = -10000000;
        int index1,index2;
        for(i=1; i<=n; i++)              // Case 2: Enumerate every edge
(i, j) (i<j), the last domino falls between key domino index1 and key
domino index2, and the time that all dominos fall is max2.
            for(j=1; j<=n; j++)
            {
                if(edge[i][j] != INF && i < j)
                {
                    if(max2 < (dist[i]+dist[j]+edge[i][j])/2.0)
                    {
                        max2 = (dist[i]+dist[j]+edge[i][j])/2.0;
                        index1 = i; index2 = j;
                    }
                }
            }

        printf("System #%d\n",tt++);    // the number of the test case
        if(max1 >= max2)                // if case 1, else case2
            printf("The last domino falls after %.1f seconds, at key
domino %d.\n",max1,index);
        else printf("The last domino falls after %.1f seconds, between
key dominoes %d and %d.\n",max2,index1,index2);
        printf("\n");
    }
    return 0;
}
```

13.5.5 106 miles to Chicago

In the movie *Blues Brothers*, the orphanage where Elwood and Jack were raised may be sold to the board of education if they do not pay $5000 in taxes at the Cook Country Assessor's Office in Chicago. After playing a gig in the Palace Hotel ballroom to earn the $5000, they have to find a way to Chicago. However, this is not so easy as it sounds, since they are chased by the police, a country band, and a group of Nazis. Moreover, it is 106 miles to Chicago, it is dark, and they are wearing sunglasses.

As they are on a mission from God, you should help them find the safest way to Chicago. In this problem, the safest way is considered to be the route that maximizes the probability that they are not caught.

Input

The input contains several test cases.

Each test case starts with two integers, *n* and *m* (2 ≤ *n* ≤ 100, 1 ≤ *m* ≤ n*(*n* – 1)/2). *n* is the number of intersections, and *m* is the number of streets to be considered.

The next *m* lines contain the description of the streets. Each street is described by a line containing three integers *a*, *b*, and *p* (1 ≤ *a*, *b* ≤ n, *a* != *b*, 1 ≤ *p* ≤ 100): *a* and *b* are the two end points of the street, and *p* is the probability in percent that the Blues Brothers will manage to use this street without being caught. Each street can be used in both directions. You may assume that there is at most one street between two end points.

The last test case is followed by a zero.

Output

For each test case, calculate the probability of the safest path from intersection 1 (the Palace Hotel) to intersection *n* (the Honorable Richard J. Daley Plaza in Chicago). You can assume that there is at least one path between intersection 1 and *n*.

Print the probability as a percentage with exactly six digits after the decimal point. The percentage value is considered correct if it differs by at most 10^{-6} from the judge output. Adhere to the format shown below and print one line for each test case.

Sample Input	Sample Output
5 7	61.200000 percent
5 2 100	
3 5 80	
2 3 70	
2 1 50	
3 4 90	
4 1 85	
3 1 70	
0	

Source: Ulm Local Contest 2005.

IDs for online judges: POJ 2472, ZOJ 2797.

Hint
The problem can be represented as an graph: intersections are represented as vertices, and streets are represented as edges. If there is a street connecting intersection *a* and intersection *b*, and *p* is the probability in percent that the Blues Brothers use this street without being caught, then edge (*a*, *b*) with weight *p*/100 is added.

If the weight of edge (*i*, *k*) is p_{ik}, and the weight of edge (*k*, *j*) is p_{kj}, then the probability that the Blues Brothers go to vertex *j* from vertex *i* through vertex *k* without being caught is $p_{ik}*p_{kj}$. We enumerate every intermedia vertex *k* in the path from vertex *i* to vertex *j*; the maximal probability that the Blues Brothers go to vertex *j* from vertex *i* through vertex *k* without being caught is max{$p_{ik}*p_{kj}$}.

Obviously, the safest path from intersection 1 to intersection *n* is the longest path intersection 1 to intersection *n*. The Floyd algorithm is used to calculate the length of the longest path between any pairs of vertices; that is, the maximal probability that they are not caught is *p*[ ][ ], where *p*[1][*n*]*100 is the solution to the problem.

13.5.6 AntiFloyd

You have been hired as a systems administrator for a large company. The company head office has *n* computers connected by a network of *m* cables. Each cable connects two different computers, and there is at most one cable connecting any given pair of computers. Each cable has a latency, measured in microseconds, that determines how long it takes for a message to travel along that cable. The network protocol is set up in a smart way so that when sending a message from computer *A* to computer *B*, the message will travel along the path that has the smallest total latency, so that it arrives at *B* as soon as possible. The cables are bidirectional and have the same latency in both directions.

As your first order of business, you need to determine which computers are connected to each other, and what the latency is along each of the *m* cables. You soon discover that this is a difficult task because the building has many floors, and the cables are hidden inside walls. So here is what you decide to do. You will send a message from every computer *A* to every other computer *B* and measure the latency. This will give you *n*(*n* − 1)/2 measurements. From this data, you will determine which computers are connected by cables, and what the latency along each cable is. You would like your model to be simple, so you want to use as few cables as possible.

Input

The first line of input gives the number of cases, *N* (at most 20). *N* test cases follow. Each one starts with a line containing *n* (0 < *n* < 100). The next *n* − 1 lines will contain the message latency measurements. Line *i* will contain *i* integers in the range [1, 10000]. Integer *j* is the amount of time it takes to send a message from computer *i* + 1 to computer *j* (or back).

Output

For each test case, output a line containing "Case #*x*:". The next line should contain *m*—the number of cables. The next *m* lines should contain three integers each: *u*, *v*, and *w*, meaning that there is a cable between computers *u* and *v*, and it has latency *w*. Lines should be sorted first by *u*, then by *v*, with *u* < *v*. If there are multiple answers, any one will do. If the situation is impossible, print "Need better measurements." Print an empty line after each test case.

Sample Input	Sample Output
2	Case #1:
3	2
100	1 2 100
200 100	2 3 100
3	

Sample Input	Sample Output
100	Case #2:
300 100	Need better measurements.

Source: Abednego's Graph Lovers' Contest 2006.

ID for online judge: UVA 10987.

Hint by the Problemsetter (http://www.algorithmist.com)
Given the length of the shortest path between every pair of vertices in an undirected graph with positive edge weights, construct such a graph, having a minimum number of edges.

Let the vertices be numbered from 1 to n, and let d_{ij} be the length of the shortest path between i and j.

The first thing is to check whether the graph actually exists, that is, do the given d_{ij}'s lengths of shortest paths exist? If there exist three (distinct) vertices i, j, k such that $d_{ik} + d_{kj} < d_{ij}$, then obviously the graph doesn't exist, because by concatenating shortest paths from i to k and from k to j we get a path from i to j, of length less than d_{ij}.

If no such vertices exist, then a complete graph K_n, with weight of edge (i, j) equal to d_{ij}, can be constructed. There are $n(n-1)/2$ edges in the graph.

Consider some pair of vertices i and j. The shortest path between them is either a single edge (i, j) of weight d_{ij} or a sequence of two or more edges, each of weight strictly less than d_{ij} (due to assumption, the weight of every edge is greater than zero). In the latter case, there exists some vertex k, different from i and j, such that $d_{ik} + d_{kj} = d_{ij}$. If no such k exists, the only option is to include the edge (i, j) in our graph. Otherwise, we can always omit edge (i, j)—there has to be a path from i to j in the final graph through a sequence of edges of length less than d_{ij} (this can be formally proven by induction on d_{ij}.)

An algorithm solving the problem is as follows:

```
Suppose n is number of vertices, array d is the shortest path lengths'
matrix.
for (i = 1; i<=n, i++)
    for (j = 1; j<=n; j++)
        for (k = 1, k<=n; k++)
            if (d[i][k]+d[k][j] < d[i][j])
                return "Need better measurements.";
for (i = 1; i<=n, i++)
    for (j = 1; j<=n; j++)
        {   flag = true;
            for (k = 1, k<=n; k++)
                if ((k != i) and (k != j) and (d[i][k]+d[k][j] == d[i][j]))
                        flag = false;
            if (flag = true)
                add edge (i, j) of weight d[i][j] to the graph;
        }
```

Chapter 14

Algorithms of Bipartite Graphs and Flow Networks

In graph theory there are some special graphs. Such graphs not only satisfy the definition of graph, but also have some special properties or satisfy some special conditions, such as bipartite graph and flow network.

Bipartite graphs and flow networks are two kinds of special graphs. Algorithms of bipartite graphs and flow networks are important in graph theory. This chapter focuses on experiments for bipartite graph and flow network.

14.1 Maximum Matching in Bipartite Graphs

A bipartite graph has a set of vertices that can be divided into two disjoint subsets such that each edge connects a vertex in one of the two subsets to a vertex in the other subset. Given a bipartite graph $G(V, E)$, a matching is a subset of edges $M \subseteq E$, if there is no common vertex for any two edges in M. A maximum matching is a matching of maximum cardinality; that is, a matching M is called a maximum matching if for any other matching M', $|M| \geq |M'|$. In this section, finding a maximum matching in a bipartite graph is discussed. A perfect matching is a matching that matches all vertices of the graph. That is, every vertex of the graph is incident to exactly one edge of the matching. A perfect matching must be a maximum matching.

For a bipartite graph, the Hungarian algorithm is used to find a maximum matching or a perfect matching.

For a complete weighted bipartite graph, the KM algorithm and minimum-cost flow algorithm can be used to find a maximum-weight matching.

The Hungarian algorithm is the foundation for all algorithms for bipartite matching.

Given a bipartite graph $G(V, E)$ and a matching M, the set of vertices with which edges in M are incident is called a cover. For matching M, an alternating path is a path where the edges belong alternatively to M and not to M, and an augmenting path is an alternating path that starts from and ends on unmatched vertices. Matching M is the maximum matching in G if there is no other matching M' in G such that $|M'| > |M|$. Steps for the Hungarian algorithm are as follows:

1. Initially matching M is empty.
2. Find an augmenting path p for M, and $M \leftarrow M \oplus p$.
3. Repeat step 2 until there is no any augmenting path in G. Matching M is a maximum matching for G.

The depth-first search (DFS) algorithm can be used to find an augmenting path. The DFS algorithm takes an unmatched vertex as the starting vertex, and it produces an augmenting path p in which the edges belong alternatively to M and not to M.

For a bipartite graph G, vertices can be divided into two disjoint sets X and Y such that every edge connects a vertex in X to one in Y. The DFS algorithm is as follows:

```
bool dfs(int i){      // Determine whether there is an augmenting path
starting from vertex i in X
  for (int j=1; j<=m; j++)
    if ((!v[j])&&(a[i][j])){       // Search all unvisited vertices which
are adjacent to vertex i
      v[j]=1;                      // visit vertex j
      if (pre[j]==0||dfs(pre[j])){      //If the precursor for j is
unmatched or there exists an augmenting path starting from the precursor
for j, then edge (i, j) is in matching, and return true
        pre[j]=i;
        return 1;
      }
    }
  return 0;                                      //return false
}
```

If *dfs*(i) returns true, then vertex i is matched. Obviously, for every vertex i, *dfs*(i) is called, and a maximum matching in a bipartite graph is obtained. Therefore, the Hungarian algorithm is as follows:

```
int ans=0;                             //Initialization
for (int i=1; i<=n; i++){              //Enumeration
  memset(v, 0, sizeof(v));
  if (dfs(i)) ans++;
}
```

Suppose there are e edges in a bipartite graph G, vertices in G are divided into two disjoint sets X and Y such that $|X| = |Y| = n$, and M is a matching in G. The time complexity of finding an augmenting path is O(e). In order to get a maximum matching, at most n augmenting paths are required for calculation. Therefore, the time complexity of the Hungarian algorithm is O($n*e$).

14.1.1 Conference

M representatives of country A and N representatives of country B (M and $N \leq 1000$) were sent to the upcoming conference. The representatives were identified with 1, 2, ..., M for country A and 1, 2, ..., N for country B. Before the conference, K pairs of representatives were chosen. Every such pair consists of one member of delegation *A* and one of delegation *B*. If there exists a pair in which both member #*i* of *A* and member #*j* of *B* are included, then #*i* and #*j* can negotiate. Everyone attending the conference was included in at least one pair. The CEO of the congress center wants to build direct telephone connections between the rooms of the delegates, so that everyone is connected with at least one representative of the other side, and every connection is made between people that can negotiate. The CEO also wants to minimize the amount of telephone connections. Write a program that, given *M*, *N*, *K*, and *K* pairs of representatives, finds the minimum number of needed connections.

Input

The first line of the input contains *M*, *N*, and *K*. The following *K* lines contain the chosen pairs in the form of two integers p_1 and p_2; p_1 is member of *A* and p_2 is member of *B*.

Output

The output should contain the minimum number of needed telephone connections.

Sample Input	Sample Output
3 2 4	3
1 1	
2 1	
3 1	
3 2	

Source: Bulgarian Online Contest, September 2001.

ID for online judge: Ural 1109.

Analysis

Because there are *M* representatives of country *A* and *N* representatives of country *B*, and *K* pairs of representatives were chosen such that for each pair, one member is from delegation *A* and the other member is from delegation *B*, the problem can be represented as a bipartite graph *G*, in which vertices in *G* are divided into two disjoint sets *X* and *Y* such that *M* representatives of country *A* are set *X* and *N* representatives of country *B* are set *Y*. The CEO wants to minimize the number of telephone connections. The edge cover number in a bipartite graph is required for computation.

Everyone is connected with at least one representative of the other side. And *N* + *M* is the number of vertices in *G*. Based on graph theory, the edge cover number plus the number of edges in a maximum matching equals *N* + *M*.

For the problem, the number of edges in a maximum matching *ans* is computed first. Then the edge cover number is obtained: *N* + *M* − *ans*. It is the minimum number of needed telephone connections.

Program

```
#include<iostream>
using namespace std;
const int V=1100;              // number of vertices
```

```
int n, m, k, x, y, pre[V];        // vertices are divided into two disjoint
   sets X and Y such that |X|=n and |Y|=m, k is the number of edges,
   vertex i is an endpoint of an edge (pre[i], i) in the matching
bool v[V], a[V][V];               //a is the adjacency matrix for a
   bipartite graph. If vertex i in Y has been visited, v[i]=true.
bool dfs(int i){                  // Determine whether there is an augmenting
   path strating from i
      for (int j=1;j<=m;j++) if ((!v[j])&&(a[i][j])){
            v[j]=1;
            if (pre[j]==0||dfs(pre[j])){
                  pre[j]=i;
                  return 1;
            }
      }
      return 0;
}
int main(){
      cin>>n>>m>>k;                      //Input N, M, K
      memset(a,0,sizeof(a));     // Initialization of the adjacency
matrix for a bipartite graph
      memset(pre,0,sizeof(pre));       // Initialization of the matching
      for (int i=1;i<=k;i++){             // Pairs of representatives
         cin>>x>>y;
         a[x][y]=1;
      }
      int ans=0;
      for (int i=1;i<=n;i++){             // Enumeration
            memset(v,0,sizeof(v));
            if (dfs(i)) ans++;
      }
      cout<<n+m-ans<<endl;       // the minimum number of needed
telephone connections
}
```

Sometimes edges' weights need to be considered in bipartite matching. We are required to find a maximum-weight matching. For example, there are n persons and m tasks. There is a fixed benefit for a person finishing a task. We need to find a maximum benefit. There are two methods to find a maximum-weight matching: the KM algorithm (its time complexity is $O(n*e^2)$) and the minimum-cost flow algorithm (its time complexity is $O(n*e)$).

14.2 Flow Networks

In this section experiments for flow networks are introduced. First, definitions and theory for flow networks are reviewed.

A flow network $G = (V, E)$ is a directed graph where each arc $(v_i, v_j) \in E$ has a nonnegative capacity $C_{ij} > 0$, and if there is no arc from vertex v_i to vertex v_j, $C_{ij} = 0$. Two vertices are distinguished in a flow network: a source vertex s and a sink vertex t. A flow in G is a real-valued function $f: V \times V \rightarrow R$.

A flow network satisfies the following properties:

1. Capacity constraint: For each $(u, v) \in E$, $0 \le f(u, v) \le C(u, v)$.

2. Flow conservation: For $\forall u \in V - \{s, t\}$,

$$\sum_{x \in V} f(x,u) - \sum_{x \in V} f(u,x) = 0$$

Given a flow network $G(V, E)$ with source vertex s and sink vertex t, and a flow f, we are normally required to find the maximum flow from the source s to the sink t. The key to the problem is to find an augmenting path p from s to t. Methods to find an augmenting path are as follows:

1. Depth-first search (DFS)
2. Breadth-first search (BFS)
3. Labeling algorithm

The above algorithms add flow each time for an augmenting path. If the maximum flow is a, and the time finding an augmenting path is m, the time complexity for computing the maximum flow is $O(a*m)$.

14.2.1 Power Network

A power network consists of nodes (power stations, consumers, and dispatchers) connected by power transport lines. A node u may be supplied with an amount $s(u) \geq 0$ of power, may produce an amount $0 \leq p(u) \leq p_{max}(u)$ of power, may consume an amount $0 \leq c(u) \leq \min(s(u), c_{max}(u))$ of power, and may deliver an amount $d(u) = s(u) + p(u) - c(u)$ of power. The following restrictions apply: $c(u) = 0$ for any power station, $p(u) = 0$ for any consumer, and $p(u) = c(u) = 0$ for any dispatcher. There is at most one power transport line (u, v) from a node u to a node v in the net; it transports an amount $0 \leq l(u, v) \leq l_{max}(u, v)$ of power delivered by u to v. Let $Con = \sum_u c(u)$ be the power consumed in the net. The problem is to compute the maximum value of Con.

u	type	s(u)	p(u)	c(u)	d(u)
0	power	0	4	0	4
1	Station	2	2	0	4
3		4	0	2	2
4	consumer	5	0	1	4
5		3	0	3	0
2	dispatcher	6	0	0	6
6		0	0	0	0

Figure 14.1 shows an example. The label x/y of power station u shows that $p(u) = x$ and $p_{max}(u) = y$. The label x/y of consumer u shows that $c(u) = x$ and $c_{max}(u) = y$. The label x/y of power transport line (u, v) shows that $l(u, v) = x$ and $l_{max}(u, v) = y$. The power consumed is $Con = 6$. Notice that there are other possible states of the network, but the value of Con cannot exceed 6.

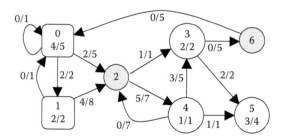

Figure 14.1 A power network.

Input

There are several data sets in the input text file. Each data set encodes a power network. It starts with four integers: $0 \le n \le 100$ (nodes), $0 \le n_p \le n$ (power stations), $0 \le n_c \le n$ (consumers), and $0 \le m \le n^2$ (power transport lines). Follow m data triplets $(u, v)z$, where u and v are node identifiers (starting from 0) and $0 \le z \le 1000$ is the value of $l_{max}(u, v)$. Follow n_p doublets$(u)z$, where u is the identifier of a power station and $0 \le z \le 10,000$ is the value of $p_{max}(u)$. The data set ends with n_c doublets $(u)z$, where u is the identifier of a consumer and $0 \le z \le 10,000$ is the value of $c_{max}(u)$. All input numbers are integers. Except the (u, v, z) triplets and the (u, z) doublets, which do not contain white spaces, white spaces can occur freely in input. Input data terminate with an end of file and are correct.

Output

For each data set from the input, the program prints on the standard output the maximum amount of power that can be consumed in the corresponding network. Each result has an integral value and is printed from the beginning of a separate line.

Sample Input	Sample Output
2 1 1 2 (0,1)20 (1,0)10 (0)15 (1)20	15
7 2 3 13 (0,0)1 (0,1)2 (0,2)5 (1,0)1 (1,2)8 (2,3)1 (2,4)7	6
(3,5)2 (3,6)5 (4,2)7 (4,3)5 (4,5)1 (6,0)5	
(0)5 (1)2 (3)2 (4)1 (5)4	

Source: ACM Southeastern Europe 2003.

IDs for online judges: POJ 1459, ZOJ 1734, UVA 2760.

Hint: The input in the table contains two data sets. The first data set encodes a network with two nodes, power station 0 with $p_{max}(0) = 15$ and consumer 1 with $c_{max}(1) = 20$, and two power transport lines with $l_{max}(0, 1) = 20$ and $l_{max}(1, 0) = 10$. The maximum value of Con is 15. The second data set encodes the network from Figure 14.1.

Analysis

The problem can be represented as a flow network. A source vertex s and a sink vertex t are added into the power network. For each power station, an arc from s to it with a nonnegative capacity p_{max}

is added. For each consumer, an arc from it to t with a nonnegative capacity c_{max} is added. For each (u, v, z) triple, there is an arc from u to v with a nonnegative capacity z. Obviously, the maximum amount of power *Con* is the maximum flow for the network.

Adjacency matrices are used to store flows ($f[n][n]$) and capacities ($c[n][n]$) in the network. The BFS algorithm is used to find an augmenting path by using a queue $q[n]$ and an array $fa[n]$ recording the augmenting path, where n is the number of vertices and $fa[i]$ is the precursor of vertex i in the augmenting path. Because a source vertex s and a sink vertex t are added, $n \leq 102$.

The time complexity of the flow network algorithm finding augmenting paths is $O(n^2 t)$, where n is the number of vertices and t is times of augmenting.

Program

```
#include <stdio.h>
#include <math.h>
#include <memory.h>
int n, np, nc, m, s, t;              //n: number of vertices, s: source
vertex, t: sink vertex, np: number of power stations, nc: number of
consumers, m: number of power transport lines
int fa[104], q[104], f[104][104], c[104][104];     // fa[ ]:augmenting
path, where fa[j] is the precursor for j, if it is a positive number, the
edge is forward edge, if it is a negative number, the edge is a backward
edge; q[ ]: a queue; f[ ][ ] and c[ ][ ] are adjacent matrixes, and are
used to store flow and capacity, respectively.
void proc( )                                    // find the maximum flow
{
    int qs, qt, d, d0, i, j, ans = 0;
    fa[t] = 1;
    while (fa[t] != 0)                          // augmenting path exists
    {
        qs = 0; qt = 1;
        q[qt] = s;                      // source vertex into the rear of the queue
        memset(fa, 0, sizeof(fa));      //                  initialize the
augmenting path
        fa[s] = s;
    while (qs < qt && fa[t] == 0)        //queue isn't empty and sink
vertex isn't found in the augmenting path
        {
            i = q[++qs];                        // get the front of the queue
            for (j = 1; j <= t; j++)            // vertex j that isn't on
augmenting path is enumerated
                if (fa[j] == 0)
                    if (f[i][j] < c[i][j]) // forward edge or backward edge
                    {
                        fa[j] = i;
                        q[++qt] = j;
                    }
                    else
                        if (f[j][i] > 0)
                        {
                            fa[j] = -i;
                            q[++qt] = j;
                        }
        }
}
```

```
      if (fa[t] != 0)                      // an augmenting path is found, adjust
the current flow
      {
          d0 = 10000000;
          i = t;
          while (i != s)                   //backward to the source
          {
             if (fa[i] > 0) // forward edge
             {
                if ((d = c[fa[i]][i] -f[fa[i]][i]) < d0)
                   d0 = d;
             }
             else            //  backward edge
                if (f[i][-fa[i]] < d0)
                   d0 = f[i][-fa[i]];
             i = abs(fa[i]);
          }
          ans += d0;             // The total flow increases d0
          i = t; // backward from the sink to adjust the flow on the
augmenting path
          while (i != s)
          {
             if (fa[i] > 0)       // forward edge, flow increases d0
                f[fa[i]][i] += d0;
             else              // backward edge, flow decreases d0
                f[i][-fa[i]] -= d0;
             i = abs(fa[i]);
          }
      }
   }
   printf("%d\n", ans);                     // output the maximum flow
}
int main( )
{
   int i, u, v, cc;
   while (scanf("%d%d%d%d", &n, &np, &nc, &m) == 4) // Input numbers of
vertices, power stations, consumers and power transport lines
   {
      s = n + 2; t = n + 1;                 // set source vertex and sink vertex
      memset(f, 0, sizeof(f));
      memset(c, 0, sizeof(c));
      for (i = 1; i <= m; i++)         //(u,v), capacity cc
      {
         while (getchar() != '(');
         scanf("%d,%d)%d", &u, &v, &cc);
         c[u + 1][v + 1] = cc;
      }
   for (i = 1; i <= np; i++)
   //from source vertex to every power station, there is an arc with
capacity cc
      {
         while (getchar() != '(');
         scanf("%d)%d", &u, &cc);
         c[s][u + 1] = cc;
```

```
        }
      for (i = 1; i <= nc; i++)
      //from every consumers to sink vertex, there is an arc with capacity
cc
      {
         while (getchar() != '(');
         scanf("%d)%d", &u, &cc);
         c[u + 1][t] = cc;
      }
      proc();                         // find the maximum flow
   }
}
```

The above problem is to find the maximum flow for the network.

The shortest path faster algorithm (SPFA) can be used to calculate the augmenting path from the source vertex *st* to the sink vertex *en*.

Flows are as edges' weights. The shortest path from the source vertex *st* to the sink vertex *en* can be calculated. The path is the augmenting path. Suppose *h* is the queue; pointers for the front and rear for *h* are *l* and *r*, respectively; and *pre* is the precursor pointer in the augmenting path.

```
int spfa( st, en, n){                 // calculate the length of the
shortest path from the source vertex st to the sink vertex en
         memset(h, 0, sizeof(h));     // Initialization: queue is empty
         memset(v, 0, sizeof(v));
         memset(d, ∞, sizeof(d));
         memset(pre, 0, sizeof(pre));
         l=1, r=1;
         h[1]=st;                     //source vertex into the queue
         v[st]=1;
         d[st]=0;                     // the length of the shortest path from the
source vertex st to itself
         while (1){
               int x=h[l];
               for (int t=fi[x], y=a[t].y; t; t=a[t].next, y=a[t].y) //all
arcs (x, i) from x are enumerated
                   if (a[t].c>0&&d[y]>d[x]+a[t].b){     // if (x, i) exists and
can be relaxed, then relax
                         d[y]=d[x]+a[t].b;
                         pre[y]=t;                 // the arc is recorded in the
augmenting path
                         if (!v[y]){               // vertex y isn't in the
queue
                               v[y]=1;
                               r=r+1; h[r]=y;
                         }
               }
               if (l==r) break;          // if the queue empty
               v[x]=0;                   // x is popped from the queue
               l=l+1;
         }
         return d[en];                // return the length of the shortest
path from the source vertex st to the sink vertex en
}
```

The algorithm calculating the new increased flow in the augmenting path from the source vertex *st* to the sink vertex *en* constitutes two steps:

1. From the sink vertex *en*, the new increased flow *a* in the augmenting path is calculated.
2. From the sink vertex *en*, the flow for edges in the augmenting path is adjusted.

```
int aug( st, en, n){                    // calculating the new increased flow
in the augmenting path
        int a=∞, ans=0, k=pre[en];      // Initialization; get the
precursor for the sink vertex
        while (k){                      // From the sink vertex en, the new
increased flow a in the augmenting path is calculated
                a=min(a, a[k].c);
                k=pre[a[k].x];          //backward
        }
        k=pre[en]; // From the sink vertex en, the flow for edges in the
augmenting path is adjusted
        while (k){
                ans+=a[k].b*a;
                a[k].c-=a;              // adjust
                a[a[k].op].c+=a;
                k=pre[a[k].x];          //backward
        }
        return ans;                     //Return
}
```

The algorithm calculating minimum-cost maximum flow is as follows:

Augmenting paths from the source vertex *st* to the sink vertex *en* are repeatedly calculated until the augmenting path doesn't exist. The new increased flows are accumulated every time.

```
int costflow( st, en, n){           // calculating minimum-cost maximum flow
        int ans=0, temp=spfa(st, en, n);    // Initialization
        while (temp<∞){     // while augmenting path exists, new increased
flow are accumulated
                ans+=aug(st, en, n);
                temp=spfa(st, en, n);       //calculating the augmenting path
        }
        return ans;                         //return total cost
}
```

An algorithm calculating minimum-cost maximum flow can be used to calculate maximum matching in bipartite graphs.

First, a bipartite graph is transferred into a network. Suppose there is a bipartite graph $G(X, Y; E)$, where X and Y are complementary sets of vertices, E is a subset of $X \times Y$, and $w(e)$ is the weight of e. A network D is constructed as follows:

1. The source vertex s and the sink vertex t are added.
2. For each $i \in X$, an arc (s, i) whose capacity is 1 and flow is 0 is constructed.
3. For each $j \in Y$, an arc (j, t) whose capacity is 1 and flow is 0 is constructed.
4. For each arc $(i, j) \in E$, where $i \in X$ and $j \in Y$, the capacity and the flow for the arc (i, j) are 1 and 0, respectively.

The minimum-cost maximum flow for D is the maximum matching in bipartite graph $(X, Y; E)$.

14.2.2 Trash

You were just hired as CEO of the local junkyard. One of your jobs is dealing with the incoming trash and sorting it for recycling. The trash comes every day in N containers, and each of these containers has a certain amount of each of the N types of trash. Given the amount of trash in the containers, find the optimal way to sort the trash. Sorting the trash means putting every type of trash in a separate container. Each of the given containers has infinite capacity. The effort for moving one unit of trash from container i to j is 1 if $i \neq j$; otherwise, it is 0. You are to minimize the total effort.

Input

The first line contains the number N ($1 \leq N \leq 150$), and the rest of the input contains the descriptions of the containers. The $(1 + i)$th line contains the description of the ith container, and the jth amount ($0 \leq \text{amount} \leq 100$) on this line denotes the amount of the jth type of trash in the ith container.

Output

You should write the minimal effort that is required for sorting the trash.

Sample Input	Sample Output
4	650
62 41 86 94	
73 58 11 12	
69 93 89 88	
81 40 69 13	

ID for online judge: Ural 1076.

Analysis

A weighted bipartite graph is given, and the maximum matching (perfect match) for the weighted bipartite graph is required for calculation. For this problem, minimum-cost maximum flow can be used. The SPFA is used to calculate the minimum-cost maximum flow most efficiently.

Modeling is the key to the problem. The problem requires that the amount of moved garbage be the least. Because the amount of garbage in each container is given, the amount of garbage that isn't moved out from containers is required as much as possible. Based on it, a flow network $G(V, E)$ is constructed, where $|V| = 2*n + 2$; source s is represented as vertex 1; sink t is represented as vertex $2*n + 2$; a set of vertices X is used to represent containers, where container X_i is represented as vertex $i + 1$, $1 \leq i \leq N$; and a set of vertices Y is used to represent trash, where Y_j is represented as vertex $N + j + 1$, $1 \leq i, j \leq n$. If the amount of trash Y_j in container X_i is a, then there is an arc from X_i to Y_j whose weight is the amount of all trash in container $X_i - a$, and there is an arc from Y_j to X_i whose weight is $-$(the amount of all trash in container $X_i - a$); from s to X_i there is an arc whose capacity is 1, and from trash Y_j to t, whose capacity is 1.

Obviously, the minimum-cost maximum flow for G is the minimal effort that is required for sorting the trash.

Program

```
#include<iostream>
#define maxn 500        //upper limit of vertices
#define maxq 10000      // upper limit of the length of the queue
#define mx 1000000      // Infinity
using namespace std;
long c[maxn][maxn]={0},g[maxn][maxn]={0},d[maxn]={0};  // adjacency
matrix for directed graph g[ ][ ], matrix for the flow length of the
shortest path c[ ][ ], the length of the shortest path d[ ]
int q[maxq]={0},pre[maxn]={0};
bool vis[maxn]={0};     // initially vertices aren't in the augmenting
path
bool b=1;              //mark for the augmenting path exists
long n,s=1,t;          // n is the number of containers and types of
trash, source vertex s=1,sink vertex t
long p=0;              // the minimal effort that is required for sorting
the trash is initialized 0
void augment()         //the flow on the augmenting path is adjusted
{
    int i=t;           // from sink vertex, calculate the flow on the
augmenting path a
    long a=mx;         // flow a is initialized Infinity
    while (i>s)
    {
        if (c[pre[i]][i]<a)a=c[pre[i]][i];
        i=pre[i]; //backward
    }
    i=t;               // the flow on the augmenting path is adjusted
from sink vertex
    while (i>s)
    {
        c[pre[i]][i]-=a;c[i][pre[i]]+=a;
        i=pre[i];      //backward
    }
}
void SPFA()                        //SPFA algorithm: calculate the
augmenting path from the source vertex to sink vertex
{
    memset(q,0,sizeof(q));         // Initialization: queue is empty
    memset(vis,0,sizeof(vis));     // Initialization: vertices aren't
visited
    memset(pre,0,sizeof(pre));     // Initialization: vertices aren't on
the augmenting path
    int l=1,r=1;                   // Initialization: front and rear for the queue
    for(int i=1;i<=t;++i)d[i]=mx;
    d[s]=0;q[1]=s;vis[s]=1;        // source vertex is pushed into the
queue,
    while (l<=r)                   // queue isn't empty
    {
        if (l==1 && r==maxq) break; // the queue is full, break
        long x=q[l];               // the front of the queue x
        for (int i=1;i<=t;++i)      // all edges (x, i) are enumerated
```

```
        if(d[x]+g[x][i]<d[i]&&c[x][i]>0) // if (x, i) exists and can be
relaxed, then relax
        {
            d[i]=d[x]+g[x][i];
            pre[i]=x;                //(x, i) is stored in the augmenting path
            if (!vis[i])                     // vertex i isn't in the queue
               {vis[i]=1;++r;if (r>maxq) r=1;q[r]=i;}
        }
        vis[x]=0;                         // x is popped from the queue
        ++l;if (l>maxq) l=1;
    }
    if (d[t]!=mx)                  // sink vertex t is reached, the
augmenting path exists
        { p+=d[t];augment();b=1;return;}
    b=0;                        // mark that the augmenting path doesn't exists
}
int main(void)
{
    cin>>n;
    t=2*n+2;                             // sink vertex
    for (int i=1;i<=n;++i)               // every container in X is
enumerated
    {
        int s=0;
        for (int j=1;j<=n;++j)           //every type of trash in Y is
enumerated
        {
            c[1+i][1+n+j]=1;
            cin>>g[1+i][1+n+j];          // the amount of trash j in
container i
            s+=g[1+i][1+n+j];            // accumulation
        }
        for (int j=1;j<=n;++j)           // every type of trash in Y is
enumerated: there is an arc from container i in set X to type of trash j
in set Y; there is an arc from type of trash j in set Y to container i in
set X; set their weights
        {
            g[1+i][1+n+j]=s-g[1+i][1+n+j]; g[1+n+j][1+i]=-g[1+i][1+n+j];
        }
    }
    for (int i=1;i<=n;++i)               // there is an arc from source
vertex to every container in set X, and from every type of trash in set Y
to sink vertex, whose capacity is 1
    {
        c[1][1+i]=1;c[1+n+i][t]=1;
    }
    b=1;                // The mark that the augmenting path exists
    while(b)SPFA();   // SPFA algorithm repeats until the augmenting path
doesn't exist
    cout<<p<<"\n"; // Output the minimal effort that is required for
sorting the trash
    return 0;
}
```

14.3 Problems

14.3.1 A Plug for UNIX

You are in charge of setting up the press room for the inaugural meeting of the United Nations Internet Executive (UNIX), which has an international mandate to make the free flow of information and ideas on the Internet as cumbersome and bureaucratic as possible.

Since the room was designed to accommodate reporters and journalists from around the world, it is equipped with electrical receptacles to suit the different shapes of plugs and voltages used by appliances in all of the countries that existed when the room was built. Unfortunately, the room was built many years ago, when reporters used very few electric and electronic devices, and is equipped with only one receptacle of each type. These days, like everyone else, reporters require many such devices to do their jobs: laptops, cell phones, tape recorders, pagers, coffee pots, microwave ovens, blow dryers, curling irons, toothbrushes, and so on. Naturally, many of these devices can operate on batteries, but since the meeting is likely to be long and tedious, you want to be able to plug in as many as you can.

Before the meeting begins, you gather up all the devices that the reporters would like to use and attempt to set them up. You notice that some of the devices use plugs for which there is no receptacle. You wonder if these devices are from countries that didn't exist when the room was built. For some receptacles, there are several devices that use the corresponding plug. For other receptacles, there are no devices that use the corresponding plug.

In order to try to solve the problem, you visit a nearby parts supply store. The store sells adapters that allow one type of plug to be used in a different type of outlet. Moreover, adapters are allowed to be plugged into other adapters. The store does not have adapters for all possible combinations of plugs and receptacles, but there is essentially an unlimited supply of the ones they do have.

Input

The input consists of one case. The first line contains a single positive integer n ($1 \le n \le 100$), indicating the number of receptacles in the room. The next n lines list the receptacle types found in the room. Each receptacle type consists of a string of at most 24 alphanumeric characters. The next line contains a single positive integer m ($1 \le m \le 100$), indicating the number of devices you would like to plug in. Each of the next m lines lists the name of a device followed by the type of plug it uses (which is identical to the type of receptacle it requires). A device name is a string of at most 24 alphanumeric characters. No two devices will have exactly the same name. The plug type is separated from the device name by a space. The next line contains a single positive integer k ($1 \le k \le 100$), indicating the number of different varieties of adapters that are available. Each of the next k lines describes a variety of adapter, giving the type of receptacle provided by the adapter, followed by a space, followed by the type of plug.

Output

The output is a line containing a single nonnegative integer indicating the smallest number of devices that cannot be plugged in.

Sample Input	Sample Output
4	1
A	

Sample Input	Sample Output
B	
C	
D	
5	
laptop B	
phone C	
pager B	
clock B	
comb X	
3	
B X	
X A	
X D	

Source: ACM East Central North America 1999.

IDs for online judges: POJ 1087, ZOJ 1157, UVA 753.

Hint
(The number of devices you would like to plug in) – (The maximal number of devices that can be plugged in) = (The smallest number of devices that cannot be plugged in). Obviously, the problem requires you to calculate the maximal number of devices that can be plugged in.

A bipartite graph is used to represent such a situation, where m devices constitute a set of vertices X, and n plugs constitute a set of vertices Y. If a device can use a plug directly, or can use a plug through adapters, there is an edge connecting the two corresponding vertices. Therefore, the maximal number of devices that can be plugged in is a maximum matching in the bipartite graph. The Hungarian algorithm can be used to solve the problem.

For this problem, what we should notice are

1. Different devices may use the same type of plug. In the list of adapters there may be some plugs that aren't used by any device.
2. When the list of adapters is input, an adjacency matrix $t[\ ][\ ]$ is set up, where $t[i][j] = 1$ represents that there can be an adapter connecting plug i and plug j. Its transitive closure $t'[i][j] = 1$ represents plug i, and plug j can be connected through adapters. Based on the transitive closure t', the bipartite graph can be constructed.

14.3.2 Machine Schedule

As we all know, machine scheduling is a very classical problem in computer science and has been studied for a long time. Scheduling problems differ widely in the nature of the constraints that must be satisfied and the type of schedule desired. Here we consider a two-machine scheduling problem.

There are two machines, *A* and *B*. Machine *A* has *n* kinds of working modes, which are called *mode_0*, *mode_1*, ..., *mode_n* − 1; likewise, machine *B* has *m* kinds of working modes, *mode_0*, *mode_1*, ..., *mode_m* − 1. At the beginning they both work at *mode_0*.

For *k* jobs given, each of them can be processed in either one of the two machines in a particular mode. For example, *job* 0 can be processed in either machine *A* at *mode_3* or machine *B* at *mode_4*, *job* 1 can be processed in either machine *A* at *mode_2* or machine *B* at *mode_4*, and so on. Thus, for *job i*, the constraint can be represented as a triple (*i*, *x*, *y*), which means it can be processed in either machine *A* at *mode_x* or machine *B* at *mode_y*.

Obviously, to accomplish all the jobs, we need to change the machine's working mode from time to time, but unfortunately, the machine's working mode can only be changed by restarting it manually. By changing the sequence of the jobs and assigning each job to a suitable machine, please write a program to minimize the times of restarting machines.

Input

The input file for this program consists of several configurations. The first line of one configuration contains three positive integers: *n*, *m* (*n*, *m* < 100), and *k* (*k* < 1000). The following *k* lines give the constraints of the *k* jobs; each line is a triple: *i*, *x*, *y*.

The input will be terminated by a line containing a single zero.

Output

The output should be one integer per line, which means the minimal times of restarting the machine.

Sample Input	Sample Output
5 5 10	3
0 1 1	
1 1 2	
2 1 3	
3 1 4	
4 2 1	
5 2 2	
6 2 3	
7 2 4	
8 3 3	
9 4 3	
0	

Source: ACM Beijing 2002.

IDs for online judges: POJ 1325, ZOJ 1364, UVA 2523.

Hint

A bipartite graph is used to represent working modes for machine A and machine B. Machine A has n kinds of working modes represented as a set of vertices X, and machine B has m kinds of working modes represented as a set of vertices Y. If *job i* can be processed in either machine A at *mode_x* or machine B at *mode_y*, and the two working modes aren't *mode_0* $((x \,\&\& \,y) \neq 0)$, there is an edge connecting vertex x and vertex y.

The problem requires you to assign each job to a suitable machine and minimize the times of restarting the machines. In the bipartite graph, an edge represents a job constraint. For a maximum matching M in the bipartite graph, an edge in M represents restarting a machine one time, for there is no common vertex for any two edges in M. Therefore, the minimal times of restarting the machine is $|M|$.

14.3.3 Selecting Courses

It is well known that it is not easy to select courses in college, for there is usually conflict among the time of the courses. Li Ming is a student who loves study every much, and at the beginning of each term, he always wants to select courses as much as possible. Of course, there should be no conflict among the courses he selects.

There are 12 classes every day and 7 days every week. There are hundreds of courses in the college, and teaching a course needs one class each week. To give students more convenience, though teaching a course needs only one class, a course will be taught several times in a week. For example, a course may be taught as both the 7th class on Tuesday and the 12th class on Wednesday; you should assume that there is no difference between the two classes and that students can select any class to go to. On different weeks, a student can even go to different classes as he or she wishes. Because there are so many courses in the college, selecting courses is not an easy job for Li Ming. As his good friends, can you help him?

Input

The input contains several cases. For each case, the first line contains an integer n $(1 \le n \le 300)$, the number of courses in Li Ming's college. The following n lines represent n different courses. In each line, the first number is an integer t $(1 \le t \le 7^*12)$, the different times when students can go to study the course. Then come t pairs of integers p $(1 \le p \le 7)$ and q $(1 \le q \le 12)$, which mean that the course will be taught at the qth class on the pth day of a week.

Output

For each test case, output one integer, which is the maximum number of courses Li Ming can select.

Sample Input	Sample Output
5	4
1 1 1	
2 1 1 2 2	

(Continued)

Sample Input	Sample Output
1 2 2	
2 3 2 3 3	
1 3 3	

Source: POJ Monthly.

ID for online judge: POJ 2239.

Hint

A bipartite graph is used to represent selecting courses. The set of time is represented as a set of vertices X, the qth class on the pth day of a week is represented as vertex i, where $i = (p-1)*12 + q$ $(1 \leq p \leq 7, 1 \leq q \leq 12)$; and the set of courses is represented as a set of vertices Y, where course j is represented as vertex j in Y $(1 \leq j \leq n)$. If course j can be taught at time i, there is an edge connecting vertex i and vertex j. Therefore, a maximum matching M in the bipartite graph is the maximum number of courses Li Ming can select.

14.3.4 Software Allocation

A computing center has 10 different computers (numbered 0–9) on which applications can run. The computers are not multitasking, so each machine can run only one application at any time. There are 26 applications, named A–Z. Whether an application can run on a particular computer can be found in a job description (see below).

Every morning, the users bring in their applications for that day. It is possible that two users bring in the same application; in that case, two different, independent computers will be allocated for that application.

A clerk collects the applications, and for each different application, he makes a list of computers on which the application could run. Then, he assigns each application to a computer.

Remember: The computers are *not* multitasking, so each computer must handle at most one application in total. (An application takes a day to complete, so that sequencing, that is, one application after another on the same machine, is not possible.)

A job description consists of:

1. One uppercase letter A–Z, indicating the application
2. One digit 1–9, indicating the number of users who brought in the application
3. A blank (space character)
4. One or more different digits 0–9, indicating the computers on which the application can run
5. A terminating semicolon
6. An end of line

Input

The input for your program is a text file. For each day, it contains one or more job descriptions, separated by a line containing only the end-of-line marker. The input file ends with the standard end-of-file marker. For each day, your program determines whether an allocation of applications to computers can be done and, if so, generates a possible allocation.

The output is also a text file. For each day, it consists of one of the following:

- Ten characters from the set {'A' ... 'Z', '_'}, indicating the applications allocated to computers 0–9. Respectively. if an allocation was possible. An underscore ('_') means that no application is allocated to the corresponding computer.
- A single character '!' if no allocation was possible.

Sample Input	Sample Output
A4 01234;	AAAA_QPPPP
Q1 5;	!
P4 56789;	
A4 01234;	
Q1 5;	
P5 56789;	

ID for online judge: UVA 259.

Hint
A flow network D is constructed as follows.

Ten different computers are represented as vertex 0 .. vertex 9, 26 applications are represented as vertex 10 .. vertex 35, source st is represented as vertex 36, and sink en is represented as vertex 37.

For a job description, if the application's name is x and the number of users who brought in the application is f, then there is an arc (st, x) from source st to x, and its capacity is f. If application x can run on computer y, then there is an arc (x, y) and its capacity is 1. For every computer y, there is also an arc (y, en) and its capacity is 1.

The total flow from the source is $sum = \sum f_{st,x}$. It is the number of runs for all applications.
The maximum flow f for flow network D is calculated.
If $f \neq sum$, then some applications don't run; else, a possible allocation is generated.

14.3.5 Crime Wave

Nieuw Knollendam is a very modern town. This becomes clear when looking at the layout of its map, which is just a rectangular grid of streets and avenues. Being an important trade center, Nieuw Knollendam also has a lot of banks. Almost on every crossing a bank is found (although there are never two banks at the same crossing). Unfortunately, this has attracted a lot of criminals. Bank holdups are quite common, and often on 1 day several banks are robbed. This has grown into a problem, not only for the banks, but also for the criminals. After robbing a bank, the robber tries to leave the town as soon as possible, most of the time chased at high speeds by the police. Sometimes two running criminals pass the same crossing, causing several risks: collisions, crowds of police at one place, and a larger risk to be caught.

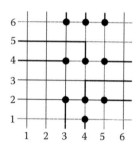

Figure 14.2 A grid.

To prevent these unpleasant situations, the robbers agreed to consult together. Every Saturday night they meet and make a schedule for the week to come: Who is going to rob which bank on which day? For every day they try to plan the getaway routes such that no two routes use the same crossing. Sometimes they do not succeed in planning the routes according to this condition, although they believe that such a plan should exist.

Given a grid of ($s \times a$) and the crossings where the banks to be robbed are located, find out whether or not it is possible to plan a getaway route from every robbed bank to the city bounds without using a crossing more than once (Figure 14.2).

Input

The first line of the input contains the number of problems p to be solved.

The first line of every problem contains the number s of streets ($1 \leq s \leq 50$), followed by the number a of avenues ($1 \leq a \leq 50$), followed by the number b ($b \geq 1$) of banks to be robbed. Then b lines follow, each containing the location of a bank in the form of two numbers, x (the number of the street) and y (the number of the avenue). Evidently, $1 \leq x \leq s$ and $1 \leq y \leq a$.

Output

The output file consists of p lines. Each line contains the text "possible" or "not possible." If it is possible to plan noncrossing getaway routes, this line should contain the word "possible". If this is not possible, the line should contain the words "not possible".

Sample Input	Sample Output
2	possible
6 6 10	not possible
4 1	
3 2	
4 2	
5 2	
3 4	
4 4	
5 4	

Sample Input	Sample Output
3 6	
4 6	
5 6	
5 5 5	
3 2	
2 3	
3 3	
4 3	
3 4	

Source: ACM Northwestern European Regionals 1996.

ID for online judge: UVA 563.

Hint

1. A flow network D is modeled.

   Nieuw Knollendam is a grid of $(s \times a)$. All grids are numbered from top to bottom and from left to right.

   Noncrossing getaway routes mean different getaway routes can't pass through the same grid. Every grid is represented as two vertices: one entrance vertex and one exit vertex. That is, for grid (i, j), its entrance vertex is $label[i][j][0]$, its exit vertex is $label[i][j][1]$, and there is an arc from the entrance vertex to the exit vertex whose capacity is 1. There are a source st and a sink en.

   In order to ensure that thieves can escape from the town, for every grid (i, j) on the boundary, $((1 \le i \le s, 1 \le j \le a)$ && $(i == 1 \| j == 1 \| i == s \| j == a))$, there is an arc $(label[i][j][1], en)$ whose capacity is 1.

   From every grid (i, j) to its adjacent grids, there are arcs whose capacities are 1. That is, there are arcs $(label[i][j][1], label[i + 1][j][0])$, $(i + 1 \le s)$; $(label[i][j][1], label[i - 1][j][0])$, $(1 \le i - 1)$; $(label[i][j][1], label[i][j + 1][0])$, $(j + 1 \le a)$; and $(label[i][j][1], label[i][j - 1][0])$, $(1 \le j - 1)$.

   In order to ensure that thieves can enter all banks, for every grid (i, j) in which there is a bank, there is an arc $(st, label[i][j][0])$ whose capacity is 1.

2. Calculate the maximum flow f for the flow network D.

3. If f is equal to the number of robbed banks b, then thieves can plan noncrossing getaway routes and output "possible"; else, output "not possible."

14.3.6 Pigs

Mirko works on a pig farm that consists of M locked pig-houses, but he can't unlock any pig-house because he doesn't have the keys. Customers come to the farm one after another. Each of them has keys to some pig-houses and wants to buy a certain number of pigs.

All data concerning customers planning to visit the farm on that particular day is available to Mirko early in the morning so that he can make a sales plan in order to maximize the number of pigs sold.

More precisely, the procedure is as follows: the customer arrives and opens all pig-houses to which he has the key, Mirko sells a certain number of pigs from all the unlocked pig-houses to him, and if Mirko wants, he can redistribute the remaining pigs across the unlocked pig-houses.

An unlimited number of pigs can be placed in every pig-house.

Write a program that will find the maximum number of pigs that he can sell on that day.

Input

The first line of input contains two integers M and N, $1 \le M \le 1000$, $1 \le N \le 100$, the number of pig-houses, and the number of customers. Pig-houses are numbered from 1 to M, and customers are numbered from 1 to N.

The next line contains M integers, for each pig-house's initial number of pigs. The number of pigs in each pig-house is greater than or equal to 0 and less than or equal to 1000.

The next N lines contain records about the customers in the following form (the record about the ith customer is written in the $(i + 2)$th line).

A K_1 K_2 ... K_A B means that this customer has the key to the pig-houses marked with the numbers $K_1, K_2, ..., K_A$ (sorted nondecreasingly) and wants to buy B pigs. Numbers A and B can be equal to 0.

Output

The first and only line of the output should contain the number of sold pigs.

Sample Input	Sample Output
3 3	7
3 1 10	
2 1 2 2	
2 1 3 3	
1 2 6	

Source: Croatia OI 2002 Final Exam—First Day.

ID for online judge: POJ 1149.

Hint

The key to the problem is to model a flow network. A flow network D is constructed based on the problem description. Each customer is represented as a vertex, where customer i is represented as vertex i, $1 \le i \le n$. The source st is as vertex $n + 1$. The sink en is as vertex $n + 2$.

For each customer, there are two ways to bug pigs:

■ Buying pigs in pig-houses which he has keys and he is the first customer who opens the pig-houses

■ Buying pigs that the previous customers can supply

When the information for customer i is inputted, if there are pig-houses which he has keys and he can open the pig-houses first, there is an arc from st to the customer, whose capacity is the total number of pigs that he can buy from these pig-houses. There is an arc from customer $i - 1$ to customer i ($i \geq 2$), whose capacity is infinity, for he can buy pigs from unlocked pig-houses that previous customers open. There is also an arc from customer i to *en*, whose capacity is the number of pigs that he wants to buy. Obviously, the maximum flow *f* for the flow network *D* is the number of sold pigs.

14.3.7 Drainage Ditches

Every time it rains on Farmer John's fields, a pond forms over Bessie's favorite clover patch. This means that the clover is covered by water for a while and takes quite a long time to regrow. Thus, Farmer John has built a set of drainage ditches so that Bessie's clover patch is never covered in water. Instead, the water is drained to a nearby stream. Being an ace engineer, Farmer John has also installed regulators at the beginning of each ditch, so he can control at what rate water flows into that ditch.

Farmer John knows not only how many gallons of water each ditch can transport per minute, but also the exact layout of the ditches, which feed out of the pond and into each other and the stream in a potentially complex network.

Given all this information, determine the maximum rate at which water can be transported out of the pond and into the stream. For any given ditch, water flows in only one direction, but there might be a way that water can flow in a circle.

Input

The input includes several cases. For each case, the first line contains two space-separated integers, N ($0 \leq N \leq 200$) and M ($2 \leq M \leq 200$). N is the number of ditches that Farmer John has dug. M is the number of intersection points for those ditches. Intersection point 1 is the pond. Intersection point M is the stream. Each of the following N lines contains three integers, S_i, E_i, and C_i. S_i and E_i ($1 \leq S_i, E_i \leq M$) designate the intersections between which this ditch flows. Water will flow through this ditch from S_i to E_i. C_i ($0 \leq C_i \leq 10{,}000{,}000$) is the maximum rate at which water will flow through the ditch.

Output

For each case, output a single integer, the maximum rate at which water may emptied from the pond.

Sample Input	Sample Output
5 4	50
1 2 40	
1 4 20	
2 4 20	
2 3 30	
3 4 10	

Source: USACO 93.

ID for online judge: POJ 1273.

Hint

A flow network D is constructed based on the problem description. N ditches are represented as N arcs, and M intersection points for those ditches are represented as vertices, where vertex i corresponds to intersection point i, $1 \leq i \leq M$. Intersection point 1 is as the pond, and intersection point M is as the stream. That is, vertex 1 is as the source, and vertex M is as the sink. For ditch k ($1 \leq k \leq n$), water is from intersection point x to intersection point y, and the capacity of water that the ditch can transport is f; then there is an arc (x, y) whose capacity is f. The maximum flow for the flow network D is the maximum rate at which water may emptied from the pond.

14.3.8 Mysterious Mountain

A group of M people is chasing a very strange animal. They believe that it will stay on a mysterious mountain T, so they decide to climb it and have a look. The mountain looks ordinary (Figure 14.3).

That is, the outline of the mountain consists of $N + 1$ segments. The end points of them are numbered $0 .. N + 1$ from left to right. That is, $x[i] < x[i + 1]$ for all $0 \leq i \leq n$. Also, $y[0] = y[n + 1] = 0$, $1 \leq y[i] \leq 1000$ for all $1 \leq y \leq n$.

According to their experience, the animal is most likely to stay at one of the N end points numbered $1 .. N$. And funny enough, they soon discover that $M = N$, so each of them can choose a different end point to look for the animal.

Initially, they are all at the foot of the mountain (i.e., at $(s_i, 0)$) For every person i, he is planning to go left or right to some place $(x, 0)$ (where x is an integer—they do not want to take time to work out an accurate place) at the speed of w_i and then climb directly to the destination along a straight line (obviously, no part of the path that he follows can be over the mountain—they can't fly) at the speed of c_i. They don't want to miss it this time, so the team leader wants the latest person to be as early as possible. How fast can this be done?

Input

The input will contain no more than 10 test cases. Each test case begins with a line containing a single integer N ($1 \leq N \leq 100$). In the following $N + 2$ lines, each line contains two integers x_i and y_i ($0 \leq x_i, y_i \leq 1000$), indicating the coordinate of the ith end points. In the following N lines, each line contains three integers, c_i, w_i, and s_i, describing a person ($1 \leq c_i < w_i \leq 100$, $0 \leq s_i \leq 1000$)— the climbing speed, walking speed, and initial position. The case containing $N = 0$ will terminate the input and should not be regarded as a test case.

Output

For each test case, output a single line containing the least time that these people must take to complete the mission, and print the answer with two decimal places.

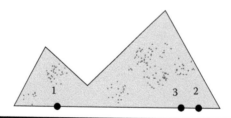

Figure 14.3 Mountain T and three people.

Sample Input	Sample Output
3	1.43
0 0	
3 4	
6 1	
12 6	
16 0	
2 4 4	
8 10 15	
4 25 14	
0	

Source: OIBH Online Programming Contest 1.

Note: In this example, person 1 goes to (5, 0) and climbs to end point 2, person 2 climbs directly to end point 3, and person 3 goes to (4, 0 and climbs to end point 1 (Figure 14.4).

ID for online judge: ZOJ 1231.

Hint
1. Calculate the positions on the ground where people can climb the mountain.

   Suppose $can[x][i]$ is the flag that end point i can be reached from $(x, 0)$. Because coordinates for $n + 2$ end points are listed from left to right, the interval for the x coordinate is $[p[0].x, p[n + 1].x]$, where $p[i]$ is the coordinate for end point i.

   For each x coordinate in the interval, all end points that can be reached from it are obtained.

   End points are enumerated in ascending order for its sequence number. For end point i, if end point j ($0 \leq j \leq i - 1$) that can be reached from $(x, 0)$ is the nearest end point, and the vector product $p_{(x, 0)} * p[i] \geq 0$, then end point i can be reached from $(x, 0)$; that is, $can[x][i] =$ true.
2. Calculate the shortest time that each person arrives at each reachable end point.

   Suppose $reach[i][j]$ is the shortest time that person i arrives at end point j. If from $(x, 0)$ end point j can be reached ($can[x][j] ==$ true), for person i, it takes

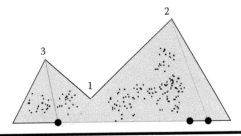

Figure 14.4   The solution to the example.

$$\frac{\sqrt{(x - p[j]x)^2 + p[j]^2}}{c_i} + \frac{x - s_i}{w_i}$$

for climbing and walking. Obviously,

$$reach[i][j] = \min_{p[0].x \le x \le p[n+1].x} \left\{ \frac{\sqrt{(x - p[j]x)^2 + p[j]^2}}{c_i} + \frac{x - s_i}{w_i} \middle| can[x][j] == true \right\}$$

3. Construct a flow network *D*.

The number of vertices is $2*n + 2$, where source *st* is represented as vertex $2*n + 1$; sink *en* is represented as vertex $2*n + 2$; persons are represented as vertex 1 .. vertex *n*; and end points are represented as vertex $n + 1$ .. vertex $2n$; end point *i* is as vertex $n + i$; $1 \le i \le n$.

From source *st* to each person *i* there is an arc (*st*, *i*) whose capacity is 1 and cost is 0, $1 \le i \le n$.

From each end point *j* to sink *en* there is an arc (*j*, *en*) whose capacity is 1 and cost is 0, $n + 1 \le j \le 2*n$.

From each person *i* to each end point *j* there is an arc (*i*, *j*) whose capacity is 1 and cost is *reach[i][j]*.

Obviously, the problem can be modeled as the maximum flow for the flow network *D*. The time that the latest person comes to the end point as early as possible is the minimum cost for *D*. The algorithm is to calculate the minimum-cost maximum flow for *D*.

4. Calculate minimum-cost maximum flow:

```
Suppose d[i] is the maximum cost for arcs that vertex i is the rear;
    double spfa( st, en){
        Queue h[ ];
        Push source st into h, d[st]=0;
        while (1){
                Pop the front of the queue x;
                Enumerate every arc from x (x, y):
                    { if (the flow of (x, y)>0&&d[y]>max(d[x], the cost of
(x,y)))){
                        d[y]= max(d[x], the cost of (x,y));
                        pre[y]=x;
                        if (y isn't in the queue){
                            y is pushed into h;
                        };
                }
                if(h is empty ) break;
                };
        };
        return d[en];
}
```

While an augmenting path from *st* to *en* is found, the flow is adjusted based on flow conservation and to calculate the maximum cost for arcs. Suppose *aug[i]* is the maximum cost for arcs for the *i*th finding augmenting path; the solution is

$$ans = \max_{1 < i \leq k}\{ang[i]\}$$

Program

```cpp
#include<iostream>
#include<math.h>
#include<cstdio>
#include<cstring>
#include<algorithm>
using namespace std;
const int V=300;                    //The upper limit of the number of vertices
const int E=30000;   // The upper limit of the size of adjacency list
const double big=1e10;   //Infinity
struct edge{      //the type of elements in adjacency list
      int x, y, next, f, op;
      double c;
  // (x, y): arc; f: flow; next: the pointer pointing to the next arc;
c: the cost of the arc
};
struct point{ //The peak coordinate
      int x, y;
};
edge a[E]; // adjacency list a[ ]
int tot, n, m, st, en; //tot: length of adjacency list; n: the number of
persons; st: source, en: sink
int fi[V], pre[V], h[V+10]; // fi[i]: the first arc for vertex i; pre[i]:
  in augmenting path the arc whose tail is i; h[]: the queue
double d[V];
bool v[V];
//The problem requires to calculate a maximum matching in a bipartite
graph, and minimize the maximum flow
void Add(int x, int y, int f, double c, int op){
//an new arc (x, y) is added into the adjacency list, f: flow, c:
   capacity, op: forward or back
      a[++tot].x=x;
      a[tot].y=y;
      a[tot].f=f;
      a[tot].c=c;
      a[tot].op=tot+op;
      a[tot].next=fi[x];
      fi[x]=tot;
}
void add(int x, int y, int f, double c){
//add a new forward arc (x, y) with capacity f, and a back arc (y, x)
with capacity 0 into adjacency list
      Add(x, y, f, c, 1);
      Add(y, x, 0, -c, -1);
}
double spfa(int st, int en){
// Calculate minimum cost maximum flow
      memset(h,0,sizeof(h));
      memset(v,0,sizeof(v));
      memset(d,66,sizeof(d));
      memset(pre,0,sizeof(pre));
      int l=1, r=1, md=V+5;
```

```
            h[1]=st; // st is added into queue h[ ]
            v[st]=1;
            d[st]=0;
            while (1){
                    int x=h[l]; // for the front in h[ ] x, enumerate all arcs
(x, y)
                    for (int t=fi[x], y=a[t].y; t; t=a[t].next,y=a[t].y)
                    if (a[t].f>0&&d[y]>max(d[x],a[t].c)){
        // the arc is added into the augmenting path
                            d[y]=max(d[x],a[t].c);
                            pre[y]=t;
                            if (!v[y]){
                                    v[y]=1;
                                    r=(r+1)%md;
                                    h[r]=y;
                            }
                    }
                    if (l==r) break;
                    v[x]=0;
                    l=(l+1)%md;
            }
            return d[en];
}
double aug(int st,int en){
  // adjust the flow in the augmenting path
        int maxf=16000000,k=pre[en];
        double ans=0;
        while (k){
                maxf=min(maxf,a[k].f);
                k=pre[a[k].x];
        }
        k=pre[en];
        while (k){
                ans=max(a[k].c,ans);
                a[k].f-=maxf;
                a[a[k].op].f+=maxf;
                k=pre[a[k].x];
        }
        return ans;
}
double costflow(int st, int en){
        double ans=0;
        double temp=spfa(st, en);
        while (temp<big){
                ans=max(ans,aug(st, en));
                temp=spfa(st, en);
        }
        return ans;
}
int N, vc[110], vw[110], sx[110];
point p[110];
bool can[1100][110];
double reach[110][110], dis[1100][110];
int cross(int X1, int Y1, int X2, int Y2){
```

```
        return X1*Y2-X2*Y1;
}
double go(int i, int x, int j){ // the i-th person go to peak j,
calculate the cost time
        double dx=abs(x-sx[i]);
        return dx/vw[i]+dis[x][j]/vc[i];
}
void build( ){ //Construct a flow network D
        memset(fi, 0, sizeof(fi));
        tot=0;
        st=2*N+1;
        en=2*N+2;
        n=2*N+2;
        for (int i=1; i<=N; i++) add(st, i, 1, 0);
        for (int i=1; i<=N; i++) add(i+N, en, 1, 0);
        for (int i=1; i<=N; i++)
          for (int j=1; j<=N; j++)
                add(i, j+N, 1, reach[i][j]);
}
int main( ){
        cin>>N;
        while (N){
                for (int i=0; i<=N+1; i++) cin>>p[i].x>>p[i].y;
                for (int i=1; i<=N; i++) cin>>vc[i]>>vw[i]>>sx[i];
                int left=p[0].x,right=p[N+1].x;
            // reachable relationships are calculated
                for (int x=left; x<=right; x++){
                        point limit;
                        limit.x=x;
                        limit.y=1000;
                        for (int i=1; i<=N; i++) if (p[i].x>=x){
                                if (cross(p[i].x-x,p[i].y,limit.x-x,limit.y)>=0){
                                        can[x][i]=1;
                                        limit=p[i];
                                } else can[x][i]=0;
                        }
                }
                for (int x=right; x>=left; x--){
                        point limit;
                        limit.x=x;
                        limit.y=1000;
                        for (int i=N; i>=1; i--) if (p[i].x<x){
                                if (cross(p[i].x-x,p[i].y,limit.x-x,limit.y)<=0){
                                        can[x][i]=1;
                                        limit=p[i];
                                } else can[x][i]=0;
                        }
                }
                memset(reach, 66, sizeof(reach));
                for (int x=left; x<=right; x++)
                for (int j=1; j<=N; j++) dis[x]
[j]=sqrt((x-p[j].x)*(x-p[j].x)+p[j].y*p[j].y);
                for (int i=1; i<=N; i++)
                for (int j=1;j<=N;j++)
```

```
        for (int x=left; x<=right; x++) if (can[x][j]) reach[i]
[j]=min(reach[i][j],go(i,x,j));
        build( ); //Construct a flow network D
        printf("%.2lf\n", costflow(st, en));
        cin>>N;
    }
}
```

SUMMARY OF SECTION IV

Graphs are used to represent objects and relationships between objects in the real world. There are two kinds of storage methods to store a graph: adjacency list and adjacency matrix.

First, experiments for BFS and DFS were introduced, for many graph algorithms are based on BFS and DFS. Then experiments for some classical graph algorithms, such as the Kruskal algorithm, Prim algorithm, Dijkstra's algorithm, and SPFA, were introduced. Finally, experiments for algorithms of bipartite graphs and flow networks were given.

Bibliography

Wu Yonghui, Wang Jiande. *Data Structure Experiment: For Collegiate Programming Contest and Education* [Chinese version]. China Machine Press, 2012.

Wu Yonghui, Wang Jiande. *Data Structure Experiment: For Collegiate Programming Contest and Education* [Traditional Chinese version]. Gotop Information, Inc., 2012.

Wu Yonghui, Wang Jiande. *Solutions and Analyses to ACM-ICPC World Finals (2004–2011)* [Chinese version]. China Machine Press, 2012.

Wu Yonghui, Wang Jiande. *Algorithm Design Experiment: For Collegiate Programming Contest and Education* [Chinese version]. China Machine Press, 2013.

Wu Yonghui, Wang Jiande. *Programming Strategies Solving Problems* [Chinese version]. China Machine Press, 2015.

Index